BIOFUELS - ECONOMY, ENVIRONMENT AND SUSTAINABILITY

Edited by **Zhen Fang**

Biofuels - Economy, Environment and Sustainability
http://dx.doi.org/10.5772/50478
Edited by Zhen Fang

CBS, Edition **2016**

Contributors

Stephen Hughes, Pantaleo, Nilay Shah, Rosa, Krzysztof Biernat, Artur Malinowski, Malwina Gnat, Minerva Singh, Shonil Bhagwat, Estelvina Rodriguez-Portillo, Jose Ricardo Duarte Ojeda, Sully Ojeda De Duarte, Anthony Basco Halog, Nana Awuah Bortsie-Aryee, Annelies Zoomers, Lucía Goldfarb, Suseno Budidarsono, Lílian Lefol Nani Guarieiro, Aline Guarieiro, Ada Rispoli, Davide Barnabè, Renzo Bucchi, Claudia Letizia Bianchi, Pier Luigi Porta, Daria Camilla Boffito, Gianni Carvoli, Carlo Pirola, Cristian Chiavetta, James A. Dyer, Raymond L. Desjardins, Suren Kulshreshtha, Brian G. McConkey, Xavier P.C. Vergé, Marcelo Sthel, Aline Rocha, Maria Castro, Victor Haber Perez, Helion Vargas, Marcelo Gomes, Georgia Mothe, Wellington Silva, Juliana Rocha, Flavio Couto

Published by InTech

Janeza Trdine 9, 51000 Rijeka, Croatia

Publishing Process Manager Iva Simcic

Technical Editor InTech DTP team

Cover InTech Design team

Additional hard copies can be obtained from orders@intechopen.com

Biofuels - Economy, Environment and Sustainability, Edited by Zhen Fang
p. cm.
ISBN 978-953-51-0950-1

INTECH

open science | open minds

Contents

Preface

Biofuels are gaining public and scientific attention driven by high oil prices, the need for energy security and global warming concerns. There are various social, economic, environmental and technical issues regarding biofuel production and its practical use. This book is intended to address these issues by providing viewpoints written by professionals in the field and the book also covers the economic and environmental impact of biofuels.

This text includes 14 chapters contributed by experts around world on the economy, evironment and sustainability of biofuel production and use. The chapters are categorized into 3 parts: Feedstocks, Biofuels, Environment

Section one, focuses on the sustainability and economy of feedstock production. Chapters 1 and 2 discuss the sustainability and biodiversity of land use for biofuel crops. Chapter 3 gives a case study on rapid expansion of soy production in a region of Argentina. Chapter 4 assesses biofuel feedstock production in Canada by farm energy analysis. Chapter 5 analyzes the processes of biofuel production using waste carbon dioxide and solar energy. Chapter 6 presents a case study on social and economic development caused by oil palm plantation in Indonesia.

Section 2, (Chapters 7-9) analyzes biofuel systems. Chapter 7 evaluates the sustainability of biofuels via life cycle and integrated sustainability modeling and analysis with consideration to temporal and spatial dimensions. Chapter 8 overviews the logistics of bioenergy systems, with particular attention to the economic and sustainability implications of the different transport, processing and energy conversion systems for heat and power generation. Chapter 9 discusses efficiently converting biomass to biofuels and value-added co-products.

Section 3, (Chapters 10-14) gives environmental analyses of biofuels. Environmental consideration and assessment of biofuels are given in Chapters 10 and 11. Evaluation of gaseous emissions by the use of biofuels is presented in Chapter 12. Energy policies in Brazil related to climate change and CO2 emission abatement are overviewed in Chapter 13. Finally, vehicle emissions from biofuel combustion are commented in Chapter 14

This book overviews social, economic, environmental and sustainable issues by the use of biofuels. It should be of interest for students, researchers, scientists and technologists in biofuels.

I would like to thank all the contributing authors for their time and efforts in the careful construction of the chapters and for making this project realizable. It is certain to inspire many young scientists and engineers who will benefit from careful study of these works and that their ideas will lead us to develop and recognize biofuel systems that are economic, sustainable and respectful of our environment.

I am grateful to Ms. Iva Simcic (Publishing Process Manager) for her encouragement and guidelines during my preparation of the book.

Finally, I would like to express my deepest gratitude towards my family for their kind cooperation and encouragement, which help me in completion of this project.

Prof. Dr. Zhen Fang
Leader of Biomass Group
Chinese Academy of Sciences
Xishuangbanna Tropical Botanical Garden, China

Section 1

Feedstocks

Chapter 1

Land Use Change Impacts of Biofuels: A Methodology to Evaluate Biofuel Sustainability

D. Barnabè, R. Bucchi, A. Rispoli, C. Chiavetta, P.L. Porta, C.L. Bianchi, C. Pirola, D.C. Boffito and G. Carvoli

Additional information is available at the end of the chapter

http://dx.doi.org/10.5772/52255

1. Introduction

Biofuel is a type of fuel whose energy derives from biological carbon fixation. Biofuels include fuels deriving from biomass conversion, solid biomass, liquid fuels and various biogases.

Despite the intent of biofuels production as an alternative to fossil fuel sources, its sustainability has been often criticized. In this context, land use change is a major issue. Indeed, considering traditional energy crop yields, vast amounts of land and water would be needed to produce enough biomass to significantly reduce fossil fuel dependency. There is also a wide debate on increasing biomass demand for the energy market which could result in a dangerous competition with the food requirements by humankind, as well as in increasing food prices. Second and third generation sources of feedstock, as well as improved sustainable production of biofuels of first generation such those from non-edible crop, are some of the fields or research handled to fight negative impact of biofuels production on land use.

Agronomic management determines which and how crops are grown: it can have far-reaching impacts on soil quality, water quality, climate change, and biodiversity. The importance of the agronomic management may be magnified as farmers, prompted by high energy-crop prices, would attempt to increase productivity of lands, enlarge the total amount of land under cultivation and expand cultivation into less productive lands.

Among biofuels, biodiesel is one of the main alternative energy sources.

In recent years, the authors have been studying innovative solutions for the field phase of feedstock production as well as for the industrial phase of transformation to produce a more sustainable biodiesel. From the agricultural point of view, the study has been focusing on alternative feedstock and good management practices to increase biomass yields keeping a high soil quality or even rescuing soils not suited anymore for edible crops.

In this context more than other, to accurately balance environmental impacts of biofuels production, it is important to consider agricultural practices applied to grow the biomass and their direct and indirect effects on soil quality. The evaluation of biofuels impacts on soil should not consider only the type of land converted, but also the trend of quality of arable land. Currently, this is still a critical aspect of life cycle analysis (LCA) tools to evaluate biofuels impacts on land use change.

Sustainability analysis of oil production for biofuel should assess the different impact on land use of intensive and extensive cultivation, should consider the not linearity in production yield and in generated impacts and should express the complex equilibrium that guarantees the biodiversity conservation. The authors are studying soil quality parameters and how these parameters could be integrated in a unique indicator able to add additional information to evaluate land use change in a LCA perspective.

The development of this innovative approach aims to improve the evaluation of biofuels impact on land use, allowing taking into account the impact of management practices on soil quality.

In particular, the authors are studying agricultural practices and their influence on soil quality related to biomass culture on marginal soils. The study is focused on agricultural practices which influence measurable parameters and which can describe soil quality trends following a biomass production process.

A methodology which can differentiate impacts of different arable land uses could be not only the base for the development of a powerful tool used by farmers to select the suitable crop and the best management practice in relation to soil type, but also a tool to describe the sustainability of different biofuel production processes in the perspective of new politic regulations and economic incentives.

2. Sustainable profile of biofuels

Biofuels offer a potentially attractive solution reducing the carbon intensity of the transport sector and addressing energy security concerns. General concern for pollution and environmental impact of energy consumption based on fossil sources has led to more and more study on the sustainability profile of available energy sources, traditional and alternative ones.

Among alternative sources, biofuels are those whose energy is derived from biological carbon fixation such as biomass, as well as solid biomass, liquid fuels and various biogases. According to this classification, also fossil fuels could be included (because of their origin in ancient carbon fixation), but they are not considered biofuels as carbon they contain has been "out" of the carbon cycle for a very long time.

Even if demand for biofuels continues to grow strongly, some biofuels have received considerable criticism as a result of:

- rising food prices;
- relatively low greenhouse gas (GHG) abatement, or even increases in some cases, based on full life-cycle assessments;
- the continuing need for significant government support and subsidies to ensure that biofuels are economically viable;
- direct and indirect impacts on land use change and the related greenhouse gas emissions;

2.1. Edible and non-edible raw materials

Biofuels currently available or in development are shared into three, sometimes also four, groups designed as "generations".

As the term "generation" indicates, biofuels are classified according to their progressive introduction on the market during the last 20-30 years[1].The final goal will combine higher energy yields, lower requirements for fertilizer and land, and the absence of competition with food together with low production costs offering a truly sustainable alternative for transportation fuels.

2.1.1. First generation biofuels

First generation biofuels are based on feedstocks that have traditionally been used as food such as corn or sugar cane for ethanol production and edible vegetable oils and animal fat for biodiesel production. The technology to produce these kinds of biofuels exists and it's quite consolidated. These fuels are currently widespread and considering production costsfor feedstocks, first generation biofuels have nearly reached their maximum market share in the fuels market.

Rising of food prices and doubts on greenhouse gases emission saving improvement are some of the hot spots on their sustainability debate.

2.1.2. Second generation biofuels

Facing the main concerns in first generation biofuels, advanced technical processes have been developing to obtain biofuels, for example ethanol and, in some cases, related alcohols such as butanol by non-edible feedstocks such as cellulose from cell wall of plant cells (rath-

1 The transesterification process of vegetable oil was first tested in 1853 by E. Duffy and J. Patrick. In 1893 Rudolf Diesel's projected the first vehicle biodiesel-powered. Only in 1990's France launched the local production of biodiesel fuel obtained by the transesterification of rapeseed oil.

er than sugar made from corn or sugar cane).Other researches are trying to find non-edible oil crops for biodiesel such as some brassicaceae (e.g., *B. carinata* and *B. juncea*), *Nicotiana tabacum, Ricinus communis, Cynara cardunculus* [1].

Even if some issues are still challenging, second generation biofuels make wider the feedstock portfolio for biofuels avoiding competition with food. Nevertheless, feedstock costs remain high (not necessarily due to the feedstock retrieval, but almost due to processing) and GHG emission savings still need to be ascertained by properly analysis of possible emission from land use change [2].

2.1.3. Third generations biofuels

Third generation biofuels, as well as second generation biofuels, are made from non-edible feedstocks, with the advantage that the resulting fuel represents an equivalent replacement produced from sustainable sources (for example fast-growing algae or bacteria) for gasoline, diesel, and aviation fuel. These alternative biofuels are anyway in developing and several technological and economic challenges still need to be faced to bring them on the market.

2.1.4. Fourth generations biofuels

Fourth generation biofuels are those which result in a negative carbon impact in the atmosphere. These fuels will be obtained from genetically engineered crops that release a lesser amount of carbon dioxide during combustionthan that absorbed from the atmosphere for their growth [3].

2.2. Land use issues

2.2.1. Demand for land

Since biofuels are derived from biomass conversion, demand for land for agro-fuel production has increased significantly over the past few years. Growing demand for land is a sensitive point in biofuels sustainability since, directly or indirectly, it influences all the three sustainability pillars: social, economic and environmental[2].

According to the so called RED directive (Renewable Energy Directive)[3], European countries have established targets for the mandatory blending of traditional transport fuels with biodiesel and bioethanol. Developing countries searching for new profitable markets, have increasingly invested in biofuel production for both domestic use and export. In general, all countries at a global level are attracted by this big demand and market, so they are targeting vast tracts of land to produce raw materials for biofuels, often with no concern for the conversion of areas of high biodiversity and high carbon stock.

---

2 Art.2 and Art.5 from "Treaty Of Amsterdam Amending The Treaty On European Union, The Treaties Establishing The European Communities And Related Acts", Official Journal C 340, 10 November 1997.

3 Directive 2009/28/EC of 23 April 2009 on the promotion of the use of energy from renewable sources and amending and subsequently repealing Directives 2001/77/EC and 2003/30/EC.

http://dx.doi.org/10.5772/52255

On one side first and second generation biofuels are still strictly dependent on a field phase of feedstock production, while on the other side, third and fourth generation biofuels are not ready to replace them as alternative source of energy. These market drivers, in consideration of the recent food crisis [4] and the financial crisis [5] causes great alarm for the growing of biofuels demand bringing to the debate often referred to as the "food or fuel dilemma" (in 2007 and 2008 cereals and protein crop drastically increased their prizes) [6]. In addition, the drought currently recorded in the USA threatens to cause a new global catastrophe driven by a speculator amplified food price bubble [7].

2.2.2. *Land Use Change (LUC)*

Currently land use is a prerogative of first and second generation biofuels so that land use change should always be taken into account in biofuel sustainability evaluation.

Cultivating biomass feedstock needs land, which might cause LUC regarddirect effect on the site of the farm or plantation and indirect effects through leakage (i.e. displacement of previous land use to another location where direct LUC could occur).

Two kind of land use change are usually described: direct land use change (dLUC) and indirect land use change (iLUC). The definition of dLUC is straightforward: direct land use change is the conversion of land, which was not used for crop production before,into land used for a particular biofuel feedstock production. The emissions caused by the conversion process can be directly linked to the biofuel load and thus be allocated to the specific carbon balance of that biofuel.

iLUC is a market effect that occurs when biofuel feedstocks are increasingly planted on areas already used for agricultural products. This causes a reduction of the area available for food and feed production and therefore leads to a reduction of food and feed supply on the world market. If the demand for food remains on the same level and does not decline, prices for food rise due to the reduced supply. These higher prices create an incentive to convert formerly unused areas for food production since the conversion of these areas becomes profitable at higher prices. This is the iLUC effect of the biofuel feedstock production. The iLUC effect of biofuels happens only through the price mechanism of the global or regional food market. Therefore iLUC in this context is always direct land use change (dLUC) for food production incentivised by the cross-price effects of an increased production of biofuel feedstocks which then translates into an additional demand for so far unused land areas [8].

From a global perspective which takes into account all land use from all production sectors of biomass, increasing biomass feedstock production has only direct LUC effect, as all interaction of markets, changes of production patterns and the respective conversion of land from one (or none) use to another will be accounted for. Thus it's a problem of scope, when the system boundaries for an analysis are reduced, "blindness" to possible impact outside of the scope is the consequence [9].

The primary risk for indirect land use change is that the use of crops for biofuels might displace other agricultural production activities onto land with high natural carbon stocks like forests, resulting in significant greenhouse gas emissions from land conversion.

The environmental profile of biofuels has to take into account the GHG emissions balance from land use. Indeed most prior studies claimed biofuels environmental benefits mainly on the base of the carbon sequestration that occurs through the growth of agronomic raw materials. These findings missed to consider in the GHG balance, the emissions that could derive from indiscriminate land use change (direct and indirect) from of high value lands to land for biofuels feedstocks production.

Currently most authors are evaluating this "carbon debt" also to calculate the so called "payback period", the time required for biofuels to overcome their carbon debt depending on the specific ecosystem involved in the land use change event [10, 11].

2.2.3. *Land Use impact assessment for agronomic system*

In relation to biofuels, land use translates not only into land occupation for a certain time, but also in possible perturbation of soil quality trend. The concept of soil quality is linked to the ability of soil to function effectively in a variety of roles. The primary measures of this effectiveness supply information on biological productivity, environmental quality, and human and animal health.

Because of its consequences on human health and environment quality, degradation of soil quality as consequence of intensive agronomic system is a major global concern. So this factor needs to be properly evaluated in the environmental assessment of agro-forestry systems involved in production of raw material for biofuels.

First methodologies for land use impact assessment in LCA don't respond to the perturbation on soil quality, giving an indication about land use impact in terms of hectare or hectare per year. Currently new methods in LCA studies and furthers indicators need to be developed to describe the aspects typical of land use impacts of agricultural systems, among these: soil quality status and its trend following to the use change, application of different types of managements, non-linear output of production [12].

2.3. Legislation on environment and renewable energy

Acid rain, air pollution, global warming, ozone depletion, smog, water pollution, and forest destruction are just some of the environmental problems that we currently have to face globally and which require long-term potential actions for sustainable development to achieve solutions.

2.3.1. *Global agreements*

To face the global environment issue, in 1979 the first World Climate Conference (WCC) took place although, only in 1992, countries joined for the first time an international treaty, the United Nations Framework Convention on Climate Change (UNFCCC), to cooperatively consider what they could do to limit average global temperature increases and the resulting climate change, and to cope with whatever impacts were, by then, inevitable. Since 1995, annually, the Conference of the Parties (COP) takes place and in 1997, with the occasion, the

Kyoto Protocol was formally adopted. In 2005, due to a complex ratification process, Kyoto Protocol entered into force introducing the operational provisions agreed by the countries to stabilize and then reduce GHG emissions [13]. The targets cover emissions of the six main greenhouse gases: carbon dioxide (CO_2), methane (CH_4), nitrous oxide (N_2O), hydrofluorocarbons (HFCs), perfluorocarbons (PFCs), and sulphur hexafluoride (SF_6).

Commitments of countries are based, since 1990, on the scientific contribution of the Intergovernmental Panel on Climate Change (IPCC) which periodically publish the Assessment Reports (AR) of the state of the knowledge on climate change [4] [14].

2.3.2. European legislation

Reduction of pollution of the atmosphere, water and soil, as well as the quantities of waste arising from industrial and agricultural installations are issues faced by the European Union (EU) in the "IPPC directive" (Integrated pollution prevention and control)[5]. This Directive defines the obligations with which industrial and agricultural activities, with an high pollution potential, must comply. It establishes a procedure for authorizing these activities and sets minimum requirements to be included in all permits, particularly in terms of pollutants released, to ensure a high level of environmental protection.

The IPPC directive requires industrial and agricultural activities with a high pollution potential to have a permit. This permit can only be issued if certain environmental conditions are met, so that the companies themselves bear responsibility for preventing and reducing any pollution they may cause.

Briefly the following are the basic obligations:

- use all appropriate pollution-prevention measures, namely the best available techniques (which produce the least waste, use less hazardous substances, enable the substances generated to be recovered and recycled, etc.);
- prevent all large-scale pollution;
- prevent, recycle or dispose of waste in the least polluting way possible;
- use energy efficiently;
- ensure accident prevention and damage limitation;
- return sites to their original state when the activity is over.

In addition, the decision to issue a permit must contain a number of specific requirements, including:

- emission limit values for polluting substances (with the exception of greenhouse gases if the emission trading scheme applies);

4 Four Assessment Reports have been completed in 1990, 1995, 2001 and 2007. All completed Assessment Reports are available on IPCC website: The IPCC Fifth Assessment Report (AR5) is scheduled for completion in 2013/14.

5 IPPC Directive (Directive 96/61/EC) recently been codified by Directive 2008/1/EC.

- any soil, water and air protection measures required;
- waste management measures;
- measures to be taken in exceptional circumstances (leaks, malfunctions, temporary or permanent stoppages, etc.);
- minimization of long-distance or transboundary pollution;
- release monitoring;
- all other appropriate measures.

In regard of IPPC themes, renewable energy resources appear to be the one of the most efficient and effective solutions. That is why there is an intimate connection between renewable energy and sustainable development, synergistically approached by energy scientists, engineers and policy makers [15].

The European Union recently updated issues on renewable energy and sustainable developments, which comprises biofuels matter, enacting the Directive 2009/28/EC on renewable energy (RED: Renewable Energy Directive). The ambitious aim of this directive is the EU reaching a 20% share of energy from renewable sources by 2020 and a 10% share of renewable energy specifically in the transport sector. National action plans have to establish pathways for the development of renewable energy sources, create cooperation mechanisms to help achieving the targets cost effectively and establish the sustainability criteria for biofuels[6]. The RED requires that all biofuels supplied to the EU market comply with the sustainability criteria. The Directive 2009/28/EC sets out sustainability criteria for biofuels in its articles 17, 18 and 19. These criteria are related to greenhouse gas savings, land with high biodiversity value, land with high carbon stock and agro-environmental practices. In order to receive government support, this compliance has to be ensured by the economic operators selling fuel on the market. Even if third countries that play a significant role in providing feedstock for EU consumed biofuels are not required to implement the requirements of the RED, the compliance with the biofuel sustainability requirements must be guaranteed by the EU Member States who count imported biofuels towards their national renewable energy targets, where such fuels are counted towards renewable energy obligations and where they receive financial support. For this situation voluntary schemes may be used as a proof of compliance with the EU sustainability criteria[7] [16].

3. Soil quality and agronomic management practices in biofuels production

The authors are involved in a three years study about the feasibility of sustainable biodiesel production in Italy[8]. This phase aims at the characterization of an innovative agronomic sol-

6 Art. 4 to the Directive 2009/28/EC.

7 EC decision 19 July 201.

ution that may positively affect the energy and GHG balance, achieving a high level of sustainability in the oilseeds production.

One of the relevant points in the evaluation of sustainability is land use impact assessment. The authors made a preliminary research on issues related to land use impact assessment such as soil quality, management practices and land use change indicators suitable to describe the agronomic solutions proposed and their impact on land use especially in terms of soil quality trend.

3.1. Soil quality

The terms "Soil Quality" and "Soil Science" were first introduced in the 1970s when it was established that the concept of soil quality should encompasses the following points [17]:

- Land resources are being evaluated for different uses;
- Multiple stakeholder groups are concerned about resources;
- Priorities of society and the demands on land resources are changing;
- Soil resources and land use decisions are made in a human or institutional context.

From a pragmatic point of view anyway the most concise definitions express soil quality as "fitness for use" [18] or as "the capacity of a soil to function" [19] or rather "the ability of the soil to perform the functions necessary for its intended use".

In the beginning, soil quality was only discussed to control soil erosion and minimizing the effects of soil loss on productivity [20]. Only in 1990s, in addition to the productivity factor, some authors began to think in terms of soil quality dependency to management practices and proposed a quantitative formula for assessing soil quality [21, 22]. Indeed soil condition, response to management, or resistance to stress imposed by natural forces or human uses, began to be taken into account as factors able to describe soil quality [23, 24].

3.1.1. Soil functions

According to the most pragmatic definitions, soil quality depends on its intended uses. Although soils cover a wide range of needs, the following are here summarized as general capabilities of soils [19]:

1. sustaining biological activity, diversity, and productivity;
2. regulating and partitioning water and solute flow;
3. filtering, buffering, degrading, immobilizing, and detoxifying organic and inorganic materials, including agricultural, industrial and municipal by-products and atmospheric deposition;

8 SUSBIOFUEL project ("Studio di fattibilità per la produzione di biocarburanti da semi oleosi di nuove specie e da sottoprodotti o materiali di scarto" – D.M. 27800/7303/09), financially supported by the Ministry of Agricultural, Food and Forestry Policies – Italy.

4. storing and cycling nutrients and other elements within the biosphere;

5. providing support of socioeconomic structures and protection for archaeological treasures associated with human habitation.

3.1.2. Soil indicators

Soil quality can be viewed in two ways: as inherent soil quality, which is regulated by the soil's inherent properties as determined by the five soil-forming factors, and as dynamic soil quality, which involves changes in soil properties influenced by human use and management.

These qualities together determine the capability of soil to function.

Inherent soil quality is independent (or slightly influenced) by land use or management practices so that is described by use-invariant properties rather linked to the soil's genesis over millennia and remain constant during the time (Figure 1). These properties include soil texture, depth to bedrock, type of clay, CEC, drainage class, and depend on the five soil-forming factors [25]:

- climate (precipitation and temperature),
- topography (shape of the land),
- biota (native vegetation, animals, and microbes),
- parent material (geologic and organic precursors to the soil),
- time (time that parent material is subject to soil formation processes).

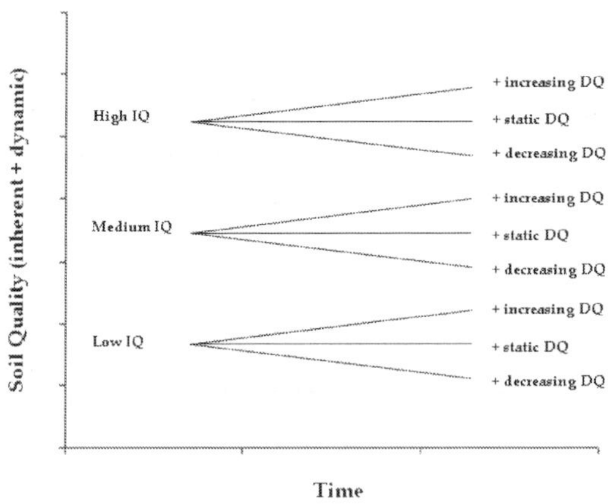

Figure 1. Trends of soil quality according to inherent properties and possible changes in dynamic properties. IQ: inherent quality; DQ: dynamic quality.

Dynamic soil quality depend on land use and management practices and it's also described through use-dependent properties among which organic matter, soil structure, infiltration rate, bulk density,water and nutrient holding capacity, biological factors (micro and macro organisms). Land management practices together with inherent soil quality characterize the trend of soil quality (Figure 1).

Soil quality is a complex matter, with inherent and dynamic properties of soil networking to determine the quality profile of a soil depending of the intended use to be evaluated. So, in order to evaluate the quality, considering the difficulty in measuring functions directly, soil properties are considered indicators to characterize soil quality and to plan the best management practices in order to avoid degradation of soils. Soil properties are usually classified as chemical, physical, and biological characteristics even if stringent classification of many indicators would not be advisable since a soil property can be ascribed to multiple categories:

- Biological indicators give a measurement of the biological activity of the soil. Soil microorganisms and macro organisms such as fungi, bacteria, earthworms and aggregation of them such as mycorrhizae, influence nutrient cycling by decomposing soil organic matter. Their movements into the soil and the results of their biological activity (e.g., cast, mucilage and hyphae growth) also influence the physical status of soil improving aggregation of soil particles, increasing water infiltration and plant root penetration;
- Physical indicators can be inherent (e.g., texture) or dynamic properties able to respond to different management practices. These indicators rely on plant roots, water and air movements into the soil;
- Chemical indicators include mineral solubility, nutrient availability, soil reaction (pH), cation exchange capacity, and buffering action. Chemical properties are determined by the amounts include and types of soil colloids (clays and organic matter).

In Table 1 a list of the main soil quality indicators is presented.

Indicator:			
Category	**Name**	**Description**	**Influence on:**
Physical	aggregate stability	ability of aggregates to resist disintegration when disruptive forces associated with tillage and water or wind erosion are applied	organic matter infiltration root growth resistance to water and wind erosion
	available water capacity	maximum amount of available water for plant uptake. The difference between the Field Capacity and the Permanent Wilting Point	organic matter water storage runoff and nutrient leaching

Indicator:			
Category	**Name**	**Description**	**Influence on:**
	bulk density	refers to soil compaction and indicate the dry weight of soil divided by its volume (g/cm^3)	organic matter structural support water, solute movement aeration
	infiltration	refers to the rate of water infiltration, the velocity at which water enters the soil (space/time)	organic matter water, solute movement and storage
	slaking	refers to the breakdown of large, air-dry soil aggregates ("/>2-5 mm) into smaller sized microaggregates (<0.25 mm) when they are suddenly immersed in water	organic matter stability of soil aggregates resistance to erosion water, solute and air movement in wet condition
	soil crusts	thin, dense, somewhat continuous layers of non-aggregated soil particles on the surface of tilled and exposed soils.	organic matter water, solute and air movement and storage salt content of soil
	soil structures and macropores	refers to the manner in which primary soil particles are aggregated. Pores exist between aggregates (macropores are larger "/>0.08 mm)	organic matter biological productivity water, solute and air movement and storage
	electrical conductivity	gives a measurement of soil salinity. It indicates the ability of a solution to be conductive.	organic matter water availability
Chemical	soil nitrate	indicates the nutrients direct available for plant roots uptake.	organic matter nutrient cycling pollution potential
	soil reaction (pH)	refers to the degree of soil acidity or alkalinity.	biological activity and productivity
Biological	earthworms	population of earthworms are measured by counting the number of earthworms/m^2	organic matter physical structure of soil plant residue depletion water, solute and air movement cycling and distribution in to the soil
	respiration	refers to carbon dioxide (CO_2) release from the soil surface	organic matter biological activity and productivity

Indicator:			
Category	**Name**	**Description**	**Influence on:**
	soil enzymes	from viable cells or stabilized soil complexes, increase the reaction rate at which plant residues decompose and release plant available nutrients.	organic matter nutrient cycling
	total organic carbon	the carbon stored in soil organic matter expressed as percentage of carbon per 100 g of soil	organic matter nutrient cycling

Table 1. Soil quality indicators. Principal source: USDA.

3.1.3. Agricultural management practices: The starting point to improve soil quality

The RED criteria basically determine only the types of ecosystems allowed for conversion into biofuel feedstock production and do not set any requirements on how the feedstock is produced. However, to pursue the sustainability of renewable energy production, especially for biofuels, agricultural choices have a significant effect in short, medium and long term on soil quality, influencing dynamic properties of soil and so modifying the trend of soil quality indicators.

Farmers' production strategy is a key point in sustainable agriculture, since interactions among possible crops, soil types and land uses are complex and strictly dependent on the situation, resulting in a variable response of soil quality to the same agronomic practice.

Table 1 shows that most indicators of soil quality are, in some way (directly or indirectly), correlated to the organic matter content of soil. A positive trend in organic substance results in an improvement of soil structure, an enhanced water and nutrient holding capacity, protection of soil from erosion and compaction, and a good biodiversity of soil organisms. As a consequence, complex relationships which describe soil quality and how it can be improved or at least maintained, could be simplified through the analysis of agronomic practices that influence organic matter.

Tillage has been reported to reduce organic matter concentrations and increase organic matter turnover rates to a variable extent [26]. The negative effect of tillage on soil organic matter originally depends on the fact that organic matter can be physically stabilized, or protected from decomposition, through microaggregation [27]. The periodical perturbation of soil structure by tillage may be the major factor increasing organic matter decomposition rates by exposing the organic matter, otherwise physically protected in microaggregates, to biodegradation [28, 29]. In addition, other tillage dependent factors contribute to reduce the organic matter content (e.g., increase in soil erosion, perturbation of helpful organisms' habitat, soil compaction).

Pest management, in some cases, could have a negative effect on soil quality due to soil organic matter deterioration. Chemical strategy of defence has an undoubted useful effect on agricultural productions, anyway plant protection productsneed to be efficiently managed in order to avoid adverse effects on non-target organisms (pollute water and air). Indeed, soil organic matter dynamics are governed largely by the decomposition activity of soil born organisms which include the decomposition of organic materials, mineralization of nutrients, nitrogen fixation, as well as suppression of crop pests and protection of roots. Chemical strategy should be limited, whenever possible, promoting the introduction of non-chemical approaches (e.g., crop rotations, cover crops, and manure management).

Nutrient management, as described above for pest management, if mismanaged can influence soil quality through adverse effect on soil biodiversity with consequence on organic matter.

Compaction has been reported to cause serious implications for the quality of the soil and the environment. Soil compaction leads to soil degradation enhancing harmful physical, chemical and biological changes in soil properties [30]. First of all, compaction reduces the amount of air, water, and space available to roots and soil organisms. Since deep compaction by heavy agricultural equipment is difficult or impossible to remedy, prevention results strategic.

Uncovered ground leads to increased wind and water erosion, drying and crusting and impoverishment of soil carbon. So crop residues and cover crops play a dual role maintaining resource quality by providing ground cover to prevent wind and water erosion and carbon input to enhance soil quality [31]. A good management of residues and cover crops should prevent delayed soil warming in spring, diseases, and excessive build-up of phosphorus at the surface.

Diversify cropping systems means diversifying cultural practices with the possibility to minimize unavoidable negative practices and maximize virtuous management practices. Different crops provide soil with different root sizes and types, contributing to improved soil structure, varied diet for soil beneficial organisms, improved pest control and organic matter.

In summary the following good agricultural practices, directly related to physical, chemical and biological soil properties (improving or stabilizing them), represent a simple but powerful handbook:

1. Avoid excessive tillage to loosen surface soil, prepare the seedbed, and control weeds and pests.

2. Use an integrated pest management approach (chemical and non-chemical), accompanied by the monitoring of pest, by the respect of application threshold and by the sustainable use of chemicals according to plant protection product labels.

3. Avoid unnecessary use of chemical fertilizers, and use properly organic ones.

4. Prevent soil compaction by repeated traffic, heavy traffic, or traveling on wet soil. Minimize soil disturbance when soil is wet.

5. Keep the ground covered through a good management of crop residues or cover crops.

6. Promote biodiversity across the landscape using buffer strips, small fields, or contour strip cropping. Promote biodiversity over time by using long crop rotations.

4. Biofuels sustainability evaluation: An overview on land use impact assessment

Biofuels are often considered the best solution to face problems connected to the growing use of fossil fuels like global warming or raw material depletion, although currently there is not yet a unique and recommended methodology to assess their environmental sustainability.

An example of a simple method to roughly evaluate a process, mostly from an economic point of view, is to calculate the Net Energy Balance (NEB) that measures the difference between the amount of energy available after the transformation process and the total energy used to produce the fuel. This method provides a quick and simple result that can give useful information about the process, but it can't be considered exhaustive to describe it. It is also used to evaluate the variation in performance of a process in a temporal horizon [32].

To have a more comprehensive and accurate result the most used methodology is the Life Cycle Assessment (LCA).

Thanks to its standardized methodology (ISO14040 and ISO14044) and the increase in quality and number of database available, LCA has recently grown in importance as one of the most complete and reliable methodology to environmental sustainability of biofuels.

Defining the goal and scope of the study, its system boundary and the functional unit (FU) to witch all the study refers, LCA allows to report all input, from raw materials to energy, and output, for example emissions and wastes, related to a process.

Furthermore LCA, considering the entire life cycle, the so called "Cradle to Grave" approach, avoids problems related to the shifting of impacts from a phase to another.

Methods more and more reliable have been developed and offer a vast and diversified range of indicators capable to fully defy impacts both on the environment and on the ecological and human dimensions making LCA a good instrument for decision makers to compare different solutions. Indicators of social and economic impacts have also been developed with the aim to give a result responding to the three pillars of sustainability.

LCA has more and more frequently been used to analyse production and use of biofuels giving indications, recognizing strengths and weaknesses, allowing a continuous improving of the system. Anyway some major challenges for applying LCA on biofuels have been identified in a recent work of McKone et al. (2011) [33]. First of all, there are uncertainties related for example to the large number and type of input used to produce biofuel. This variability is not only linked to the chosen crop but also to the site, to the agricultural practices, the yield, not forgetting the weather. Standardize such a complicated system requires a huge amount of data and parameters can vary seasonally. Another problem is connected to the composition of biofuel that can deeply vary considering different crops, the treatments applied, the technology used to produce biofuel (in addition new technologies and practices to produce biofuel are still under revision and possible improving in the final yield is not yet predictable), such that many effects on the environment and on the human health are not yet well defined. Lastly, but likewise important, the problem strictly connected to the agricultural phase. The use of land to produce crop for biofuel can have multiple involvements, by a side it could led to a change in the use of the soil, for example from forestry to crop, or the use of the harvest could change from food to raw material for biofuel. All these factors can cause emission to air and to water, soil depletion, or an increase in the agricultural areas to face the growing demand of biomass.

Even in cases in which land use aspects are considered extremely important such as in biofuels production, these aspects are not generally assessed in LCA [34].

Many approaches exist, providing suggestions for indicators, which are suitable to model land use impacts in LCA, but few of them provide detailed instructions on how to calculate quantified indicators. The most interesting approaches can be divided in land use quantification using biodiversity and land use quantification using soil functionality. Even if some promising studies on biodiversity within land use have been proposed(see for instance [35, 36]),the functionality approach seems to show more links towards an application in practice. However, biodiversity is an important issue and should be part of the land use impact assessment.

The first Life Cycle Impact Assessment methodologies assessed the use of land by recording the amount of land used (ha or ha* year) as an indication of the impacts [37].

It was common practice especially in LCA of agronomic system to evaluate land use as m^2year, meaning that less impact is linked to less use of land. This approach doesn't consider several aspects in soil quality and it doesn't allow differentiating different impacts due to same occupation but with different intensities (i.e. extensive or intensive cultivation).

However today is acknowledged that changes in the quality of land should also be assessed in LCA.

Land use and associated factors such as ecosystem services and biodiversity are likely either not be addressed or captured only by a crude measure of area. Even when considered, these measures typically reported provide no practical help in our environmental management efforts; nothing that usefully informs about choices and decisions in product development or supply chain management. This leaves a gaping hole in the supposedly holistic picture by a life cycle approach. However it is still not common practice to include land use impacts in LCA studies and an agreed coherent and consistent method has yet to be defined, in the last years some interesting approaches have been proposed.

http://dx.doi.org/10.5772/52255

To date the ILCD Handbook identified (see Table 2) three midpoint methods and underlying models for land use and suggested the use of the one based on Soil Organic Matter (SOM) developed by Milà i Canals. Also five endpoint methods are selected, but all of them are considered too immature by the ILCD Handbook to be recommended.

Midpoint method	Underlying model	Reference
ReCiPe	Not based on a specific model	De Schryver and Goedkoop (2009)
Milàl Canals	Based on Soil Organic Matter	Milài Canals(2007)
Baitz	Based on seven quality indicators	Baitz (2002); Bos, Wittstock (2008)
Endpoint method	**Underlying model**	**Reference**
EPS 2000	Base on species diversity loss and production of wood	Jarvinen and Miettinen (1987)
Ecoindicator 99	Based on species diversity loss	Koller (2000), Goedkoop and Spriensma (2000)
ReCiPe	Based on species diversity loss	De Schryver, Goedkoop (2009)
LIME	Based on species diversity loss and production of wood	Itsubo et al (2008)
Swiss Ecoscarcity	Based on species diversity loss	Koller (2001), Koller and Scholz (2008)

Table 2. Selected midpoints and endpoints methods. Source: ILCD Handbook [38].

ReCiPe: it takes into account the surface area occupied or transformed without any further characterization. In that sense, ReCiPe is not a characterization model but rather a selection of LCI parameters.

Baitz (2002): based on the method proposed by Baitz and further developed by Bos and Wittstock. This method describes the impacts related to land occupation and transformation using an inventory of seven indicators:

- erosion stability
- filter buffer and transformation function for water
- groundwater availability and protection
- net primary production
- water permeability and absorption capacity
- emission filtering absorption and protection
- ecosystem stability and biodiversity.

All indicators are calculated as elementary flows and until now, the different indicators cannot be combined or weighted at the midpoint level.

Milà i Canals (2007): this method considers Soil Organic Matter (SOM) as a soil quality indicator. SOM is qualified as a keystone soil quality indicator, especially for assessing the impact on the fertile land use. It influences properties like buffer capacity, soil structure and fertility. Evaluation of change in one indicator is interrelated to changes in other indicators: the loss of organic matter reduces soil fertility and degrades soil structure.

The LCA practitioners is expected to know the location, the timeframe and the SOM values before and after the land occupation, the SOM value of the reference land system, the relaxation rate and associated SOM values. Based in this, the LCA practitioners are expected to calculate the characterisation factor for the foreground system.

The choice of the method developed by Milà i Canals is based on general scientific criteria and on stakeholder acceptance and applicability to LCI datasets.

The scientific criteria used in the ILCD Handbook are: completeness of scope, environmental relevance, scientific robustness and certainty, transparency, reproducibility and applicability. Degree of stakeholder acceptance and suitability for communication in a business and policy contexts has been also evaluated. Each criterion has been specified through a number of sub criteria.

According to these criteria the method of Milà i Canals resulted the best one but it reached the level III that means the method is recommended but should be applied with caution (level I: recommended and satisfactory; level II: recommended but in need of some improvement).

The method developed by Milà i Canals takes into account both occupation and transformation process of land as function of the area used, the time (duration of occupation and transformation process) and the quality of land before, during and after the land use. Occupation process refers to the use of a land for a certain purpose, assuming no intended transformation of the land properties during this use. In contrast, a transformation process implies the change quality of a land area according to the requirements of a given new type of occupation process. SOM is the indicator for quality definition of a land, but this methodology is easily adjustable to express impact of land occupation and transformation using different quality indicators.

The method defines formulas for occupation and transformation impact and also data sources for SOM and a calculation model to obtain SOM from Soil Organic Carbon (SOC) measurement. The authors provide also considerations about the reference to measure occupation and transformation impact differentiating between attributional and consequential LCA studies. If the LCA is aiming at describing the system's impacts (attributional approach), the study should focus on determining all the impacts caused by the studied activity relative to a situation where this activity is not undertaken. Thus the adequate reference situation for attributional LCA studies is natural relaxation (natural recover of the land quality). On the other hand, if the study aims at evaluating the consequences of changes in land use (consequential LCA), only the changes in land use impacts directly due to the studied system respect an alternative system are considered. Therefore the alternative system becomes the reference. This reference situation should be derived from statistical time series for land use [39].

Milà i Canals method to evaluate land use impact is suitable to be used and improved measuring land quality with different indicators.

For instance, adapting the method of Biatz (2002) to the framework on land use impact assessment set up by Milà i Canals et al.(2007), at the Department for Life Cycle Engineering of Fraunhofer Institute in Germany a tool called LANCA® has been developed to calculate land use indicator values based on ecosystem functions.

The PE-Gabi database (2011) includes several land quality parameters as inventory flows based on this approach.

The LANCA® method requires the user determine five soil quality parameters based on an extensive amount of site-specific soil parameters in order to calculate land quality in different time steps [40].

Quality alteration is defined to be the change in quantifiable land characteristics. Occupation [$m^{2*}y$] is defined as the occupation of the area during the time of its use. Transformation [m^2] is the irreversibly affected area of a land use [41].

To represent land quality and their calculation some of the parameters proposed by Baitz (2002) are used:

- *Erosion resistance*: input data required are soil texture, declination, summer precipitation, type of land use, skeletal content humus content, kind of surface.
- *Mechanical filtration:* inputs needed are soil texture, distance surface to groundwater.
- *Physicochemical filtration:*for its calculation the effective cation exchange capacity and the type of land use are needed.
- *Ground water replenishment:* input data required are soil texture, type of land use, precipitation, evapotranspiration, distance surface to groundwater and declination
- *Biotic production:* depends on declination, spoil texture, skeleton content, nutrient supply, water supply, mean annual temperature and erosion sensibility.

An example from LANCA® method report follows. Figure 2 shows a possible quality alteration due to a defined land use: starting at a quality A in t1, an hypothetic land use change leads to a quality deterioration represented by the situation B in t2. During use, it is assumed, that the quality is constant. After the end of the use, the land quality can recover until reaching the situation C in t3.

After the use the land is able to increase its quality via renaturation or succession from B to C. Accordingly C displays the land quality after regeneration and is thus the reference situation for the calculation of occupation. Transformation is the quality difference of the land after use (C) and before the use (A).

These quality values are inventory flows for the Life Cycle Assessment. To characterize the inventory flows and to adapt them to the characterization of emission-based impact categories, their absolute values are multiplied by the characterization factor c=1,-1 respectively accordingly to display the difference between negative and positive effects of the increase of the land quality parameters values.

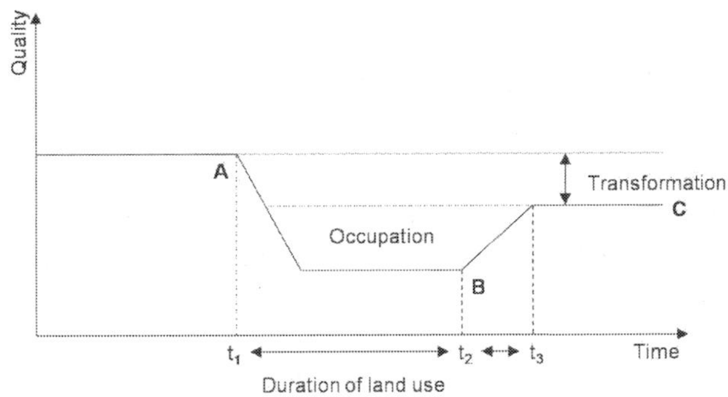

Figure 2. Land occupation and transformation. Source: LANCA® method report figure 2-1.

As a conclusion, to identify land use impacts LANCA® quantifies changes of different aspects of land quality using the influence of the land use on different ecosystem functions.

Same as in all LCIA methods, simplifications of established methods had to be made for being able to adapt them to LCA requirements. For instance differentiations between land use types such as conventional and organic farming are not possible yet.

5. SUSBIOFUEL project: A case-study

In relation to land use change and soil quality the authors present here preliminary results of a three year study "SUSBIOFUEL" (2010-2013) about the feasibility of using new oilseed species for biodiesel production in Italy [42]. The intent is to propose an innovative agronomic solution that may affect the energy and the GHG emission balance in order to achieve a high level of sustainability in the oilseeds production.

As previously discussed in paragraph 2.2, beside GHG emission saving, land use is a critical point in biofuels sustainability evaluation. To set up an agronomic proposal in compliance with the project objectives and the current needs of sustainability in this field, the authors studied a feedstock sustainable production plan facing the issues which follow:

1. WHERE to produce the feedstock for biodiesel?
2. WHAT are we going to produce?
3. HOW are we going to produce?

4. Which soil tillage?
5. Which pest management?
6. Which nutrient management?
7. Which irrigation management?
8. Which cropping system?
9. HOW can we evaluate the sustainability of the biodiesel?

5.1. Agronomical aspects

The agronomic issues listed above were faced and for each of them the authors proposed a solution taking into account that the aim is to produce biodiesel, in Italy, and according to all the sustainability pillars.

5.1.1. Land choice

Marginal lands have received an increased attention by the bioenergy industry as an alternative to cropland for feedstock supply that could help to address the food *versus* fuel debate challenging the industry's further development [43].

The marginality of soils could be ascribed to several different factors so that the term "Marginal land" expresses a wide variety of soil constituting a concept with faint boundaries rather than a definition. For example, aproduction oriented definition establishes that a soil is considered marginal when the ratio of agricultural production to the inputs required to achieve that is low. According to this definition, the combination crop/land needs to be evaluated in order to decide if a soil should or not be considered marginal for a specific crop.

In the context of SUSBIOFUEL project, as well as crop/land peculiarities, the authors evaluated the agronomic management to assess the oilseeds productivity potential of a promising energy non-edible crop.

The authors identified soils rendered marginal by nematode high pest pressure as a good candidate for sustainable production of feedstock for biodiesel market. Using these lands to grow energy crops, even though the lands are less productive, can provide some additional environmental benefits, including restoration of degraded land and carbon sequestration.

5.1.2. Oil crop choice

To face the ethical and economic problem of using edible crops for biodiesel production purposes, the authors made a selection of the most promising crops to be introduced in the Mediterranean zone among the non-edible ones, taking into account that currently the Mediterranean basin comprises also slightly-arid lands [1].

A promising non-edible energy crop seems to be the tobacco (*Nicotiana tabacum*), which currently exists both in the non-GMO and GMO version for improved oilseed yield and resistance factors against herbicides and insects [44]. In addition, from the climatic point of view its taproot system, widely branched, make it able to survive also in arid condition with limited water needing. Considering all these characteristics, its high oil yield makes it very competitive in front of mainstream oil crops as rapeseed, sunflower and soybean.

The remaining meal revealed to be relevant for combustion or to be used as a protein source for livestock. In addition, the presence of consolidate agricultural practices and know-how make clear the advantage of using a well-known species as tobacco as alternative feedstock for biodiesel. The research on "Energy Tobacco" has also found new economies for the agronomic management and practices which currently are under development [45].

5.1.3. Crop rotation system

The large biodiversity of Brassicaceae reveal incipient species, among which *Brassica juncea* and *Brassica carinata*. Besides the potential as raw material for biodiesel, their high content of glucosinolates (GSL) make them able to recover soils made marginal by soil-borne pests as nematodes (e.g. galling nematodes from the *Meloidogyne* genus and cist nematodes from *Heterodera* and *Globodera* genera) [46, 47]. Many researchers also report weed-suppressive effects of Brassicaceae [48, 49] as well as filtering-buffering effects against heavy metals pollution [50].

Considering the characteristics of tobacco, about high adaptability to hard pedo-climatic conditions, the authors tested the possibility to produce tobacco oilseeds for the biodiesel market on marginal soils. According to a sustainable agriculture approach, the harvest should be achieved in full compliance and in an attempt to restore the soil quality. The authors set up a crop rotation between a cover crop with naturally biocidal effects (*B. juncea* or *B. carinata*) and the tobacco oilseed crop. The cultivation and green manuring of the Brassicaceae is expected to improve soil quality, providing soil pest control and organic matter to land. This crop rotation would substitute chemical approaches with highly toxic products (e.g. methyl bromide fumigation[9]; 1,3-dichloropropane) yet under the regulation of law which restrict their uses[10].

Thanks to this practice the soil could be rapidly good enough to produce oilseeds with satisfying yields for industrial destination. Furthermore a reduction in inputs of fertilizers is also expected due to preservation of organic matter content of soil. This practice offers the possibility to rescue soils availability for food production. Indeed, after some cycles of this rota-

9 Methyl bromide is readily photolyzed in the atmosphere to release elemental bromine which contributes to stratospheric ozone depletion. Due to this highly toxic effect, this substance is subject to phase-out requirements of the 1987 Montreal Protocol on Ozone Depleting Substances.

10 COMMISSION DECISION of 20 September 2007 concerning the non-inclusion of 1,3-dichloropropene in Annex I to Council Directive 91/414/EEC and the withdrawal of authorisations for plant protection products containing that substance, Official Journal of the European Union, 25 September 2007.

tion, the pest control and the progressive increase of organic matter should make the soil eligible again for quality productions.

5.1.4. Tillage

Besides the energetic and economic point of view, conventional tillage is reported to have negative long term influence on soil quality. In relation to some *B. napus* cultivars, some studies showed that although the amount of yield was the highest at conventional tillage, it may be more agronomically sustainable to plant under no-tillage or minimum tillage [51].

Considering the crop rotation between a Brassicaceae (*B. juncea* or *B. carinata*) and tobacco to produce tobacco oilseeds, the authors decided to follow a minimum tillage approach. For the Brassicaceae, as pre-sowing land operation, the authors choose to apply only one low input tillage technique among those suggested by published official local specifications for integrated production. At Brassicaceae flowering time, the green manuring of this crop was tested to evaluate the possibility of exploiting this operation to also prepare the soil surface for successive transplant of tobacco plantlets.

5.1.5. Pest management

Soil born pest, and nematode in particular, are the main issues of marginal soils chosen for the agronomic system to be tested by the authors. Nematodes are worm-like invertebrates known since a long time but the development of plant protection products effective against these parasites is still a challenge of research and development for agrochemical industries. From one side the agrochemicals dedicate low budget for this field of research compared to other sectors such as insecticides and fungicides, but from the other side, researchers have to face some hot pointspeculiar to nematicides development which can be summarized as follow [52]:

1. they live confined to soil or within plant roots, so that the delivery of a chemical to the immediate surroundings is difficult,
2. the outer surface of nematodes is a poor biochemical target and is impermeable to many organic molecules,
3. the delivery of a toxic compound by an oral route is nearly impossible because most phytoparasitic species ingest material only when feeding on plant roots.

For all these reasons, nematicides have tended to be broad-spectrum toxicants possessing high volatility, resulting in highly toxic compounds for the environment (e.g. ozone depletion) and biodiversity subjected to progressive withdrawal of authorizations worldwide.

Some selectivity improvement is being achieved by using agrochemicals with a less wide spectrum, for example fungicides against nematodes but anyway currently the management of plant-parasitic nematodes through alternative strategies seems to become more and more pressing. Among the non-chemical alternatives, biofumigation and solarization are outstanding, and so are crop rotation, use of resistant varieties, and grafting, which are effective means of control when included in an integrated crop management system. According to this school of thought, the authors tested the possibility to halt the marginalization of contaminated soils introducing a crop rotation system between a Brassicaceae, able to fight nematodes and improving soil organic matter at the same time, and a promising oilseeds non-edible crop, the tobacco plant.

5.1.6. Nutrient and irrigation management

Soil fertility can be improved by managing nutrient stocks and flows. A range of intervention strategies are available to farmers. Land users tend to purchase and use fertilizer nutrients in areas with good market access and higher agricultural potential. Combining manures with inorganic fertilizers can result in significant synergy and increased nutrient and water use efficiencies [53].

The authors decided to exploit the green manure as partial source of nutrients, complementing the nutrient needs of the successive oilseeds crop with organic poultry manure. To optimize the agronomic system (crop/soil/management) from the nutrient point of view, the authors also tested the possibility to apply inorganic fertilizer, sharing the total dose rate of application on the two crops: half on the oilseed crop (the crop which bring the harvest) and the other half on the Brassicaceae aiming at increasing its biomass production to maximize the biofumigant effect of the crop.

The Brassicaceae/tobacco crop rotation, taking into consideration the climate of the Mediterranean basin and in particular those of experimental trial sites chosen by the authors, should not need high input of water. This characteristic depends on the peculiarities of crops involved in the crop rotation and it is in favour of a sustainable agronomic management. Indeed, the Brassicaceae take advantage of the water naturally supplied by the winter season of growing, while the tobacco plant due to itstaproot system, widely branched, is able to survive also in arid condition with limited water needing. The authors tested the production supplying only emergency irrigation for the tobacco crop.

5.1.7. Experimental details of field trials

The agronomic rotation Brassicaceae/tobacco was tested under a wide range of situations. Three field triallocations were chosen for seasons 2010 and 2011, taking into account Italy's wide latitudinal distribution (two locations in the north and one location in the south)[11].After two years of experimentation, the author decided to maintain two of these locations, in order to concentrate the attention on the most representative sites[12].Experimental design was thought to produce oilseeds from *N. tabacum* and from traditional oilseed crops (sunflower, soybean, rapeseed in 2010-2011 and soybean in 2012), used as comparison to validate themethodology. For the third year of experimentation, the authors decided to dedicate more land to the tobacco, limiting the space available for the traditional oilseed crops. They

11 Altedo (BO), Vaccolino (FE) and Santa Margherita di Savoia (FG).

12 Altedo (BO), Santa Margherita di Savoia (FG).

chose to compare tobacco only with soybean, since physic-chemical characteristics of its oil is the most comparable. Each field was divided into two parts and the Brassicaceae (*B. Junceain* 2010 or *B. carinata* in 2012-2012, depending on the sowing time) were sown only in one halfof the field. To maximize the biofumigant effect, green manuring of the Brassicaceae biomass was carried out when the crop reached flowering. After this, sowing of soybeanas well as the transplant of tobacco plantlets took place in both parts of the field. In order to make the proposal as flexible as possible, four different fertilization treatments on the oilseed crops were used in 2010-2011: low input (30 kg/ha of chemical fertilizer[13]), medium input (90 kg/ha of chemicalfertilizer), high input (140 kg/ha of chemical fertilizer) or organic input (10000 kg/ha of poultry manure). In 2012, to test the possibility of increase the biofumigat effect of the Brassicaceae, the author decided to split the total amount of fertilization dosage on both crops in rotation: a half on the Brassicaceae and the other half on the oilseed crop (tobacco or soybean).In all the field trials, untreated plots were set up as control. All field tests were conducted under Good Experimental Practices (GEP).

Yet in the first year of experimentation the authors assured that the green manure of Brassicaceaedo not increase the sulphur content of the successive crop and its oil [1], which is therefore suitable for biodiesel production[14]. To evaluate the effect of the green manure of Brassicaceaeon nematode infection, countings of *Meloidogyne* spp. were carried out on soil samples taken from both sides of the field while effects on yield of crops grown in succession were monitored recording the fresh weight per hectare of plant biomass from both sides of the field (or when possible the seed yield). The authors also checked the weed-suppressive effects of Brassicaceae.

5.1.8. Results and discussion on agronomical aspects

Research on alternative biofuel faces the increasing demand for energy requirements by means of a more sustainable energy supply. From this point of view, greenhouse gases saving are expected from biofuels.

The first year of experimentation showed that the use of *B. juncea* as green manure does not influence the sulphur content in sunflower seeds and oil, suggesting no sulphur accumulation occurs in succeeding crops. Theplants grown in succession of *B. juncea* resulted in higher biomass. This could be due either to the increase in the organic matter content or to the pest control. Indeed, counting of nematodes revealed a strong effect of the green manure of *B. juncea* on nematode control. These data trends were confirmed in the second year also for *B. carinata*: the authors observed that the positive effects on biomass correspond to a similar effect on seed yields. In addition the weed suppressive effect of the green manure with *B. carinata* was also observed and reported in Figure 3.

The third year of experimentation will end in 2013, so data from this season are not available.

---

13 NPK fertilizer was composed of 46% urea; 48% P2O5; 50% K.

14 The contents of this element in the final product must be under 10 ppm (UNI EN 14214 - Automotive fuels. Fatty acid methyl esters (FAME) for diesel engines. Requirements and test methods).

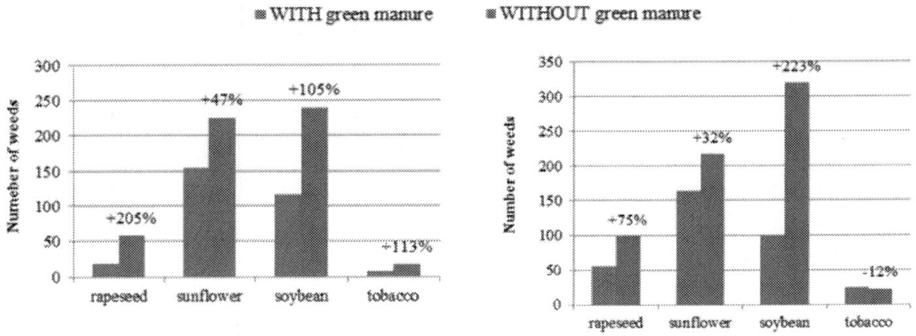

Figure 3. Field trials carried out in Altedo (BO) on the left and in Vaccolino (FE) on the right. Assessment of weeds.

5.2. Evaluation of sustainability aspects

In the SUSBIOFUEL project LCA has been chosen as sustainability evaluation and decision making tool.

Several scenarios have been evaluated to assess the environmental burdens related to different feedstock and agronomic management system.

In this chapter, two main scenarios have been selected to compare the production of biofuel using two different crops. Furthermore the possible improving in soil quality due to the agronomic practices proposed will be evaluated.

Both scenarios consider a crop rotation between a cover crop with naturally biocidal effects (*B. juncea* or *B. carinata*) and the oilseed crop, soybean or tobacco. For tobacco crop, a preliminary greenhouse phase was considered. Authors choose to set tillage, nutrient and irrigation management at the lowest level suggested by the good agricultural practices as explained in paragraph 5.1. The use of the rotation with brassica has been considered sufficient and no other pesticides were added in the model.

The functional unit is 1 litre of oilseed, system boundaries goes from the seed preservation to the oil production.

Data used in this study are both collected directly on the experimental fields and from good agricultural practices (GAP) vade mecum [54], data from Ecoinvent Database were also used.

The method chosen to evaluate the potential impacts of the system is the CML 2001 (updated in November 2009) using GaBi LCA software. The following impact categories have been assessed:

- Abiotic Depletion (ADP), expressed in kg Sb-Equiv

http://dx.doi.org/10.5772/52255

- Acidification Potential (AP), expressed in kg SO2-Equiv.
- Eutrophication Potential (EP) expressed in kg Phosphate-Equiv.
- Global Warming Potential (GWP 100 years), expressed in kg CO2-Equiv.
- Ozone Layer Depletion Potential (ODP, steady state), expressed in kg R11-Equiv.
- Photochem. Ozone Creation Potential (POCP), expressed in kg Ethene-Equiv.

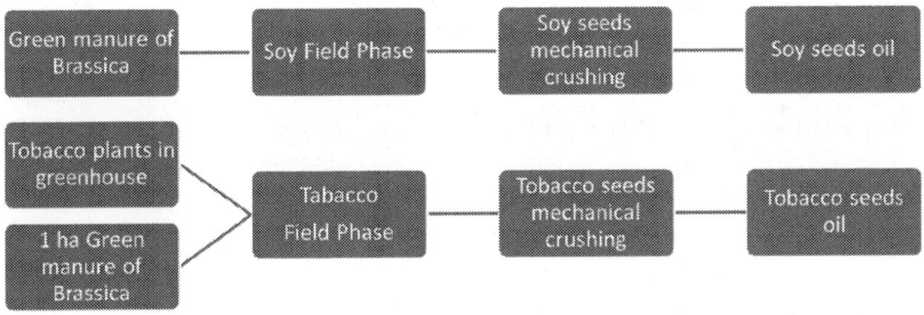

Figure 4. Flowcharts of soy and tobacco field phase of production.

In order to observe the holistic aim of Life Cycle Assessment, the authors did a special effort trying to include considerations about Land Use impact in their analysis aware of the capital environmental importance of this issue in biofuels sustainability evaluation.

The method of Milà i Canals has been considered as the most consistent with the scope of the study. The ultimate goal of the project is the sustainable production of biodiesel with a contemporary rescue of marginal soils. The Soil Organic Matter indicator chosen by Milà i Canals constitutes a trade-off between an easy to measure and a representative indicator of soil quality.

This indicator could confirm the expected increase in SOM in marginal soils after the crop rotation and management system defined in the SUSBIOFUEL project. SOM evaluation could help also in GHG emission assessment (biofumigant green manure practice already showed important saving in CO_{eq} emissions, calculated with a simplified LCA approach [55].

Measures of SOM in the marginal soil before the crop rotation for each scenario represent the reference situation for the land use indicator, while measures of SOM during and after the SUSBIOFUEL tests constitutes the quality value needed for occupation and transformation impacts calculations.

In order to have information useful for decision making between different project options, the authors recurred to the Soil Conditioning Index.

The Soil Conditioning Index (SCI) is a Windows-Excel based model developed by NRSC (Natural Resource Conservation Service) US Department of Agriculture to estimate soil carbon trends. This tool can predict the consequences of cropping system and tillage practices on the status of soil organic matter in a field[56]. SCI estimates the combined effect of three variables on trends in organic matter, as a simple weighted average.

The soil conditioning index formula is:

$$SCI = (OM \times 0.4) + (FO \times 0.4) + (RF \times 0.2) \tag{1}$$

Where OM accounts for organic material returned to the soil (as a function of biomass produced), FO represents tillage and field operation effects and ER is the sorting and removal of surface soil material by sheet, rill and wind erosion.

Controlling erosion and building organic matter do not guarantee good soil quality, but in most cropping situations they are prerequisites to improving and protecting soil quality and productivity. The SCI is a quick way to characterize the organic matter dynamics of a farming system and can help assess good soil management. The following information is needed about the field to calculate the SCI:

- Soil texture
- Climate data
- All crops in the crop rotation
- Typical yield for each crop
- Additional applications of organic matter or removals of organic matter
- All field operations (tillage, fertilizer and manure application, harvest)
- Rates of erosion

The SCI can predict if a particular management system will have a positive or negative trend in SOM. If the SCI value is negative, soil organic matter is predicted to be declining under a given production system, and corrective measures should be planned. If the SCI value is zero or positive, soil organic matter is predicted to be stable or increasing.

The Soil Conditioning Index represents a support to plan and design conservation crop rotation and residue management practices when low organic matter, surface crusting or erosion are identified as concerns and it helps producers in changes in SOM monitoring or prediction.

The use of this semi-quantitative tool allows running several what-if scenarios which results could be useful to drive decisions taken in the project.

Understanding processes that affect soil quality can guide in management decisions and practices that will maintain or improve the soil resource.

Appropriate management strategies can significantly reduce the payback period and enhance greenhouse gas benefits associated with biofuel production system.

5.2.1. Results and discussion on sustainability evaluation

Results obtained represent the comparison between the two scenarios which don't include yet land use impact category, since further researches are needed on this topic. As shown in Figure 5 for all impact categories selected the tobacco oil production, despite the greenhouse phase, generate less than 30% of potential impacts in comparison with soya oil. Results have been normalizedto find which impact categories are the more important. Figure 5 shows that these processes have a great effect on global warming potential, nevertheless, the other impact categories, apart from Ozone Layer Depletion Potential, have anyhow to be considered.

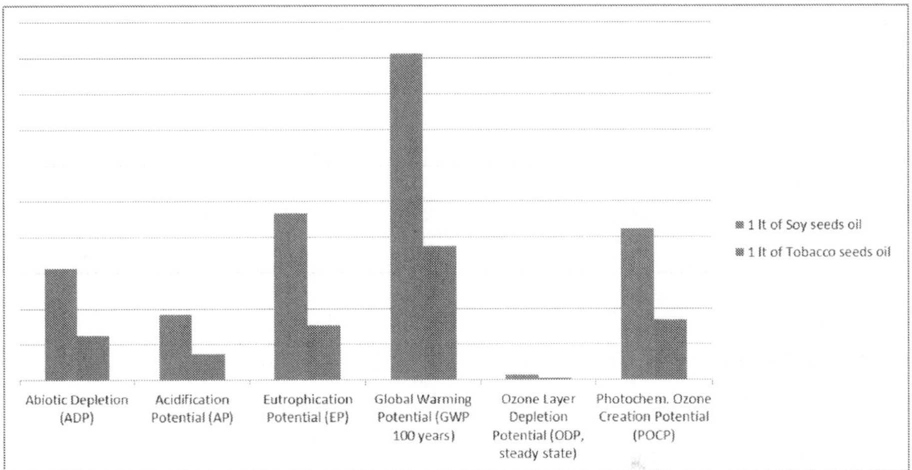

Figure 5. Scenario results normalization[15].

Even if these results have to be considered preliminary, they give the indication about the validity of the use of tobacco as non-edible oilseed crop.

6. Conclusion

The use of oilseeds derived from non-edible crops represents a first step in order to increase the sustainable profile of biodiesel production.

Beyond the use of non-edible crops, to face the land use change (one of the main criticisms-about biodiesel sustainability) the authors propose to set up the non-edible system of pro-

15 For confidentiality reasons, unit of measures and numerical results are detailed on SUSBIOFUEL internal report.

duction onmarginal soils. According to the state of the art on soil quality properties and indicators, soil organic matter is outstanding, so each phase of production was thought to respect, and if possible improve, this property of the soil. In this scenario the aim of the production is not only the oilseeds harvest, but also the maintenance and if possible the rescue of the full soil functionality. Taking into account these considerations, the authors analysed which oilseed crop would have been the most suitable, which kind of marginal soil, which the best agronomic practices to follow in this particular situation. In this publication the authors present a case study which contributes significantly to a wider portfolio of land-use strategy. Tobacco was individuated as the most promising non-edible oilseed crop and the possibility to produce tobacco oilseeds from soils rendered marginal by nematode infestation was analysed.

The authors verified that the green manure of *B. juncea*or *B. carinata* (depending on the sowing period) resulted in nematode infestation drastically decreasing and improved soil quality reflected in higher seeds yield of crops in agronomic succession. In addition weed-suppressive effect of this agronomic practice was shown, avoiding chemical herbicide applications for this agronomic system. The restoring of soil fertility avoiding the fumigant usage, and in the meantime the generation of income from non-edible vegetable oils, assure the ethical, economic and environmental sustainability of the solution. It should be also considered that food production from marginal soils would worsen soil depletion and nematodes infestation.

Preliminary results, according to the traditional LCA, confirmed that tobacco is a promising non-edible oilseed crop according to the agronomic practices applied, for those soils rendered marginal by nematode infestation.

This study reports the impact of cover crops and their green manures on the density and damage of root-knot and lesion nematodes to oilseed crops, as well as those of tillage, soil amendments, crop rotation, and cover crops on oilseeds yield and root rot severity. The incidence and severity of root diseases is an indirect assessment of soil health for specific commodity/soil use [57]. In order to evaluate the sustainability of this scenario through the LCA methodology, it would be relevant to estimate the benefits on soil quality of the agronomic system proposed. For this reason the authors are studying how to complete the information supplied by the traditional land use indicator.

Policy strategies will be needed to increasingly shift abandoned or low biodiversity value marginal lands to this kind of ecologically-friendly practices.

Acknowledgements

The authors gratefully acknowledge financially support by Italian Ministero delle Politiche Agricole, Alimentari e Forestali (project SUSBIOFUEL – D.M. 27800/7303/09).

Author details

D. Barnabè[1], R. Bucchi[1], A. Rispoli[1], C. Chiavetta[2], P.L. Porta[3], C.L. Bianchi[4], C. Pirola[4], D.C. Boffito[4] and G. Carvoli[5]

1 Agri 2000 Soc. Coop., Bologna, Italy

2 University of Bologna – DICAM, Bologna, Italy

3 ENEA – UTVALAMB, Bologna, Italy

4 Università degli Studi di Milano – Dipartimento di Chimica, Milano, Italy

5 Khemistar S.r.l, Novara, Italy

References

[1] Bianchi C.L. et al. Non Edible Oils: Raw Materials for Sustainable Biodiesel. Biodiesel – Feedstocks and Processing Technologies: InTech; 2011.

[2] Searchinger T. Use of U.S. Croplands for Biofuels Increases Greenhouse Gases Through Emissions from Land-Use Change. Science 2008;319(5867): 1238-1240.

[3] Kalita D. Hydrocarbon plant – New source of energy for future. Renewable andSustainable Energy 2008;12: 455-471.

[4] Food and Agriculture Organization. FAO: Food price index data from the UN's. http://www.fao.org/ (accessed 12 July 2012).

[5] International Monetary Fund. IMF: World Economic Outlook: Crisis and Recovery, April 2009. http://www.imf.org/external/index.htm (accessed 12 July 2012).

[6] Azar C., Emerging scarcities - Bioenergy-food competition in a carbon constrained world. In: Scarcity and growth revisited: natural resources and the environment in the new millennium. John Hopkins University Press; 2005. p359-377.

[7] Lagi M., Bar-Yam Yavni, Bar-Yam Yaneer, UPDATE July 2012 — The Food Crises: The US Drought, July 23, 2012.

[8] Delzeit R., Klepper G., Lange M. Review of IFPRI study "Assessing the land use consequences of European biofuel policy and its uncertainties", Kiel Institute for World Economy. http://www.ebb-u.org/EBBpressreleases/Review_iLUC_IfW_final.pdf (accessed20 July 2012).

[9] Fritsche U.R., Hennenberg K., Hunecke K. Sustainability Standards for internationally traded Biomass The "iLUC Factor" as a Means to Hedge Risks of GHG Emissions

from Indirect Land Use Change - Working Paper -. Öko-Institut, Darmstadt Office 2010. http://www.oeko.de/oekodoc/1030/2010-082-en.pdf (accessed 20 July 2012).

[10] HyungtaeK., Seungdo K., Bruce E. Dale. Biofuels, Land Use Change, and Greenhouse Gas Emissions: Some Unexplored Variables. Environmental Science Technology 2009;43 (3) 961–967.

[11] Searchinger T., Heimlich R., Houghton R. A., Dong F., Elobeid A., Fabiosa J., Tokgoz S., Hayes D., Yu T.H. Use of U.S. Croplands for Biofuels Increases Greenhouse Gases Through Emissions from Land-Use Change. Science 2008;319(5867) 1238-1240.

[12] Geyer R., Lindener J.P., StomsD.M.,Davis W.F., Wittstock B. Coupling GIS and LCA for biodiversity assessment of land use. Part 2: Impact assessment. International Journal of LCA 2010; 15:692-703.

[13] United Nations Framework Convention on Climate Change. UNFCCC: History. http://unfccc.int/2860.php (accessed 12 July 2012).

[14] Intergovernmental Panel on Climate Change. IPCC: Fourth Assessment Report (AR4). http://www.ipcc.ch/(accessed 12 July 2012).

[15] Ibrahim D. Renewable energy and sustainable development: a crucial review. Renewable and Sustainable Energy Reviews 2000;4(1) 157–175.

[16] European Commission. EU: Renewable Energy: http://ec.europa.eu/energy/renewables/biofuels/sustainability_schemes_en.htm (accessed 12 July 2012).

[17] Warkentin B.P., Fletcher H.F. Soil quality for intensive agriculture: conference proceedings, 1977, Proc. Int. Sem. on Soil Environ. andFert. Manage. in Intensive Agric. Soc. Sci. Soil and Manure, Tokyo, Japan, Natl. Inst. of Agric. Sci.

[18] Larson W.E., Pierce F.J. Conservation and enhancement of soil quality. In: Dumanski J., Pushparajah E., Latham M., Myers R. (ed.) Evaluation for Sustainable Land Management in the Developing World Vol. 2. Technical Papers, p175– 203; 1991: conference proceedings, Sept. 15-21, 1991, Int. Board for Soil Res. and Management, Bangkok, Thailand.

[19] Karlen D.L., Mausbach M.J., Doran J.W., Cline R.G., Harris R.F., Schuman G.E. Soil quality: A concept, definition, and framework for evaluation. Soil Sci. Soc. Am. J. 1997;61 4-10.

[20] Pierce F.J., Larson W.E., Dowdy R.H., Soil loss tolerance: maintenance of long-term soil productivity. Journal of Soil and Water Conservation 1984;39 136–138.

[21] Larson W.E., Pierce F.J. Conservation and enhancement of soil quality. In: Dumanski J., Pushparajah E., Latham M., Myers R. (ed.) Evaluation for Sustainable Land Management in the Developing World Vol. 2. Technical Papers, p175– 203; 1991: conference proceedings, Sept. 15-21, 1991, Int. Board for Soil Res. and Management, Bangkok, Thailand.

[22] Larson, W.E., Pierce, F.J., The dynamics of soil quality as a measure of sustainable management. In: Doran J.W. (ed.) Defining Soil Quality for a Sustainable Environment. Soil Science Society of America Special Publication No. 35. Soil Science Society of America and American Society of Agronomy, Madison, WI; 1994. p37– 51.

[23] Arshad M.A., Coen G.M. Characterization of soil quality: physical and chemical criteria. American Journal of Alternative Agriculture 1992;7 25– 31.

[24] Haberern J. Coming full circle—the new emphasis on soil quality. American Journal of Alternative Agriculture 1992;7 3– 4.

[25] Hans J. et al., editor. Factors of Soil Formation: A System of Quantitative Pedolog. Mineola, N.Y.: Dover; 1941.

[26] Kemper W. D., Schneider N. N., Sinclair T. R. No-till can increase earthworm populations and rooting depths. Journal of Soil and Water Conservation 2011;66(1) 13A-17A.

[27] Six J., Conant R. T., Paul E. A., Paustian K. Stabilization mechanisms of soil organic matter: Implications for C-saturation of soils. Plant and Soil 2002;241(2) 155-176.

[28] Reicosky D.C., Kemper W.D., Langdale G.W., Douglas C.L., Rasmussen P.E. Soil organic matter changes resulting from tillage and biomass production. Journal of Soil and Water Conservation 1995;50(3) 253-261.

[29] Balesdent J., Chenu C., Balabane M. Relationship of soil organic matter dynamics to physical protection and tillage. Soil and Tillage Research 2000;53(3–4) 215–230.

[30] Soane B.D., van Ouwerkerk C., Implications of soilcompaction in crop production for the quality of the environment. Soil and Tillage Research 1995;35(1–2) 5–22.

[31] Reicosky D.C., Forcella F. Cover crop and soil quality interactions in agroecosystems.

[32] Bureau, J.-C., Disdier, A.C., Gauroy, C., Tréguer, D. A quantitative assessment of the determinants of the net energy value of biofuels (2010) Energy Policy, 38 (5), pp. 2282-2290.

[33] McKone, T.E., Nazaroff, W.W., Berck, P., Auffhammer, M., Lipman, T., Torn, M.S., Masanet, E., Lobscheid, A., Santero, N., Mishra, U., Barrett, A., Bomberg, M., Fingerman, K., Scown, C., Strogen, B., Horvath, A. Grand challenges for life-cycle assessment of biofuels (2011) Environmental Science and Technology, 45 (5), pp. 1751-1756.

[34] Cherubini F., Stromman A.H. Life cycle assessment of bioenergy system: state of the art and future challenges. Bioresource Technology 2011; 102(2):437-451.

[35] Brentrup F., Hanninghof J., Lammel J. andKuhlmann H. Life Cycle Impact Assessment of Land Use based on the Hemeroby concept. International Journal of LCA 2002; 7 (6) 330-348.

[36] Bare J. Recommendation for land use impact assessment: first step into framework, theory and implementation. Clean Techn Environ Policy 2011; 13:7-18.

[37] Milà i Canals L., Romanya' J., Cowell S.J. Method for assessing impacts on life supportfuction (LSF) related to the use "of fertile land" in life cycle assessment (LCA). Journal of Cleaner Production 2007; 15 1426-1440.

[38] European Commission-Joint Research Centre - Institute for Environment and Sustainability: International Reference Life Cycle Data System (ILCD) Handbook- Recommendations for Life Cycle Impact Assessment in the European context.

[39] Milà i Canals L., Bauer C., Depestele J., Dubreuil A., FreiermuthKnuchel R., Gaillard G., Michelsen O., Muller-Wenk R., RYdgren B. Key elements in a framework for land use impact assessment within LCA. International Journal of LCA 2007; 12 (1): 5-15.

[40] Mattila T., Helin T., Antikainen R. Land use indicators in life cycle assessment. A case study on beer production. International Journal of LCA 2012; 17:277-286.

[41] Beck T., Bos U., Wittstock B., Baitz M., Fischer M., Sedlbauer K., LANCA – land use indicator value calculation in life cycle assessment. Fraunhofer, Stuttgart 2010.

[42] SUSBIOFUEL project ("Studio di fattibilità per la produzione di biocarburanti da semi oleosi di nuove specie e da sottoprodotti o materiali di scarto„ – D.M. 27800/7303/09), financially supported by the Ministry of Agricultural, Food and Forestry Policies – Italy.

[43] Milbrandt A., Overend R.P. Assessment of Biomass Resources from Marginal Lands in APEC Economies 2009. APEC#209-RE-01.4.

[44] Fogher, C. Sviluppo di un Ideotipo di Tabacco per la Produzione di Seme da Usare a Fini Energetici. Presentedat the workshop "La Filiera del Tabacco in Campania: Ristrutturazione e/o Riconversione. La ricerca come motore per l'innovazione tecnologica, la sostenibilità e la competitività della filiera", Portici, Italy, February, 2008.

[45] University of Kentucky, College of Agriculture. UKAg: Planning And Constructing An Outside Float Bed For Tobacco Transplant Production. http://www.uky.edu/Ag/TobaccoProd/pubs/Float%20basics.pdf (accessed 16 July 2012).

[46] Romero E., Barrau C., Romero, F. Plant Metabolites Derived from Brassica spp. Tissues as Biofumigant to Control Soil Borne Fungi Pathogens. Proceedings of the 7th International Symposium on Chemical and non- Chemical Soil and Substrate Disinfestation, ISBN 9789066056237, Leuven, Belgium, September 2009.

[47] Curto G., Lazzeri L. Brassicacee, un Baluardo Sotterraneo Contro i Nematodi. Agricoltura, 2006;34(5) 110-112.

[48] Al-Khatib K., Libbey C., Boydston R. Weed Suppression with Brassica Green Manure Crops in Green Pea. Weed Science, 1997;45(3) 439-445. ISSN 1939-7291.

[49] Krishnan G., Holshauser D. L., Nissen S. J. Weed Control in Soybean (Glycine max) with Green Manure Crops. Weed Technology, 1998;12(1) 97-102.

[50] Palmer C.E., Warwick S., Keller W. Brassicaceae (Cruciferae) Family, Plant Biotechnology, and Phytoremediation. International Journal of Phytoremediation, 2001;3(3) 245–287. ISSN: 1549-7879.

[51] Torabi H., Naghdibadi H.A., Omidi H., Amirshekari H., Miransari M. Effects of soil tillage, canola (Brassica napus L.) cultivars and planting date on canola yield, and oil and some biological and physical properties of soil. Follow Archives of Agronomy and Soil Science 2008;54(2) 175-188.

[52] David J. Chitwood, USDA-ARS. Nematicides. Beltsville, Maryland. Pag. 1104 http://naldc.nal.usda.gov/download/43874/PDF(accessed 13 July 2012).

[53] Piha M. I., Optimizing Fertilizer Use and Practical Rainfall Capture in a Semiarid Environment with Variable Rainfall. Experimental Agriculture 1993;29 405–415.

[54] Ribaudo F. Prontuario di Agricoltura per il geometra, il perito agrario e l'agrotecnico. Bologna, Italy: Ed. Edagricole. 1993.

[55] Lazzeri L., D'Avino L., Gies D. 2010. Additional Benefits of the Efficacy in Containing Soilborne Pest and Pathogens with Biofumigation Plants and Materials. Proceedings of the VIIth International Symposium on Chemical and Non-Chemical Soil and Substrate Disinfestation Leuven – Belgium. Acta Horticulture 883, 323-329.

[56] Natural Resources Conservation Service, US Department of Agriculture. Guide to using the Soil Conditioning Index 2002.

[57] Abawi G.S, Widmer T.L. Impact of soil health management practices on soilborne pathogens, nematodes and root diseases of vegetable crops. Applied Soil Ecology, 2000;15(1) 37–47.

Chapter 2

Tropical Agricultural Production, Conservation and Carbon Sequesteration Conflicts: Oil Palm Expansion in South East Asia

Minerva Singh and Shonil Bhagwat

Additional information is available at the end of the chapter

http://dx.doi.org/10.5772/52420

1. Introduction

1.1. Biodiversity conservation and agricultural conflicts

Agricultural expansion remains one of the leading causes of deforestation, biodiversity losses and environmental degradation across the world, especially in the tropics (Angelsen et al., 1999 and Norris, 2008). From 1961 to 1993, the world population increased by 80% (Goklany, 1998). Due to the rapidly expanding human populations large increases in the supply of agricultural products are required, which may lead to the transformation of many landscapes (including biodiversity-rich tropical rainforests) to agricultural landscapes (Ewers et al., 2009). The quest for further land for agricultural production has already caused significant habitat loss and fragmentation, posing substantial threats to the world's biodiversity, forests and ecosystems (Goklany, 1998). According to Matson and Vitousek (2006), many involved in conservation believe that the twin goals of increasing agricultural production and conservation are fundamentally incompatible. Indeed representatives of developing countries (where the tropical forests and majority of the world's biodiversity reside) argue that their developmental needs are partially met by deforestation, since it provides arable land and timber export revenues (Leplay et al., 2010). Further, agricultural revenues accrued from cultivating plantation crops, such as timber, palm and coffee, are significant drivers of deforestation in many parts of the tropics (Kaimowitz and Angelsen, 1998).

Agricultural expansion is often accompanied by significant reductions in tree cover, fallow vegetation, habitat diversity and forest connectivity. Habitat loss and modification are considered to be among the most important drivers of species loss worldwide (Pimm and Raven 2000). Conservative estimates of the effects of anthropogenic land-use changes on global breeding bird populations, from a global level meta-analysis, indicate that they have caused losses of between a fifth and a quarter of pre-agricultural bird numbers, across a wide range of temperate and tropical land-use types (Gaston et al., 2003). Large-scale deforestation is also a significant driver of habitat fragmentation, i.e. breaking up of formerly contiguous habitat areas (Laurance and Peres, 2006). Forest fragmentation in turn increases the susceptibility of habitats to edge effects, isolation and disruption of vital ecological processes, followed by species losses from individual forest fragments (Lovejoy, 1986). Forest fragmentation also reduces the ability of forest-dependent species such as understory birds, herbivorous insects and dung beetles to transverse the landscape to reach more suitable habitats (Laurance et al., 2011). Thus, habitat loss, modification and fragmentation synergistically drive biodiversity losses (Hilty et al., 2006).

At a regional level, Singapore represents an extreme case of habitat loss, with a loss of 95% of its primary rainforest. This extreme habitat loss is correlated with high rates of species losses at all taxonomic levels. Extinction rates have been substantially higher for forest specialists (33%) than for species that can persist in open/modified habitats and forest edges (7%). However, they have been highest for highly habitat-specific taxa, e.g. 67% and 59% for habitat-specific birds and mammals, respectively, in Singapore (Brook et al., 2003).' Note, we have deleted the references to 'the island', because Singapore consists of 63 islands, and it is not clear whether you mean the main island or all the islands. Alternatively, you could clarify whether you mean the main islands or the islands generally. There have been considerable efforts to protect biodiversity in reserves and parks, but many of these refuges are small, fragmented, isolated or poorly protected (Harvey et al., 2008). Besides having inadequate dimensions, most protected areas are embedded within agricultural landscapes, and the buffer zones are inadequate to alleviate effects of fragmentation, contamination by agrochemicals, hunting and unsustainable or illegal logging (DeFries et al. 2005).

Biodiversity conservation in many parts of the tropics is threatened by rapid habitat loss and destruction (Sodhi et al., 2004). South East Asia, which hosts four biodiversity hotspots, is especially vulnerable and it is estimated that 42% of the region's species could be extinct by 2100 (Sodhi and Brooks, 2006). In recent decades, in addition to swidden agriculture, rapid urbanization and timber logging, the expansion of oil palm monocultures has been one of the most significant drivers of deforestation and habitat loss in South East Asia (Corlett and Primack, 2008) and other tropical countries. The coverage of oil palm plantations in Indonesia and Malaysia in 2008 is illustrated in Figure 1.

From figure 1 it can be seen that the area under oil palm plantations has risen significantly since 1990 and this expansion is a major driver for forest loss. The rapid growth of the oil palm industry can be attributed to its uses, which include food products, and other consumer goods such as cosmetics, industrial lubricant and biofuels (Corley & Tinker 2003). The demand for vegetable oil based biofuel (such as oil palm based biodiesel) is expected to rise exponentially over the coming decades. This in turn is expected to cause an increase in forest loss as natural forests are removed to make way for oil palm monocultures (Butler, 2007). This paper presents an overview of the impact of oil palm plantations on biodiversity and carbon storage. Subsequently, the paper describes the role of human modified landscapes on conservation of biodiversity and carbon stocks. Finally, the paper provides a detailed overview and description of the different strategies and landscape configurations that may allow for the reconciliation of conservation priorities with oil palm expansion.

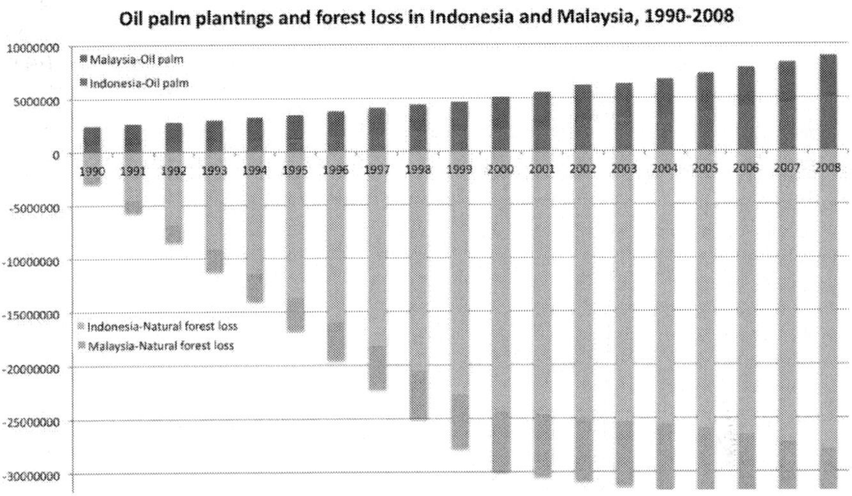

Figure 1. Total forest covered by oil palm plantations, and hence losses of natural forests, in Indonesia and Malaysia (Mongabay, 2012)

2. Oil palm expansion and associated biodiversity & carbon storage challenges

In recent decades, oil palm has become one of the most rapidly expanding crops in the world (Koh & Wilcove 2007) and occupies 13.8 million hectares in the tropics (FAO, 2009). The rapid growth of oil palm cultivation can be attributed to its value as both a source of food and as raw material for various consumer products, such as cosmetics, industrial lubricants and biofuels (Corley & Tinker 2003). This rapid oil palm monoculture expansion has been taking place at the cost of primary forests in the tropics (Corley, 2009).

In Indonesia and Malaysia, in the last 20 years deforestation has been driven by expansion of agricultural plantations, especially oil palm monocultures. The replacement of natural forest cover with oil palm monocultures has had detrimental effects on the regional biodiversity. The plantations support less biodiversity than the primary forests or even disturbed and logged forests they replace (Fitzherbert et al., 2008). The lower biodiversity value of oil palm monocultures can be attributed to the absence of major structurally complex components of forest vegetation, including forest trees, lianas and epiphytic orchids (Danielsen et al. 2009). Owing to the lack of the structurally complex flora, oil palm monocultures support less than half the vertebrate species found in the natural forests they replace (Persey and Anhar, 2010). An early exploration of these phenomena examined the impact of converting primary lowland forests to logging concessions and oil palm plantations in Sumatra on birds, primates, squirrels, tree shrews and bats (Danielsen and Heegaard (1995). Logging reportedly caused no change in the abundance of primate groups, squirrels and tree shrews, but was associated with a 38%-50% decline in the species richness of bats, while conversion of primary lowland forests to oil palm plantations led to species-poor and less diverse animal communities with fewer specialized species. For instance, only 5%-10% of primary forest bird species were recorded (and no mammalian species apart from one primate species) in oil palm plantations.

These findings have been corroborated by recent research focussing on the impact of converting primary forest to oil palm plantations on various species. Across all taxa, only 15% of species recorded in primary forest were found in oil palm plantations in a study by Fitzherbert et al. (2008). However, the ability of oil palm monocultures to support biodiversity varies across different taxa.

In Malaysia, primary forests support up to 80 mammalian species, while disturbed forests provide habitats for about 30 mammal species and oil palm plantations only support up to 12 species (WWF 2012, Clay (2004)). Corroborative research by Maddox et al. (2007) indicates that oil palm monocultures provide poor habitats for most endangered terrestrial mammal species of South East Asia. Orang-utans are a critically endangered mammal species, restricted to the islands of Sumatra and Borneo. In the recent past, expanding oil palm estates have encroached on their major habitat, the primary lowland forests, and this habitat loss has been associated with sharp decline in their population (Friends of the Earth, 2009). Further, most endangered species, including the iconic Sumatran tiger, tapirs and clouded leopard have not been detected in oil palm plantations. In addition, some species (including deer, macaques and pangolins) have shown a limited tolerance for oil palm plantations (Maddox et al., 2007).

Fukuda et al. (2009) examined effects of land use change on the diversity and distribution of bat species in an intact lowland Dipterocarp forest and surrounding land of modified use types, including secondary forests, orchards and oil palm monocultures. The intact lowland forest had the highest species diversity, and many of the specialist species such as frugivorous and insectivorous bats were absent from oil palm plantations. Oil palm plantations and orchards were only able to provide food sources for three megachiropteran species, and even these species displayed a preference for primary and secondary forests (Fukuda et al., 2009). In addition, Sheldon et al. (2010) found that only between 5% and 53% of bird species recorded in Bornean primary forest were also present in nearby oil palm plantations. Similarly, Phillipps and Phillipps (2010) reported that lowland primary forests in Borneo support more than 220 habitat specialist (often endemic and vulnerable) birds, while oil palm plantations support less than a dozen bird species, most of which are generalists? Conversion of primary forests and logged forests to oil palm plantations has also reportedly decreased the species richness of forest birds in Peninsular Malaysia, by 77% and 73%, respectively (Peh et al., 2006). Similar trends in bird species population declines have been observed elsewhere in the world. Notably, in Thailand, conversion of forests to rubber and oil palm plantations has resulted in a 60% decline in bird species richness, with insectivorous and frugivorous birds declining more rapidly than omnivorous birds (Aratrakorn et al., 2006). Furthermore, bird communities in the oil palm and rubber plantations are reportedly extremely similar. The results confirm that a high proportion of species formerly present in the region are unable to adapt to conversion of forest to oil palm and rubber plantations, resulting in large losses of bird species and the replacement of species with high conservation status by those with extensive ranges and low conservation status.

The species losses have not been restricted to birds. For instance, analyses by Koh and Wilcove (2007) indicate that conversion of primary forests and logged forests to oil palm plantations has decreased species richness of forest butterflies by 83% and 79%, respectively, in South East Asia. Further, Fayle et al. (2010) found that conversion of lowland forests to oil palm plantations in Sabah (Malaysia) causes a 64% reduction in ant species richness. Monoculture plantations such as those of oil palm also have much lower species diversity of dung beetles than primary or even logged forests (Chung et al., 2000). Similar trends in the diversity of insect species across different land use types have also been observed elsewhere in the world. For example, Barlow et al. (2007) surveyed dung beetles across different land use types, including *Eucalyptus* plantations, and found that dung beetle communities have very low species diversity in plantations. Furthermore, the depletion of dung beetle abundance is likely to have detrimental consequences for the maintenance of ecosystem services associated with them. Dung beetles have high ecological importance and are tightly linked with mammal communities (Edwards et al., 2011). Hence any alteration in their community structure can have a domino effect on higher taxonomic groups. Turner and Foster (2009) found that conversion of primary forests and logged forests to oil palm plantations in Malaysia also detrimentally affects the abundance of arthropods at all levels of the forest ecosystem. In another survey, of moth population distributions and diversity in Danum, Sabah, 67% of the species recorded in primary forest were not detected at oil palm sites and adjacent, disturbed forest (Chey, 2006). The results showed that biodiversity loss (using moths as indicator species) was most significant in oil palm plantations, the most common form of forest conversion in Sabah.

Hence, on the basis of evidence presented in the literature cited above, it may be inferred that oil palm plantations tend to have less taxonomically diverse species assemblages than more diverse habitats such as logged forests. These monocultures are poor substitute habitats for most tropical forest species, particularly those of high conservation concern, such as vulnerable, endemic or range-restricted species. Plantation assemblages are typically dominated by a few generalists (e.g. macaques and dusky munias), alien invasives (e.g. crazy ants), pests (e.g. rats), and their predators (e.g. pythons and barn owls) (Sodhi et al., 2010). This has been attributed to the fact that the extreme disturbance that follows conversion to oil palm plantation has a significantly negative impact on the flora and fauna of the area.

In addition to having detrimental effects on biodiversity, oil palm expansion also contributes to carbon emissions in South East Asia (Koh et al., 2011). Forest conversion to oil palm monoculture causes a net release of approximately 650 Mg CO_2 equivalents per hectare, and emissions resulting from peat forest conversion are even higher due to the decomposition of drained peat and consequent release of greenhouse gases (GHGs; Germer and Sauerborn, 2008). This conversion in turn creates a "biofuel carbon debt" by releasing

17 to 420 times more CO_2 than the annual GHG reductions that these biofuels would provide by displacing fossil fuels (Fargione et al., 2008). Research by Danielsen et al. (2009) indicates that it could take between 75-93 years for carbon emissions saved from the use of biofuels to compensate for the carbon released by forest conversion. CO_2 emissions from conversion of peat swamp forest, in particular, are far greater than gains from substitution of fossil fuels with palm oil (Hooijer et al., 2006).

While the carbon sequestration and conservation value of old growth and undisturbed primary forests is well established, less than pristine human modified landscapes also have considerable conservation and carbon storage potential. Research by Berry et al. (2010) indicates that more than 90% of the species recorded in primary forests are also present in logged forests. In addition to providing vital habitats for biodiversity, human-modified landscapes (such as logged forests) have significant carbon storage potential. During recovery, Berry et al. (2010) found that logged forest accumulated carbon five times more rapidly than natural forest (1.4 and 0.28 Mg C ha^{-1} $year^{-1}$, respectively), providing significant carbon storage.

Hence, in order to create palm oil monocultures that do not damage local biodiversity and carbon stocks, it is vital to include human modified/degraded landscapes in any conservation program considered for a given area.

3. Role of human-modified landscapes in biodiversity conservation

A significant body of literature (discussed in the previous section) indicates that converting land to oil palm plantations has a detrimental effect on most taxa. However, literature also indicates that the marginal and/or degraded habitats often found within or outside palm oil concessions can retain high conservation values. The fate of biodiversity within patches of lowland forests is inextricably linked to the broader landscape context, including how the surrounding agricultural matrix is designed and managed (Wallace et al. 2005). Species are more likely to persist in protected areas that incorporate local environmental heterogeneity, because this provides them with a range of conditions, thereby allowing them to adjust their local distribution in response to environmental change (Gillson and Willis, 2004). Whilst this matrix is rarely of the quality found in protected areas, using the conservation potential of unprotected lands can help overcome many of the shortfalls of the protected area system (Maddox et al., 2007) and combat the effects of habitat loss, fragmentation and edge effects. For instance, selectively logged forests can make important contributions to the conservation of tropical biodiversity, provided that they are managed in a way that maintains environmental heterogeneity (Hamer et al., 2003).

There are also other land use types that can counter the threats oil palm monocultures pose to biodiversity. While intensive oil palm monocultures offer few biodiversity conservation benefits (Koh, 2008), other plantation systems, such as cocoa agroforests and mixed fruit orchards, may provide valuable habitats (Hartley 2002). This dichotomy may be due to the vegetation structure of monocultures being too simple to support high levels of species diversity, while multi-species plantations (especially those that encompass remnants of native vegetation) have much more complex structure, with significant resemblance to primary vegetation (Lindenmayer and Hobbs, 2004). The dichotomy has important implications for both conservation strategies and the design of oil palm plantations, indicating that modified landscapes and new planning strategies may be required to reduce their impacts on local biodiversity (Maddox et al., 2007).

The problems associated with reducing the adverse effects of agricultural expansion on nature are substantial and require urgent attention. Two very different management approaches have been recommended in the literature to address these problems: land sparing, in which farm yields are increased and pressure to convert land for agriculture is reduced (at the potential cost of reducing wildlife populations on farmland); and land sharing/wildlife friendly farming, in which on-farm practises that are benign to wildlife are applied (Balmford et al., 2005).

3.1. The land sparing approach

In land sparing, homogeneous areas of farmland are managed to maximize yields, while separate reserves target biodiversity conservation. Land sparing is associated with an island model of modified landscapes, where islands of nature are seen as separate from human activities. The resulting agricultural systems are mostly homogenous and industrial in style, striving for maximum economic efficiency. They commonly rely on high inputs of fertilizers and pesticides, crop diversity is usually low, and individual fields tend to be large (Fischer et al., 2008). Indeed, according to Waggoner (1996), high yields of food crops have to be maintained if mankind is to put aside any land for biodiversity conservation. This arguably makes a compelling argument for land sparing.

Green et al. (2005) have presented a model that identifies the trade-offs involved. According to their research on a range of taxa in developing countries, high yield farming provides a means of allowing more species to persist. In accordance with the model: the abundance of imperilled bird species is reportedly 60-fold lower in fragments and 200-fold lower in oil palm plantations than in primary contiguous forest; retention of forest fragments does not increase bird abundance in adjacent oil palm plantations; and fragments have lower species richness than contiguous forest, with an avifaunal composition that is more similar to that of the plantations than contiguous forest (Edwards et al., 2010). Therefore, from a conservation perspective, any investment in the retention of fragments would be better directed toward the protection of primary contiguous forest. In contrast, censuses of birds in El Trunifo Biosphere reserve, southern Mexico, indicate that shade coffee plantations may have bird diversity levels similar to those of natural forests (Tejeda-Cruz and Sutherland, 2004). However, the coffee agricultural systems and natural forests differ in species composition, notably birds that are highly sensitive to disturbance are more abundant in primary forests. Hence, although coffee agricultural systems may play an important role in maintaining local biodiversity, their promotion as a "silver bullet" may encourage the transformation of primary forests to shade coffee plantations, leading to a loss of forest species. Similarly, in North Borneo there is reportedly little difference in butterfly diversity between primary and logged forests, but marked differences in the composition of butterfly assemblages between the two habitats. Butterfly species with higher shade preferences and narrower geographic distributions are particularly adversely affected by logging (Hamer et al., 2003). A similar, earlier survey of butterfly species distribution in Buru, Indonesia, indicated that their species richness and abundance was higher in primary forests than in other land use types. The primary forests also housed endemic species and species whose distribution is restricted to the province (Hill et al., 1995). Dunn (2004) compared the effects of logging and conversion of forest to agriculture or pasture on ant, bird, and lepidoptera species richness by combining data from 34 studies of tropical forests in Africa, Asia and the Americas, finding that forest conversion to agriculture or pasture decreased the species richness of ants and animals overall.

On the basis of the impacts that forest clearance has on the distribution different kinds of taxa, many authors have recommended using land sparing techniques as a way of conserving maximum possible biodiversity. Due to the impacts of forest clearance on the distribution of various taxa, many authors (e.g., Ghazoul et al., 2010) have recommended land sparing strategies to conserve maximum possible biodiversity. Accordingly, studies by Anderson et al. (2009) indicate that achieving both high agricultural production and biodiversity is difficult within a single land management system, and some 'land sparing' is necessary to achieve biodiversity targets. Research by Phalan (2009) in Ghana also indicates that increasing yields on existing croplands may enhance possibilities to spare natural habitats, by reducing the area of land needed for agriculture. According to these authors, the persistence of many species maybe better enhanced by enlarging or maintaining areas of natural forest cover while intensifying production on converted land to ensure that demands for plantation crops (such as oil palm) are met by maximizing yields on unforested land. This view is supported (*inter alia*) by findings presented by Green et al. (2005) and Balmford et al. (2005) that more bird diversity could be sustained by portioning land between intensive agriculture and wildlife reserves. Thus, although land sparing is a weak strategy, and can only be applied under a limited set of circumstances, it can have positive outcomes for nature conservation (Ewers et al., 2009).

Despite its proposed benefits, the land sparing approach has been heavily critiqued in a body of peer-reviewed literature. Notably, Matson and Vitousek (2006) suggest that intensive agricultural systems may fail to spare land for nature and the high use of agricultural inputs may adversely impact areas downstream from the agricultural farms. Thus, the high use of agro-chemicals, such as fertilizers, which is required for land intensification, could have detrimental effects on biodiversity, although extensive integrated land management could enhance profitability and benefit native vegetation (Moll et al., 2007). In contrast, Polasky et al. (2005) found that biodiversity conservation objectives can potentially be achieved in agricultural landscapes with thoughtful land-use planning. Due to the degree of conflict between conservation and agricultural returns, with the assumption that sensitive species survive only within and economic activity only outside reserves, land-sparing is a much less integrated approach than some options. Further, it does not recognize, or incorporate, the important contributions to the conservation of tropical biodiversity that other land use types, such as selectively logged forests, can make provided that their management maintains environmental heterogeneity (Hamer et al., 2003).

According to Norris (2008), it is important to recognize that some biodiversity is retained in agricultural landscapes, and that it is the degree of biodiversity retention that we need to understand and effectively manage. In regions such as Indonesia there are relatively few pristine areas and agriculture–forest mosaics now comprise much of the landscape (Koh et al., 2009). In this scenario, given the fragmented nature of most tropical ecosystems, agricultural landscapes should be an essential component of any conservation strategy (Perfecto and Vandermeer, 2010). In this context, we examine an alternative land management option; wildlife friendly farming.

3.2. Land sharing/wildlife friendly farming

Estimates suggest that more than half of all species exist principally outside Protected Areas, mostly in agricultural landscapes (Blann, 2006). Such species can be conserved by modifying existing agricultural landscapes through the principles of eco-agriculture and wildlife friendly farming (Scherr and McNeely, 2007).

In wildlife-friendly farming, as opposed to land sparing, conservation and production are integrated within more heterogeneous landscapes (Fischer et al., 2008), but agricultural yields tend to be relatively low (Green et al. 2005). Therefore, a larger land area is typically needed to produce the same agricultural yield. While this leaves less land for permanent preservation, more biodiversity can occur on the "wildlife-friendly" agricultural land itself (Fischer et al., 2008). Wildlife friendly farming systems apply the principle that agricultural landscapes can be designed and managed to host wild biodiversity of many types, with neutral or even positive effects on agricultural production. It may be regarded as an integrated conservation–agriculture strategy, in which biodiversity conservation is an explicit objective of agriculture and rural development, which are explicitly considered in shaping conservation and livelihood strategies (Scherr and McNeely, 2007). For instance, agroforestry systems are important for the livelihood of many families in rural areas of the tropics, especially the humid tropics where most biodiversity hotspots are located. They may also provide critical refuges for wildlife. Hence, wildlife friendly land planning strategies like these should be considered important components of integrated landscape conservation strategies (Perfecto and Vandermeer, 2008).

Harvey et al. (2008) discuss the illustrative case of Mesoamerican biodiversity hotspots, where 80% of the primary forest has been converted to agricultural landscapes and the survival of endemic species is threatened. The cited authors propose an integrated landscape management strategy, in which conservation and production units within the agricultural matrix are managed jointly for long-term sustainability and agricultural production. While this approach is unlikely to maximize agricultural productivity, traditional small-scale farms can provide both farmer livelihoods and biodiversity conservation (Harvey et al., 2008), since forested and non-forested habitats contribute to biodiversity conservation.Forest fragments, riparian forests, tree plantations and other types of remnant and introduced tree cover serve as habitats for many species, enhance landscape connectivity, and retain potential for forest regeneration and restoration (Chazdon 2003; Harvey et al. 2006*a* Furthermore, landscapes exploited for various other uses, such as diverse coffee agroforestry (Moguel & Toledo 1999; Komar 2006), cocoa agroforestry (Rice & Greenberg 2000; Harvey et al. 2006*b*) and traditional agro-ecological land systems, such as organic farming and swidden agriculture (Finegan & Nasi 2004), also retain high levels of both wild and agricultural biodiversity and offer much greater conservation value than the high-yield intensive agricultural landscapes that are proposed to replace them.

According to Bhagwat et al. (2008), agroforestry systems can alleviate resource use pressure in protected areas, enhance habitats for certain species and improve the connectivity of landscape components. Traditional agroforestry systems such as jungle rubber, shade-grown coffee and cocoa systems in the tropics resemble natural rainforests in many structural respects, and thus have been suggested to be promising wildlife-friendly land use strategies, conserving a significant proportion of tropical rainforest diversity while providing significant economic returns (Schroth et al., 2004). As primary forests shrink, fragment and are reduced to isolated habitat islands in a landscape dominated by agricultural farms/plantations, agroforestry systems could help maintain higher levels of biodiversity both within and outside protected areas. Although it has been correctly noted that agroforestry systems displace natural ecosystems, they offer greater potential for reconciling biodiversity conservation with agricultural production than intensive monocultures (Schroth, 2003).

To illustrate the potential conservation utility of agroforestry systems, For example, jungle rubber gardens of Indonesia are low-input rubber agro-forests that structurally resemble secondary forest and retain high populations of wild species. Beukema et al. (2007) compared the plant and bird diversity in Indonesian jungle rubber agroforestry systems to that in primary forest and rubber plantations. They found that species richness in the jungle rubber systems was slightly higher for terrestrial pteridophytes, similar for birds and lower for epiphytic pteridophytes, trees and vascular plants as a whole than in primary forest. For subsets of 'forest species' of terrestrial pteridophytes and birds, species richness was lower in the jungle rubber systems than in primary forest. These findings show that forest-like land use can be potentially used to support species diversity in an impoverished landscape increasingly dominated by monoculture plantations (Beukema et al., 2007). Similarly, Perfecto et al. (2005) investigated the biodiversity conservation potential of shade-grown coffee systems and concluded that due to the retention of canopy cover such agro-forests can provide both biodiversity conservation and economic returns from agricultural production. According to Harvey, (2007) transformation from near-primary forest to cocoa agroforestry in Sulawesi (Indonesia) has had little effect on overall species richness, but reduced plant biomass and carbon storage by ≈75%, and the species richness of forest-using taxa by ≈60%. In contrast, increased land use intensity in cacao agroforestry, coupled with a reduction in shade tree cover from 80% to 40%, has caused only minor quantitative changes in biodiversity and maintained high levels of ecosystem functioning while doubling farmers' net income. Hence, Harvey, (2007) concluded that low-shade agroforestry provides the best available compromise between economic forces and ecological needs. Similarly, Harvey and Villalobos (2007) examined species diversity across various agroforestry systems and concluded that heterogeneous mixed plantations of tree crops such as banana and cocoa are more hospitable to biodiversity than monoculture plantations. Round et al. (2006) compared the distributions of birds in a mixed fruit orchard and a nearby isolated forest patch on Khao Luang mountain, southern Thailand, with that in natural forest. They found the orchard was about 75% as rich in bird species as the forest, and the avifauna in the former was dominated by smaller fugivorous, nectarivorous and widespread generalist species. Sundaic birds contributed 26% of sightings in the orchard, although understory insectivores were poorly represented. However, while agricultural diversification may assist in restoring modest levels of diversity in areas already degraded or committed to human use, it should not be seen as a substitute for conventional protection of forest and wildlife. Communities in disturbed habitats are known to contain significantly more generalized species than those associated with pristine habitats (Taki and Kevan 2007), hence it is vital to identify the optimal approach for a given scenario.

4. Land sparing versus wildlife friendly farming choices

Both land sparing and land sharing approaches have advocates, and pros and cons, raising questions about which, if either, approach should be used. The optimal choice between land sparing and wildlife-friendly farming, or combining the two, to maintain production and biodiversity will differ between landscapes (Hodgson et al., 2010). The choice of which strategy to use could be informed by yield-density relations, which describe relationships between population densities of particular species and agricultural yields. Land sparing is a recommended strategy for species whose populations decline rapidly when their natural intact habitats are converted to low intensity agricultural landscapes. In contrast, wildlife friendly farming may be more appropriate for species whose densities persist even when their intact natural habitats are converted in this manner (Mattison and Norris, 2005).

Instead of taking an either-or approach, Koh et al. (2009) draw on both strategies, advocating the creation of agroforestry zones between high conservation value areas and intensive agricultural plantations (such as palm oil monocultures) to create more heterogeneous landscapes that can benefit both biodiversity and rural communities. However, before selecting any of the aforementioned approaches many other landscape level factors must be taken into account. The landscape considerations that can arise and why they need to be addressed has been illustrated by a comparison of distributions and abundance patterns of multiple species (frogs, bats and dung beetles) between a cloud forest and coffee agroforestry in Mexico (Pineda et al., 2005). The three indicator sets of species considered responded differently to land use change from montane cloud forest to coffee agroforestry. The species richness of frogs in the coffee agroforestry system was just a fifth of that in the cloud forests, and a third of the forest species were recorded in both the forest and the coffee agroforestry system. In contrast, the abundance and diversity of beetle species were higher in the coffee agroforestry system, while the composition of bat species did not significantly differ between the land use types. This variation in species distribution has been attributed to differences in the permeability of different species through the two habitats and their physiological traits. Shaded coffee agroforestry systems form a matrix around remnant patches of the cloud forest, and provide connections between the forest fragments and other habitats in the landscape.

Although such systems cannot be regarded as substitutes for the primary forest habitats they replace, they can allow the preservation of biodiversity in a fragmented, human-modified landscape (Pineda et al., 2005). Wells et al. (2007) surveyed the species assemblages of small non-volant mammals in three unlogged and three logged forests and compared the species richness, dominance and evenness of small non-volant mammals between these habitats in Sabah, Indonesia. They found species richness and diversity were higher in logged forests, and all common species were present in both habitats. However, closer analysis revealed that while habitat modification did not have a significant effect on the distribution of most of the species it had a profound impact on the rarer species. Rare and specialized species, such as arboreal mammals and Viverridae, were completely missing from logged forest sites. Hence, while the conservation value of logged forests cannot be negated, it can be inferred that the presence of logged forests may not help the conservation and preservation of habitat-specific and range-restricted species.

These case studies not only illustrate potential configurations of a designer landscape (e.g. a primary forest-coffee agroforestry or primary forest-logged forest system) but also the fact that in order to create a designer landscape, characteristics of the surrounding matrix and the ability of given species to adapt to a modified landscape have to be analysed. Many other landscape elements can also help in the creation of designer landscapes. In addition to structurally complex agroforestry systems, these include small rainforest fragments in agricultural landscapes, high conservation value forests and wildlife corridors (aspects of which are discussed in detail below). However, before deciding which configuration of landscape elements/habitats can help create a designer landscape suitable for achieving a desired set of conservation goals, it is vital to understand the theoretical foundations of designer landscapes.

5. Theoretical foundations of designer landscapes

From a biodiversity conservation perspective, maintaining vast areas of unmodified land has been considered to be desirable. However, increasingly biodiversity has to persist, as far as possible, in human-dominated landscapes that have been heavily modified for producing agricultural and forestry commodities. In such a scenario there is widespread evidence that heterogeneous, designer landscapes provide greater biodiversity benefits than intensively managed monocultures (Fischer et al., 2006). For instance, carefully designed landscapes can play important roles in offsetting the negative edge effects of intensive agricultural plantations adjoining patches of native vegetation (Koh et al., 2009).

Fischer et al. (2006) recommend several strategies that are likely to help efforts to maintain diverse species and ecological processes in production landscapes. The most important is maintenance of large patches of native vegetation, because the structural complexity of native vegetation is very important for maintaining vital ecological processes, which in turn are vital for the persistence of biodiversity. They also provide vital habitats for species that cannot persist in the absence of native vegetation. However, maintenance of structural complexity within commodity production landscapes (including designer landscapes) is also vital for conserving biodiversity. The key element of this strategy is the "matrix"; the area surrounding patches of native vegetation (Forman, 1995), which significantly influences the within-community remnant dynamics (Laurance, 1990). A matrix with complex vegetation structures similar to those of native vegetation supplies a variety of ecosystem benefits, including support for ecological processes, varied habitats for species outside patches of native vegetation or protected areas, shelter, breeding sites, additional food resources and the capacity to recover from disturbance (Chazdon et al., 2009).

"Matrix effects" may also explain why the abundance and richness of species can be maintained in non-forest habitats such as agricultural production systems, logged forests (Barlow et al., 2007) and mixed cropping systems. These modified landscapes successfully provide habitat elements and ecological processes needed for species persistence outside of a primary forest habitat. For instance, as mentioned above, jungle rubber gardens of Indonesia structurally resemble secondary forest and allow wild species of various taxa to persist Beukema et al. (2007). The cited findings, and other examples, illustrate the potential importance of the landscape matrix surrounding primary forest or primary forest patches for enabling species persistence and biodiversity conservation in non-forest habitats. A number of examples illustrate the role landscape matrix surrounding the primary forest patches or primary forests play in enabling species persistence and biodiversity conservation in non-forest habitats.

Oil palm plantations, particularly, can be made wildlife friendly by preserving forest fragments, riparian strips, etc., and/or by providing habitat complexity within them in the form of epiphytes and understory plants (Jackson et al., 2010). These elements can maintain both connectivity of habitats for animal species and vital ecological processes at landscape scale (Bennett, 2003). The following sections discuss some of these landscape configurations, and particularly how they can help the creation of more sustainable oil palm plantations.

5.1. Biodiversity in small fragmented rainforest patches

As discussed above, deforestation is occurring at an alarming rate in the tropics, leading to high rates of forest fragmentation, especially in South East Asia (Sodhi et al., 2004), and species losses resulting from these processes have been described as the greatest threats to biodiversity (Myers, 1989). It has been argued that small fragments may become the last refuges of many rainforest species on the brink of extinction, and in areas with little residual rainforest fragments can act as 'seeds' for re-establishing extensive forest (Turner and Corlett, 1996). Thus, since high rates of land conversion coincide with high levels of species richness and endemism, it is imperative to determine the capacity of small forest remnants to support biodiversity.

In a rapid evaluation of butterfly species in a single remaining forest patch in Costa Rica, Daily and Ehrlich (1995) found that even small (20-30 ha), isolated forest fragments retained a large butterfly fauna. This suggests that even heavily managed systems of largely exotic plants (such as agricultural systems) could be designed to serve as corridors for butterflies and perhaps some other groups of organisms and connect separated forest patches. Benedick et al. (2006) evaluated impacts of habitat fragmentation on the species richness and faunal composition of butterflies in tropical rainforests in Sabah, Borneo, considering both α- and β-diversity patterns to assess the relative importance of patch size, isolation and vegetation structure for the diversity and similarity of species assemblages. They concluded that variations among fragments added substantial species diversity to the butterfly assemblage in Bornean rainforests. The reported data indicate that, despite having lower species richness than contiguous primary forest, relatively small and isolated rainforest remnants make substantial contributions to regional diversity. Similar conclusions were drawn by Struebig et al. (2008) from an analysis of the biodiversity value of forest fragments in peninsular Malaysia. They estimated frequencies of insectivorous bat species at seven continuous forest sites and 27 forest fragments, then tested the effects of fragment isolation and area on the abundance, species richness, diversity, composition and nestedness of assemblages, and the abundance of the 10 most common species. The results indicate that while large tracts of contiguous forests are needed to support rare, forest specialist species and their retention should be a high conservation priority, forest fragments also perform vital conservation roles. More specifically, fragments larger than 300 ha appear to contribute substantially to landscape-level bat diversity (retaining comparable diversity to continuous forest), but small fragments (less than 150 ha) also have some conservation value, despite having more variable species composition than either larger fragments or continuous forest.

Small, isolated forest remnants are generally accorded low conservation status and given little protection, consequently they often disappear over time because of continued anthropogenic disturbance. However, the results presented by Struebig et al. (2008) indicate that the conservation value of small forest remnants' contributions to environmental heterogeneity should not be neglected. Similarly, Arroyo-Rodriguez et al. (2009) found small (<5 ha) forest patches to be important for the conservation of regional plant diversity in the tropical rainforests of Mexico. Despite their small dimensions, they reportedly contained diverse communities of native plants, including endangered and economically important species. Hence, the conservation and restoration of small patches may be essential for effective preservation of plant diversity in this strongly deforested and unique Neotropical region. Further, Estrada and Coates-Estrada (2002) examined responses of bat species' abundance and distribution to habitat destruction and deforestation by surveying a continuous forest patch, fragments and an agricultural mosaic habitat in Mexico. While most bat species were found in the primary forest patch, the co-occurrence of the three habitats aided conservation of the bats by allowing them to disperse through the landscape. However, a one-size-fits-all approach cannot be applied when trying to include isolated rainforest patches as part of a conservation strategy and results of these studies must be interpreted cautiously.

Research on the impact of forest fragmentation on leaf litter ant communities in Sabah further illustrates its detrimental effect on species abundance and composition. Bruhl et al. (2003) evaluated distributions of these ants in two forest fragments and a contiguous forest patch. The species numbers and diversity recorded in the forest fragments were less than half of those recorded in the primary forest patch. Further, the community composition was also altered by fragmentation. Similar effects, on other species, have been observed elsewhere in the world. Notably in Tanzania, protected areas are increasingly becoming habitat islands in a sea of modified landscapes (such as agricultural fields). This habitat isolation has been correlated with mammalian extinctions (Newmark, 1996). Indeed, research by Turner and Corlett (1996) indicates that in nearly all cases of tropical fragmentation there have been losses of local species, most severely for large species and species that have highly specific habitat requirements. These effects have been attributed to forest edge effects, reductions in population size and changes in community structure arising from fragmentation.

In spite of evidence of species currently persisting in isolated forest fragments, possible "extinction debts" must also be considered, i.e. the possibility that some populations may have survived extensive habitat loss but still face extinction owing to a time lag between initial habitat loss and eventual population decline (Cowlishaw, 1999). Nevertheless, the surrounding matrix can ameliorate effects of the isolation of habitat patches much more than can be readily explained by classical theories such as island biogeography theory. Thus, matrix modification may provide opportunities for reducing the isolation of individual primary habitat patches, and thus extinction risks of populations in fragments (Ricketts, 2004). Hence, in addition to other measures, management schemes such as improving the landscape matrix and habitat connectivity, e.g., by incorporating wildlife corridors may make valuable contributions to conservation landscapes, as discussed in the following section.

5.2. Improving landscape matrices and connectivity

Bhagwat et al. (2005) examined the diversity of pants, micro-fungi and birds at 58 sites (in 10 forest reserves, 25 sacred groves and 23 coffee plantations) in the Western Ghats, India. They found that the biodiversity in the sacred groves had been influenced by native tree cover in the surrounding landscape (as opposed to patch size), and recommended that the tree-covered landscape matrix should be preserved to conserve this biodiversity This is consistent with earlier findings that the quality of the matrix surrounding forest fragments plays a significant role in population persistence (Fahrig 2001). Further, matrix quality can be improved by maintaining heterogeneous landscapes with a variety of vegetated features.

Horskins (2005) has defined matrix as the landscape elements surrounding a habitat of focal interest (for instance, a primary forest fragment). One such element is a wildlife or biological corridor, described as a linear stretch of habitat, differing in structure and vegetation type to that of the matrix type, linking two habitat patches (such as rainforest patches) that would otherwise be isolated from one another (Laurance, 1990). Wildlife corridors can allow the conservation of species in fragmented habitats by increasing the connectivity between isolated habitat patches and integrating populations into single demographic units. Connecting patches of habitat with corridors has been shown to slow extinction rates of species, and species richness has been preserved for longer periods of time in connected than in disconnected habitat patches (Gilbert et al., 1998). In addition, Laurance and Laurance (1999) showed that linking primary forest patches to a large continuous primary forest provided more habitats for mammalian species and greatly facilitated their movement in Australia. They found the linking of primary forest patches to a large continuous primary forest greatly allowed for movement and habitat for mammals.

Agroforests can also play important roles in expanding habitats for a wide range of species and provide wildlife corridors that can facilitate the movement of species across a given landscape (Elevitch, 2004). In addition to supporting native species of plants and animals, agroforestry systems may contribute to the conservation of biodiversity by enhancing the connectivity of populations, communities, and ecological processes in fragmented landscapes. As mentioned above, habitats that can maintain such connectivity across landscapes are commonly referred to as biological or wildlife corridors, and their potential value is further considered below.

5.3. Wildlife corridors

Hilty et al. (2006) describe a wildlife corridor as a landscape element that connects two patches of suitable habitat by passing through a matrix of unsuitable habitat. The concept of wildlife corridors emerged from research on habitat fragmentation, and was based on the premise that fragments that are linked by a corridor of similar suitable habitat are likely to have greater conservation value than isolated fragments of similar size and could combat the adverse effects of habitat fragmentation (Diamond 1975). Protection or provision of continuous habitat corridors to link isolates such as nature reserves or old-growth forest patches is seen as more beneficial than retaining individual habitat patches (Bennett, 2003). Indeed, wildlife corridors have been shown to facilitate the migratory movements of birds and mammals (Newmark, 1993), and are currently being proposed as a means of combating the isolation of protected areas in Tanzania (Caro et al., 2009).

Extensive habitats like these provide permeable corridors for species to disperse through landscapes much more readily than through monocultures (Watts et al., 2005). Thus, the presence of wooded corridors and shrub-matrix can make intensive monoculture plantations more hospitable to the presence and dispersal of species, including endemics (Castellon and Sieving, 2006).

5.4. Riparian margins

Riparian margins are linear landscape elements that can play vital roles in offsetting the effects of habitat fragmentation and loss (Lees and Peres, 2007). They can support the persistence and dispersal of a wide assemblage of taxa, especially small vertebrate taxa (Lees and Peres, 2007) and bird species (Wild Asia, 2011) by providing structural connectivity. Notably, the ability of riparian margins to allow species dispersal is of vital importance for forest birds, which face severe population decline resulting from fragmentation in tropical countries like Costa Rica. Further, forest specialist bird species also use riparian margins for dispersing through fragmented landscapes (Gillies and St Clair, 2008). As an illustration of the benefits of riparian margins for countering adverse effects of habitat fragmentation, forest birds that depend on army ants for food respond negatively to fragmentation, and disappeared very rapidly when 1 and 1 ha fragments were isolated from contiguous forests in the Brazilian Amazon (Bierregaard et al., 1992 in Bennett (2003)). However, a fragmented patch (2 km long and 100-300 m wide) that remained connected to contiguous forest by a riparian margin retained the same species composition and abundance of understory birds as the continuous forest. The riparian margins were further found to enhance the movement of birds through the fragmented landscape and maintenance of population and species composition (Bennett, 2003). Riparian

http://dx.doi.org/10.5772/52420

margins also have significant conservation value and serve as buffers against fragmentation for other species, such as arboreal mammals and marsupials, including the tree kangaroos of tropical rainforest remnants in Australia (Laurance, 1990). This assertion is confirmed by data on populations of butterflies, dung beetles and ants in interior rainforests, rainforest edges, remnant rainforest riparian strips and arable land presented by Hill (1995). "Rainforest interior species", including the most abundant species of ants in the rainforest interior, three of the butterflies and one of the dung beetle species were found in the streamside riparian strips, but not in either the rainforest edges or arable land.

The presence of riparian margins in and around agricultural plantations allows the diversification of these habitats and maintenance of vital ecological processes, thus enabling greater persistence of species in agricultural landscapes (Tscharnkte, 2007). In 2010, a survey in an oil palm plantation in Sarawak, conducted to evaluate the variation in avian biodiversity among different habitats within the plantation, found that the presence of riparian buffer zones enhanced bird abundance and diversity in it. Although no high-conservation value species were detected in the plantation during the survey, the findings indicate that the presence of natural habitat within an oil palm plantation can help increase species diversity and aid biodiversity conservation (Wild Asia, 2011).

The importance of riparian vegetation has also been enshrined in law. For instance, Brazilian forestry legislation calls for the protection of riparian buffer zones in private land holdings (Lees and Peres, 2007). However, the importance of riparian margins, especially for protecting habitat-specific species, should not be over-estimated. While streamside vegetation and riparian margins may be helpful, research by Warkentin et al. (1995) indicates that riparian woodlands in Mexico are dominated by habitat generalists rather than pristine forest avifauna. Nevertheless, the presence of riparian margin helps in maintaining a wide range of species and species diversity in agricultural mosaics.

It is often difficult to evaluate the efficacy of tropical forest conservation programs. Hence, it is vital to develop and select appropriate metrics for assessing both the conservation value of different forest/land use types and the efficacy of conservation schemes designed to preserve them. However, more than one metric is needed to avoid biased judgement, and ideally both the conservation value (biodiversity protection and ecosystem services) and carbon sequestration offered by alternative land use types should be considered (Macdonald et al., 2011). Schemes that target the biodiversity conservation and ecosystem services potential of a land use type (e.g. Forest Stewardship Council schemes) and the carbon sequestration services provided by tropical forests (such as Reducing Emissions from Deforestation and Forest Degradation, REDD) have already been established, and need to be assessed, implemented and monitored on a case-by case basis.

6. High conservation value forests

High Conservation Value Forest (HCVF) approach emerged in 2001 as a supplementary standard within the Forest Stewardship Scheme. High Conservation Value Forest (HCVF) is a forest management standard, introduced by the Forest Stewardship Council in 2001, intended to identify and protect forest areas with critical or outstanding ecological, social and cultural values, e.g., habitats for rare threatened or endangered species, water/soil quantity and quality protection, and/or sites of cultural and/or religious significance (Levy, 2009). When a forest is designated as HCVF these attributes are identified and should be effectively managed. The concept has been transferred to the plantation sector, where progressive oil palm plantation managers use it as a means of claiming environmental responsibility. In non-forested areas, more general "HCV" approaches (Jepson et al., 2011) can play important roles in guiding conservation priorities for different land use types. For instance, demarcating a forest patch as an HCVF can make a case for directing logging pressures towards other forests and plantations, with lower conservation value, where logging/agricultural expansion may be more compatible with preservation of biodiversity (Frumhoff and Losos, 1998). The demarcation of forest areas as HCVFs can also lay foundations for designing policy mechanisms or certification schemes to prevent undesirable exploitation (Gullison, 2003). Further, the HCVF approach can play a very significant role in shifting the focus from conserving a specific type of forest (such as old growth forests) to the values that make a forest important, such as its biodiversity conservation potential (Jennings et al., 2003). It can greatly promote sustainable forest management and be beneficial for forest zoning and forest benefit compensation, as well as informing nature reserve establishment (Huafeng, 2007). The concepts encompassed in HCV can also be applied to evaluate the conservation value of non-forest landscapes, such as human-modified habitats and mixed-rural habitats. Although primary forests have the highest conservation values, application of this scheme may allow biodiversity conservation in other land use types with high conservation values, such as secondary forests and isolated forests, by bringing them under the purview of suitable land management policies (Sodhi et al., 2004).

Arguably, in this era of rapid deforestation, designing landscapes that can reconcile biodiversity conservation and agricultural production may require different combinations of the land management approaches discussed in this chapter, with various similarities and unique features among regions.

6.1. Reconciling biodiversity conservation with carbon storage

One of the obvious shortcomings of the HCV concept is the neglect of carbon storage services in its defining criteria. However, it can be used in conjugation with Reducing Emissions from Deforestation and Forest Degradation (REDD) schemes to meet carbon storage objectives together with biodiversity conservation goals. Global level analysis by Strassburg et al. (2010) indicates that species richness is strongly associated with carbon stocks. However, this relation varies significantly across regions. For instance, in South East Asia plantations of rapidly growing crops, such as acacia and pines, have low biodiversity conservation potential but can enhance carbon sequestration (Corlett and Primack, 2008). Further, vital habitats for vulnerable and endemic species may not overlap with areas of high carbon storage. Hence, measures that promote the persistence of conservation-dependent species may not necessarily assist carbon sequestration. Hence, spatial planning should ideally aim to maximize biodiversity and enhance carbon storage simultaneously (ProForest, 2011).

"Designer landscape" is a theoretical concept with important practical implications. In South East Asia oil palm monoculture –forest mosaics comprise a substantial part of many landscapes (Koh et al., 2009), in which remnants of primary, old growth forests occur as fragmented, isolated patches of various sizes in a matrix of oil palm plantations (Bennett, 2003). In this situation, more heterogeneous, designer landscapes could be created by incorporating features such as agroforestry zones between high conservation value forests and oil palm monoculture plantations. Further, isolated and fragmented habitat patches could be connected across such landscapes by retaining or creating wildlife corridors (Koh et al., 2009).

Struebig et al. (2010) argued that the designer landscape approach is highly context-dependent and constrained by socio-political and biological realities. Nevertheless, in principle there is some room for a designer approach, to develop landscapes where oil palm plantations co-exist with forests, in partnership with local communities, which could yield significant conservation benefits (Phalan, 2009). Ghazoul et al. (2010) recognise risks (wildlife conflicts, hunting and encroachment) inherent in the approach advocated by Struebig et al. (2010), but argue that a strict land-sparing and/or HCV combination would be subject to similar risks. Given that biodiversity and ecosystem services are increasingly expected to persist in a human-modified landscape, the practical realities (and feasibility) of the 'designer landscape' approach should be carefully considered. Many configurations of landscape level strategies could allow the creation of designer landscapes that meet desired goals. However, very few landscape level studies have been carried out that could form the theoretical basis for informing these landscape level strategies and the subsequent creation of designer landscapes. In addition to the on-going Stability of Altered Forest Ecosystem (SAFE) project in Danum, examining the effects of fragmentation, Laurance et al. (2011) have presented results of a 32-year study on Amazonian forest fragments. The results indicate that species in forest fragments are vulnerable to edge effects and that isolation effects decline with the recovery of secondary forests. However, there have been very few landscape level studies that could provide a sound theoretical basis for informing landscape level strategies and subsequent creation of designer landscapes. Thus, further examination of both patch level processes and wider landscape contextual issues is required to facilitate the creation of designer landscapes that meet multiple (often conflicting goals) in given areas.'

7. Conclusions

Agricultural expansion, driven by rapid increases in the human population, is a major cause of deforestation, biodiversity losses and environmental degradation globally, especially in the tropics. Accompanying searches for land to support the increases in agricultural production, and increase revenues, have caused significant losses, modifications and fragmentation of habitats, which are posing substantial threats to the world's biodiversity, forests and ecosystems. There have been considerable efforts to protect biodiversity in reserves and parks, but many of the refuges are small, fragmented, isolated, or poorly protected. In addition, most Protected Areas are embedded in agricultural landscapes, and existing buffer zones are inadequate for alleviating effects of fragmentation, contamination by agrochemicals, hunting, and unsustainable or illegal logging. In recent decades, the expansion of oil palm monocultures has been one of the most significant drivers of deforestation and habitat loss in various tropical regions, including South East Asia, which hosts four biodiversity hotspots and is particularly vulnerable to these processes. The replacement of natural forest cover with oil palm monoculture plantations has detrimental effects on the biodiversity in the region, since they support substantially less biodiversity than the primary forests or even the disturbed and logged forests they replace. This is largely due to the absence of major structurally complex components of forest vegetation (including forest trees, lianas and epiphytic orchids) and associated fauna, including numerous threatened, iconic species.

In addition to adversely affecting biodiversity, oil palm expansion also increases carbon emissions in South East Asia, reportedly resulting in a net release of approximately 650 Mg CO_2 equivalents per hectare or more, resulting in a "biofuel carbon debt" by releasing 17 to 420 times more CO_2 than the annual GHG reductions that oil palm-derived biofuels would provide by displacing fossil fuels. Estimates indicate that it could take 75-93 years for carbon emissions saved from the use of biofuels to compensate for the carbon released by forest conversion. Furthermore, less than pristine human-modified landscapes also have considerable conservation and carbon storage potential. For instance, more than 90% of the species recorded in primary forests have been found in logged forests, and during recovery logged forests may accumulate carbon five times more rapidly than natural forest, providing substantial carbon storage. In addition, marginal and/or degraded habitats often found within or nearby palm oil concessions can retain high conservation values. Thus, in order to create palm oil monocultures that do not damage the local biodiversity and carbon stocks, it is vital to include human modified/degraded landscapes in any conservation program for a given area.

The fate of biodiversity within patches of lowland forests is inextricably linked to the broader landscape context, including how the surrounding agricultural matrix is designed and managed. Species are more likely to persist in protected areas that incorporate local environmental heterogeneity, because it provides a range of conditions that allows them to adjust their local distribution in response to environmental changes. Other land use types can also counter the adverse effects of oil palm monocultures on biodiversity. Notably, other plantation systems such as cocoa agroforests and mixed fruit orchards may provide valuable habitats, because complex multi-species plantations (especially those that encompass remnants of native vegetation) bear significant resemblance to primary vegetation. This has important implications both for conservation strategies and the design of oil palm plantations, which may entail the use of modified landscapes and new planning strategies to reduce impacts on local biodiversity.

Two contrasting management approaches have been recommended for boosting agricultural production while maintaining biodiversity: land sparing, which increases farm yields while reducing pressure to convert land for agriculture (at the potential cost of reducing wildlife populations on farmland) and land sharing/wildlife friendly farming, in which on-farm practices that are benign to wildlife are applied. In land sparing, homogeneous areas of

http://dx.doi.org/10.5772/52420

farmland are managed to maximize yields, while separate reserves target biodiversity conservation. Inputs of fertilizers and pesticides are generally high, crop diversity usually low, and individual fields tend to be large. Land sparing is a weak conservation strategy, and can be applied under a limited set of circumstances, but it can have positive outcomes for nature conservation. Notably, from a conservation perspective the protection of primary contiguous forest appears to be more beneficial than investing in the retention of fragments. However, despite its proposed benefits, the land sparing approach has been heavily critiqued, since intensive agricultural systems may fail to spare land for nature and the high use of agricultural inputs may adversely impact areas downstream from the agricultural farms.

In wildlife-friendly farming, as opposed to land sparing, conservation and production are integrated within more heterogeneous landscapes, recognizing: that biodiversity is retained in agricultural landscapes; that we need to understand and effectively manage the *degree* of biodiversity retention; and that many species can be conserved by modifying existing agricultural landscapes through the principles of eco-agriculture. In wildlife-friendly farming, agricultural yields tend to be lower. Therefore, larger land areas are typically required to produce the same agricultural yield. This leaves less land for permanent preservation, but more biodiversity may be retained on the "wildlife-friendly" agricultural land itself. Hence, wildlife friendly land planning strategies should be considered important components of integrated landscape conservation strategies.

Both land sparing and land sharing have pros and cons. Further, the optimal strategy for maintaining production and biodiversity differs between landscapes (Hodgson et al., 2010). The choice of strategy could be informed by yield-density relations, i.e. relationships between population densities of particular species and agricultural yields. Land sparing may be most appropriate for species whose populations decline rapidly when their natural intact habitat is converted to low intensity agricultural landscapes, while wildlife friendly farming may be more appropriate for species whose densities are maintained when the intact natural habitat is converted to low yield agricultural landscapes (Mattison and Norris, 2005). However, a strategy combining the two approaches may be optimal in many cases, incorporating (for instance) the creation of agroforestry zones between high conservation value areas and intensive agricultural plantations to create more heterogeneous landscapes that can benefit both biodiversity and rural communities. Nevertheless, before selecting any of the abovementioned approaches numerous other landscape level factors should be considered. These include the permeability of sensitive species through habitats and their physiological traits. No heavily human-modified landscapes can be regarded as a full substitute for primary forest habitats, but they may allow the preservation of biodiversity if appropriately fragmented or designed. However, in order to design an appropriate landscape, characteristics of the surrounding matrix and the ability of sensitive species to adapt to it have to be addressed, and both the value and characteristics of potentially important elements need to be considered. Such elements may include rainforest fragments, high conservation value forests, wildlife corridors and riparian margins.

Several strategies have been recommended for establishing landscape patterns that are likely to maintain many species and ecological processes in production landscapes. The most important is the maintenance of large patches of native vegetation, which are very important for maintaining vital ecological processes and (hence) the persistence of biodiversity. In addition, maintenance of structural complexity, an appropriate (preferably native vegetation) "matrix" and suitable levels of connectivity between elements is vital. A matrix with a similar vegetation structure to that of native vegetation supplies a variety of ecosystem benefits, support for ecological processes, habitats for species outside patches of native vegetation or protected areas, shelter, breeding sites, additional food resources and the ability to recover from disturbance. Oil palm plantations, in particular, can be made wildlife friendly by the preservation of forest fragments, riparian strips, etc., and/or by providing habitat complexity within the plantations in the form of epiphytes and understory plants. "Extinction debt" must also be considered, i.e. populations that have survived extensive habitat loss may still face extinction owing to a time lag between initial habitat loss and eventual population decline.

Finally, it is essential to integrate biodiversity conservation with robust carbon storage strategies. Global level analysis indicates that species richness is strongly associated with carbon stocks, but the strength of the relationship varies significantly across regions. For instance, in South East Asia plantations of fast-growing crops, such as acacia and pines, have low biodiversity conservation potential but may have high carbon sequestration value. Further, vital habitats of vulnerable and endemic species may not overlap with areas of high carbon storage. Thus, spatial planning that maximizes biodiversity co-benefits along with carbon storage is required.

It has been argued that the designer landscape approach is highly context-dependent and constrained by socio-political and biological realities. Further, many configurations of landscape level strategies could allow the creation of designer landscapes that meet desired goals. Thus, further research on (*inter alia*) patch-level processes and wider contextual issues is required to develop robust strategies for creating designer landscapes that meet often conflicting conservation, agricultural production and carbon sequestration goals.

Author details

Minerva Singh and Shonil Bhagwat

University of Oxford, UK

References

[1] Anderson, B. J, Armsworth, P. R, Eigenbrod, F, Thomas, C. D, Gillings, S, Heinemeyer, A, Roy, D. B, & Gaston, K. J. (2009). Spatial covariance between biodiversity and other ecosystem service priorities' *Journal of Applied Ecology, 46* (4), 888-896.

[2] Angelsen, A. (1999). Agricultural expansion and deforestation: modelling the impact of population, market forces and property rights'. *Journal of Development Economics*, , 58(1), 185-218.

[3] Aratrakorn, S, Thunhikorn, S, & Donald, P. F. (2006). Changes in bird communities following conversion of lowland forest to oil palm and rubber plantations in southern Thailand'. *Bird Conservation International, 16* (1), 71-82.

[4] Arroyo-rodriguez, V, Pineda, E, Escobar, F, & Bénitez-malvido, J. (2009). Value of small patches in the conservation of plant-species diversity in highly fragmented rainforest'. *Conservation Biology*, , 23(3), 729-739.

[5] Balmford, A, Green, R. E, & Scharlemann, P. W. (2005). Sparing land for nature: exploring the potential impact of changes in agricultural yield on the area needed for crop production'. *Global Change Biology*, , 11(10), 1594-1605.

[6] Barlow, J, Gardner, T. A, Araujo, I. S, et al. (2007). Quantifying the biodiversity value of tropical primary, secondary and plantation forests'. *Proceedings of the National Academy of Sciences*, , 104(47), 18555-18560.

[7] Benedick, S, Hill, J. K, Mustaffa, N, Chey, V. K, Maryati, M, Searle, J. B, Schilthuizen, M, & Hamer, K. C. (2006). Impacts of rain forest fragmentation on butterflies in northern Borneo: species richness, turnover and the value of small fragments'. *Journal of Applied Ecology*, , 43(5), 967-977.

[8] Bengtsson, J, Angelstam, P, Elmqvist, T, Emanuelsson, U, Folke, C, Ihse, M, Moberg, F, & Nystrom, M. (2003). Reserves, resilience and dynamic landscapes'. *Ambio*, , 32(6), 389-396.

[9] Bennett, A. F. (2003). *Linkages in the landscape: the role of corridors and connectivity in wildlife conservation*, 2nd edn. IUCN, Gland, Switzerland and Cambridge, UK.

[10] Berry, N, Phillips, O, Lewis, S, Hill, J, Edwards, D, Tawatao, N, Ahmad, N, Magintan, D, Khen, C, Maryati, M, Ong, R, & Hamer, K. (2010). The high value of logged tropical forests: lessons from northern Borneo'. *Biodiversity and Conservation, 19* (4), 985-997.

[11] Beukema, H, Danielsen, F, Vincent, G, Hardiwinoto, S, & Van Andel, J. (2007). Plant and bird diversity in rubber agroforests in the lowlands of Sumatra'. *Agroforestry Systems* , 70(3), 217-242.

[12] Bhagwat, S. A, Cushalappa, C. G, Williams, P. H, & Brown, N. D. (2005). A landscape approach to biodiversity conservation of sacred groves in the Western Ghats of India'. *Conservation Biology*, , 19(6), 1853-1862.

[13] Bhagwat, S. A, Willis, K. J, Birks, H. J. B, & Whittaker, R. J. (2008). Agroforestry: a refuge for tropical biodiversity?' *Trends in Ecology and Evolution*, , 23(5), 261-267.

[14] Bhagwat, S. A, Cole, L. E. S, & Willis, K. J. (2010). Biodiversity conservation, rural livelihoods and sustainability of oil palm landscapes: problems and prospects.

[15] Blann, K. (2006). *Habitat in Agricultural Landscapes: How Much is Enough?* West Linn: Defenders of Wildlife.

[16] Broadbent, E. N, Asner, G. P, Keller, M, Knapp, D. E, Oliveira, P. C. J, & Silva, J. N. (2008). Forest fragmentation from deforestation and selective logging in the Brazilian Amazon'. *Biological Conservation,* , 141(7), 1745-1757.

[17] Brook, B. W, Sodhi, N. S, & Ng, P. K. L. (2003). Catastrophic extinctions follow deforestation in Singapore', *Nature, 424* , 420-423.

[18] Brühl, C. A, Eltz, T, & Linsemair, K. E. (2003). Size does matter- effects of tropical rainforest fragmentation on leaf litter and ant community in Sabah, Malaysia'. *Biodiversity and Conservation,* , 12(7), 1371-1389.

[19] Butler, R. mongabay.com (October 04, (2007). Biodiesel demand could destroy world's forests. http://news.mongabay.com/2007/biodiesel.htmlLast accessed: August 14, 2012

[20] Caro, T, Gardner, T. A, Stoner, C, Fitzherbert, E, & Davenport, T. R. B. (2009). Assessing the effectiveness of protected areas: paradoxes call for pluralism in evaluating conservation performance'. *Diversity and Distributions,* , 15(1), 178-182.

[21] Castellon, T. D, & Sieving, K. E. (2006). An Experimental Test of Matrix Permeability and Corridor Use by an Endemic Understory Bird'. Conservation Biology, , 20(1), 135-145.

[22] Chazdon, R. L. (2003). Tropical forest recovery: legacies of human impact and natural disturbances'. *Perspectives in Plant Ecology, Evolution and Systematics,* , 6(1), 51-71.

[23] Chazdon, R. L, Harvey, C. A, Komar, O, Griffith, D. M, Ferguson, B. G, Martínez-ramos, M, Morales, H, Nigh, R, Soto-pinto, L, Van Breugel, M, & Philpott, S. M. (2009). Beyond reserves: a research agenda for conserving biodiversity in human-modified tropical landscapes'. *Biotropica, 41* (2), 142-153.

[24] Chey, V. K. (2006). Impacts of forest conversion on biodiversity as indicated by moths'. *Malayan Nature Journal* , 57(4), 383-418.

[25] Chung, A. Y. C, Eggleton, P, Speight, M. R, Hammond, P. M, & Chey, V. K. (2000). The diversity of beetle assemblages in different habitat types in Sabah, Malaysia'. *Bulletin of Entomological Research,* , 90(6), 475-496.

[26] Clay, J. W. (2004). *World Agriculture and the Environment.* Washington, DC: Island Press.

[27] Corlett, R. T, & Primack, R. B. (2008). Tropical Rainforest Conservation: A Global perspective'. In: Carlson, W. and Schnitzer, S. (eds). *Tropical Forest Community Ecology.* Oxford: Wiley-Blackwell. , 442-457.

[28] Corley, R. H. V, & Tinker, P. B. (2003). *The Oil Palm*. 4th edition. Oxford: Blackwell Science Ltd.

[29] Corley, R. H. V. (2009). How much palm oil do we need?' *Environmental Science and Policy 12* (2), 134-139.

[30] Cowlishaw, G. (1999). Predicting the Pattern of Decline of African Primate Diversity: an Extinction Debt from Historical Deforestation'. *Conservation Biology, 13* (5), 1183-1193.

[31] Daily, G. C, & Ehrlich, P. R. (1995). Preservation of biodiversity in small rainforest patches: rapid evaluations using butterfly trapping'. *Biodiversity and Conservation, 4* (1), 35-55.

[32] Danielsen, F, & Heegaard, M. (1995). Impact of logging and plantation development on species diversity: a case study from Sumatra'. In: Sandbukt, O. (ed.) *Management of tropical forests: towards an integrated perspective.* Oslo: Centre for Development and the Environment, University of Oslo. , 73-92.

[33] Danielsen, F, Beukema, H, Burgess, N. D, Parish, F, Brühl, C. A, Donald, P. F, Murdiyarso, D, Phalan, B, Reijnders, L, Struebig, L, & Fitzherbert, E. B. (2009). Biofuel Plantations on Forested Lands: Double Jeopardy for Biodiversity and Climate'. *Conservation Biology*, , 23(2), 348-358.

[34] Defries, R, Asner, G. P, Achard, F, Justice, C, Laporte, N, Price, K, Small, C, & Townshend, J. (2005). Monitoring tropical deforestation for emerging carbon markets'. In: Moutinho, P. and Schwartzman, S. (eds.) *Tropical Deforestation and Climate Change.* Washington, DC: IPAM. , 1-27.

[35] Diamond, J. M. (1975). The island dilemma: lessons of modern biogeographic studies for the design of natural reserves'. *Biological Conservation, 7* (2), 129-146.

[36] Dunn, R. R. (2004). Recovery of faunal communities during tropical forest regeneration'. *Conservation Biology*, , 18(2), 302-309.

[37] Edwards, D. P, Fisher, B, & Boyd, E. (2010). Protecting degraded rainforests: enhancement of forest carbon stocks under REDD+'. *Conservation Letters*, , 3(5), 313-316.

[38] Edwards, D. P, Larsen, T. H, Docherty, T. D. S, Ansell, F. A, Hsu, W. W, Dehré, M. A, Hamer, K. C, & Wilcove, D. S. (2011). Degraded lands worth protecting: the biological importance of Southeast Asia's repeatedly logged forests'. *Proceedings of the Royal Society B, 278* (1702), 82-90.

[39] Elevitch, C. R. ed.). ((2004). *The Overstory Book: Cultivating Connections with Trees.* 2nd Edition. Holualoa: Permanent Agriculture Resources.

[40] Elmqvist, T, Folke, C, Nyström, M, Peterson, G, Bengtsson, J, & Walker, B. And Norberg, J. ((2003). Response diversity, ecosystem change, and resilience'. *Frontiers in Ecology and the Environment*, , 1(9), 488-494.

[41] Estrada, A, & Coates-estrada, R. (2002). Dung beetles in continuous forest, forest fragments and in an agricultural mosaic habitat island at Los Tuxtlas, Mexico'. *Biodiversity and Conservation,* , 11(11), 1903-1918.

[42] Ewers, R. M, Scharlemann, J. P. W, Balmford, A, & Green, R. E. (2009). Do increases in agricultural yield spare land for nature?' *Global Change Biology, 15* (7), 1516-1526.

[43] Fahrig, L. (2001). How much habitat is enough?' *Biological Conservation, 100* (1), 65-74.

[44] FAO (2009). *FAO statistics: Production and crops*. The Food and Agriculture Organization of the United Nations. Available at: http://faostat.fao.org/site/567/default.aspx#ancor.Accessed 12 August 2012)

[45] Fargione, J, Hill, J, Polasky, S, & Hawthorne, P. (2008). Land clearing and the biofuel carbon debt'. *Science,* , 319(5867), 1235-1238.

[46] Fayle, T. M, Turner, E. C, Snaddon, J. L, Chey, V. K, Chung, A. Y. C, Eggleton, P, & Foster, W. A. (2010). Oil palm expansion into rain forest greatly reduces ant biodiversity in canopy, epiphytes and leaf- litter'. *Basic and Applied Ecology,* 11(4), 337-345.

[47] Finegan, B, & Nasi, R. (2004). The biodiversity and conservation potential of shifting cultivation landscapes'. In: Schroth, G., da Fonseca, G.A.B., Harvey, C.A., Gascon, C., Vasconcelos, H. L. and Izac, A.M.N. (eds.). *Agroforestry and biodiversity conservation in tropical landscapes*. Washington, DC: Island Press. , 153-197.

[48] Fischer, J, Lindenmayer, D. B, & Manning, A. D. (2006). Biodiversity, ecosystem function, and resilience: ten guiding principles for commodity production landscapes' *Frontiers in Ecology and the Environment,* , 4(2), 80-86.

[49] Fischer, J, Brost, B, Daily, G. C, Ehlrich, P. R, Goldman, R, Goldstein, J, Lindenmayer, D. B, Manning, A. D, Mooney, H. A, Pejchar, L, Ranganathan, J, & Tallis, H. (2008). Should agricultural policies encourage land sparing or wildlife- friendly farming'. *Frontiers in Ecology and the Environment,* , 6(7), 380-385.

[50] Fitzherbert, E. B, Struebig, M. J, Morel, A, Danielsen, F, Brühl, C. A, Donald, P. F, & Phalan, B. (2008). How will oil palm expansion affect biodiversity?' *Trends in Ecology and Evolution,* , 23(10), 538-545.

[51] Forman, R. T. T. (1995). *Land Mosaics: The ecology of landscapes and regions.* New York, NY: Cambridge University Press.

[52] Friends of the Earth ((2009). *The Oil for Ape Scandal*. Available at: http://www.foe.co.uk/resource/reports/oil_for_ape_summary.pdfAccessed 12 August 2012)

[53] Frumhoff, P. C, & Losos, E. C. (1998). *Setting priorities for conserving biological diversity in tropical timber production forests*. Washington, DC: Center for Tropical Forest Science, Smithsonian Institute.

[54] Fukuda, D, Tisen, O. B, Momose, K, & Sakai, S. (2009). Bat diversity in the vegetation mosaic around a lowland Dipterocarp forest of Borneo'. *Raffles Bulletin of Zoology* , 57(1), 213-221.

[55] Gaston, K. J, Blackburn, T. M, & Goldewijk, K. K. (2003). Habitat conversion and global avian biodiversity loss'. *Proceedings of the Royal Society B, 270* (1521), 1293-1300.

[56] Gaveau, D. L. A, Epting, J, Lyne, O, Linkie, M, Kumara, I, Kanninen, M, & Leader-williams, N. (2009). Evaluating whether protected areas reduce tropical deforestation in Sumatra'. *Journal of Biogeography,* , 36(11), 2165-2175.

[57] Germer, J, & Sauerborn, J. (2008). Estimation of the impact of oil palm plantation establishment on greenhouse gas balance'. *Environment, Development and Sustainability, 10* (6), 697-716.

[58] Ghazoul, J, Koh, L. P, & Levang, P. (2010). Checking the reality check'. *Trends in Ecology and Evolution* , 25(1), 8-9.

[59] Gilbert, F, Gonzalez, A, & Evans-freke, I. (1998). Corridors maintain species richness in the fragmented landscapes of a microecosystem' *Proceedings of the Royal Society B, 256* (1396), 577-582.

[60] Gillies, C. S. and St. Clair, C.C. ((2008). Riparian corridors enhance movement of a forest specialist bird in fragmented tropical forest'. *Proceedings of the National Academy of Sciences, 105* (50), 19774-19779.

[61] Gillson, L, & Willis, K. J. (2004). As Earth's testimonies tell': wilderness conservation in a changing world'. Ecology Letters, , 7(10), 990-998.

[62] Goklany, I. M. (1998). Saving habitat and conserving biodiversity on a crowded planet'. *BioScience,* , 48, 941-953.

[63] Green, R. E, Cornell, S. J, Scharlemann, J. P. W, & Balmford, A. (2005). Farming and the fate of wild nature'. *Science,* 307(5709), 550-555.

[64] Gullison, R. E. (2003). Does forest certification conserve biodiversity?' *Oryx, 37* (2), 153-165.

[65] Hamer, K. C, Hill, J. K, Benedick, S, Mustaffa, N, Maryati, M, & Chey, V. K. (2003). Ecology of butterflies in natural and selectively logged forests of northern Borneo: the importance of habitat heterogeneity'. *Journal of Applied Ecology,* , 40(1), 150-162.

[66] Hartley, M. J. (2002). Rationale and methods for conserving biodiversity in plantation forests'. *Forest Ecology and Management,* 155(1-3): 81-95.

[67] Harvey, C. A, & Medina, A. Sánchez Merlo, D., Vílchez, S., Hernández, B., Saenz, J.C., Maes, J.M., Casanovas, F. and Sinclair, F.L. ((2006a). Patterns of animal diversity associated with different forms of tree cover retained in agricultural landscapes'. *Ecological Applications,* , 16(5), 1986-1999.

[68] Harvey, C. A, González, J, & Somarriba, E. (2006b). Dung beetle and mammal diversity in forests, indigenous agroforestry systems and plantain monocultures in Talamanca, Costa Rica'. *Biodiversity and Conservation* , 15(2), 555-585.

[69] Harvey, C. A. (2007). Designing agricultural landscapes for biodiversity conservation'. In Scherr, S.J. and McNeely, J.A. (eds.). Farming with nature: The science and practice of ecoagriculture. Washington, DC: Island Press. , 146-165.

[70] Harvey, C. A, & Gonalez-villalobos, J. A. (2007). Agroforestry systems conserve species-rich but modified assemblages of tropical birds and bats'. *Biodiversity and Conservation* , 16(8), 2257-2292.

[71] Harvey, C. A, Komar, O, Chazdon, R, Ferguson, B. G, Finegan, B, Griffith, D. M, Martínez-ramos, M, Morales, H, Nigh, R, Soto-pinto, L, Van Breugel, M, & Wishnie, M. (2008). Integrating agricultural landscapes with biodiversity conservation in the Mesoamerican hotspot'. *Conservation and Policy, 22* (1), 8-15.

[72] Hill, J. K, Hamer, K. C, Lace, L. A, & Banham, W. M. T. (1995). Effects of selective logging on tropical forest butterflies on Buru, Indonesia'. *Journal of Applied Ecology* , 32(4), 754-760.

[73] Hilty, J. A, Lidicker, W. Z, & Merenlender, A. M. (2006). *Corridor ecology: The science and practice of linking landscapes for biodiversity conservation*. Washington, DC: Island Press.

[74] Hodgson, J. A, Kunin, W. E, Thomas, C. D, Benton, T. G, & Gabriel, D. (2010). Comparing organic farming and land sparing: optimizing yield and butterfly populations at a landscape scale'. *Ecology Letters 13* (11), 1358-1367.

[75] Hole, D. G, Perkins, A. J, Wilson, J. D, Alexander, I. H, Grice, F, & Evans, A. D. (2005). Does organic farming benefit biodiversity?' *Biological Conservation,* , 122(1), 113-130.

[76] Holzschuh, A, Steffan-dewenter, I, & Tscharntke, T. (2008). Agricultural landscapes with organic crops support higher pollinator diversity'. *Oikos,* , 117(3), 354-361.

[77] Hooijer, A, Silvius, M, Wösten, H, & Page, S. (2006). PEAT-CO_2, Assessment of CO_2 emissions from drained peatlands in SE Asia'. *Delft Hydraulics report Q3943 (2006)*

[78] Horskins, E. (2005). *The Effectiveness of Wildlife Corridors in Facilitating Connectivity: Assessment of a Model System from the Australian Wet Tropics.* PhD thesis, Queensland University of Technology.

[79] Huafeng, D. K. D. (2007). High Conservation Value Forests Theory,Practice and Future Tendency'. *World Forestry Research,* , 2007 - 05 .

[80] Jackson, L, Van Noordwijk, M, Bengtsson, J, Foster, W, Lipper, L, Pulleman, M, Said, M, Snaddon, J, & Vodouhe, R. (2010). Biodiversity and agricultural sustainagility:

from assessment to adaptive management'. *Current Opinion in Environmental Sustainability*, 2(1-2): 80-87.

[81] Jennings, S, Nussbaum, R, Judd, N, & Evans, T. (2003). *The High Conservation Value Forest Toolkit*'. ProForest. Available at: http://lvoh.aprojekt.sk/upload/File/medzinarodny%20koncept/hcvftoolkitglobal.pdfAccessed 13 August 2012)

[82] Jepson, P, Barua, M, & Buckingham, K. (2011). What is a conservation actor?' *Conservation and Society, 9* (3), 229-235.

[83] Kaimowitz, D, & Angelsen, A. (1998). *Economic Models of Tropical Deforestation: A Review*. Bogor, Indonesia: Centre for International Forestry Research.

[84] Koh, L. P, & Wilcove, D. S. (2007). Cashing in palm oil for conservation'. *Nature*, , 448, 993-994.

[85] Koh, L. P. (2008). Can oil palm plantations be made more hospitable for forest butterflies and birds?' Journal of Applied Ecology, , 45(4), 1002-1009.

[86] Koh, L. P, Levang, P, & Ghazoul, J. (2009). Designer landscapes for sustainable biofuels'. *Trends in Ecology and Evolution, 24* (8), 431-438.

[87] Koh, L. P, Miettinen, J, Liew, S. C, & Ghazoul, J. (2011). Remotely sensed evidence of tropical peatland conversion to oil palm', *Proceedings of the National Academy of Sciences, 108* (12), 5127-5132.

[88] Komar, O. (2006). Ecology and conservation of birds in coffee plantations: a critical review'. *Bird Conservation International*, , 16(1), 1-23.

[89] Laurance, W. F. (1990). Edge effects in tropical forest fragments: Application of a model for the design of nature reserves'. *Biological Conservation, 57* (2), 205-219.

[90] Laurance, S. G, & Laurance, W. F. (1999). Tropical wildlife corridors: use of linear rainforest remnants by arboreal mammals'. Biological Conservation, 91(2-3): 231-239.

[91] Laurance, W. F, & Peres, C. A. eds.). ((2006). *Emerging Threats to Tropical Forests* Chicago: University of Chicago Press.

[92] Laurance, W. F, Camargo, J. L. C, Luizão, R. C. C, et al. (2011). The fate of Amazonian forest fragments: a 32 year investigation', *Biological Conservation, 144* (1), 56-67.

[93] Lees, A. C, & Peres, C. A. (2007). Conservation Value of Remnant Riparian Forest Corridors of Varying Quality for Amazonian Birds and Mammals'. Conservation Biology, , 22(2), 439-449.

[94] Leplay, S, Busch, J, Delacote, P, & Thoyer, S. (2010). Local and national REDD income under alternative PES designs: theory and illustration from Sumatra', *International Conference: Environment and Natural Resources Management in Developing and Transition Economies* CERDI November. Clermont Ferrand., 18-19.

[95] Levy, M. (2009). A Review of High Conservation Value Forest (HCVF) Assessments in intensively managed plantations'. OCTF, http://www.eci.ox.ac.uk/octf/events/junlevy.pdf

[96] Lindenmayer, D. B, & Hobbs, R. J. (2004). Fauna conservation in Australian plantation forests- A review', *Biological Conservation* , 119(2), 151-168.

[97] Lovejoy, T. E. Bierregaard Jr., R.O., Rylands, A.B., Malcom, J.R., Quintela, C.E., Harper, L.H., Brown Jr., K.S., Powell, A.H., Powell, G.V.H., Schubart, H.O.R. and Hays, M.B. ((1986). Edge and other effects of isolation on Amazon forest fragments'. In: Soulé, M.E. (ed.). *Conservation Biology: the Science of Scarcity and Diversity*. Sunderland, MA: Sinauer Associates Inc. , 257-285.

[98] Lovejoy, T. E, & Hannah, L. eds.). ((2005). *Climate Change and Biodiversity*. Michigan: Yale University Press.

[99] Macdonald, E. A, Collins, M, Johnson, P. J, Clayton, L. M, Malhi, Y, Fisher, J. B, Milner-gulland, E. J, & Macdonald, D. W. (2011). Wildlife conservation and reduced emissions from deforestation in a case study of Nantu National Park, Sulawesi 1. The effectiveness of forest protection- many measures, one goal'. *Environmental Science and Policy* , 14(6), 697-708.

[100] Maddox, T, Priatna, D, Gemita, E, & Salmpessy, A. (2007). *The conservation of tigers and other wildlife in oil palm plantations, Jambi Province, Sumatra, Indonesia (October 2007)*. London: Zoological Society of London.

[101] Matson, P. A, & Vitousek, P. M. (2006). Agricultural intensification: will land spared from farming be land spared for nature?' Conservation Biology, , 20(3), 709-710.

[102] Mattison, E. H. A, & Norris, K. (2005). Bridging the gaps between agricultural policy, land-use and biodiversity'. *Trends in Ecology and Evolution*, , 20(11), 610-616.

[103] Moguel, P, & Toledo, V. M. (1999). Biodiversity conservation in traditional coffee systems of Mexico'. *Conservation Biology*, , 13(1), 11-21.

[104] Moll, J, Crosthwaite, J, Dorrough, J, Shea, D, Shea, R, Moxham, C, & Straker, A. (2007). Farm businesses can profitably manage biodiversity'. A paper to the Conference: *Biodiversity Balancing Conservation and Production- Case Studies from the Real World*, Launceston, June 2007., 26-28.

[105] Mongabay (2012). http://news.mongabay.com/2012/ucs_vegetable_oil_deforestation.html>Last accessed Aug 16, 2012

[106] Myers, N. (1989). A major extinction spasm: predictable and inevitable?' In: Western, D. and Pearl, M.C. (eds.). *Conservation for the twenty-first century*, New York, NY. Oxford University Press. , 42-49.

[107] Newmark, W. D. (1993). The role and design of wildlife corridors with examples from Tanzania'. *Ambio* , 22(8), 500-504.

[108] Newmark, W. D. (1996). Insularization of Tanzanian parks and the local extinction of large mammals'. *Conservation Biology, 10* (6), 1549-1556.

[109] Norris, K. (2008). Agriculture and biodiversity conservation: opportunity knocks'. Conservation Letters, , 1(1), 2-11.

[110] Peh, K. S-H, Sodhi, N. S, De Jong, J, Sekercioglu, C. H, Yap, C. A-M, & Lim, S. L-H. (2006). Conservation value of degraded habitats for forest birds in southern Peninsular Malaysia'. *Diversity and Distributions* , 12(5), 572-581.

[111] Perfecto, I, Vandermeer, J, Mas, A, & Soto-pinto, L. (2005). Biodiversity, yield, and shade coffee certification'. *Ecological Economics, 54* (4), 435-446.

[112] Perfecto, I, & Vandermeer, J. (2008). Biodiversity conservation in tropical agroecosystems: A new paradigm'. *Annals of the New York Academy of Science, (The Year in Ecology and Conservation Biology 2008)* 1134: 173-200.

[113] Perfecto, I, Vandermeer, J, & Wright, A. (2009). *Nature's matrix: Linking agriculture, Conservation and food sovereignty*. London: Earthscan Press.

[114] Perfecto, I, & Vandermeer, J. (2010). The agricultural matrix as an alternative to the land-sparing/agricultural intensification model: facing the food and biodiversity crises'. *Proceedings of the National Academy of Sciences, 107* (13), 5786-5791.

[115] Persey, S, & Anhar, S. (2010). Biodiversity Information for Oil Palm'. *2nd International Conference on Oil Palm and Environment 2010: Measurement and mitigation of environmental impact of palm oil production.* Bali, INDONESIA February. Proceeding ICOPE 2010., 23-25.

[116] Phalan, B. (2009). Does farming have to cost the earth?' Department of Zoology, University of Cambridge. Available at: http://www.srcf.ucam.org/awtbi/documents/Doesfarminghavetocosttheearth.pdfAccessed 14 August 2012)

[117] Phillipps, Q, & Phillipps, K. (2010). *Phillipps' Field Guide to the Birds of Borneo: Sabah, Sarawak, Brunei and Kalimantan.* Midpoint Trade Books.

[118] Pimm, S. L, & Raven, P. (2000). Biodiversity: Extinction by numbers'. *Nature,* , 403, 843-845.

[119] Pineda, E, Moreno, C, Escobar, F, & Halffter, G. (2005). Frog, bat, and dung beetle diversity in the cloud forest and coffee agroecosystems of Veracruz, Mexico'. Conservation Biology, , 19(2), 400-410.

[120] Polasky, S, Nelson, E, Lonsdorf, E, Fackler, P, & Starfioeld, A. (2005). Conserving species in a working landscape: land use with biological objectives'. *Ecological Applications,* , 15(4), 1387-1401.

[121] Puechmaille, S. J, Soisook, P, Yokobul, M, & Piyapan, P. Ar Gouilh, M., Khin, M.M., Khin, K.K., Mackie, I., Bumrungsri, S., Dejtaradol, A., Tin, N., Si Si, H.B., Satasook, C., Bates, P. and Teeling, E. ((2009). Population size, distribution, threats and conser-

vation status of two endangered bat species *Craseonycteris thonglongyai* and *Hipposideros turpis'. Endangered Species Research, 8(1-2): 15-23.*

[122] Rice, R. A, & Greenberg, R. (2000). Cacao cultivation and the conservation of biological diversity'. *Ambio* , 29(3), 167-173.

[123] Ricketts, T. H. (2004). Tropical forest fragments enhance pollinator activity in nearby coffee crops'. *Conservation Biology, 18* (5), 1262-1271.

[124] Round, P. D, Gale, G. A, & Brockelman, W. Y. (2006). A comparison of bird communities in mixed fruit orchards and natural forest at Khao Luang, southern Thailand'. *Biodiversity and Conservation, 15* (9), 2873-2891.

[125] Scherr, S. J. and McNeely, J.A (eds). ((2007). *Farming with Nature: the Science and Practice of Ecoagriculture.* Washington, DC: Island Press.

[126] Schroth, G. (2003). Decomposition and nutrient supply from biomass'. In: Schroth G, Sinclair FL (eds.) *Trees, crops and soil fertility: concepts and research methods.* Wallingford: CAB International. , 131-150.

[127] Schroth, G. da Fonseca, G.A.B., Harvey, C.A., Gascon, C., Vasconcelos, H. L. and Izac, A.M.N. (eds.) ((2004). *Agroforestry and biodiversity conservation in tropical landscapes.* Washington, DC: Island Press.

[128] Sheldon, F. H, Styring, A, & Hosner, P. A. (2010). Bird species richness in a Bornean exotic tree plantation: A long-term perspective'. *Biological Conservation* , 143(2), 399-407.

[129] Sodhi, N. S, Koh, L. P, & Prawiradilaga, D. M. Darjono, Tinulele, I., Putra, D.D. and Tan, T.H.T. ((2004). Land use and conservation value for forest birds in Central Sulawesi (Indonesia)'. *Biological Conservation* , 122(4), 547-558.

[130] Sodhi, N. S, & Brooks, B. W. (2006). *Southeast Asian Biodiversity in Crisis.* Cambridge: Cambridge University Press.

[131] Sodhi, N. S, Posa, M. R. C, Lee, T. M, Bickford, D, Koh, L. P, & Brook, B. W. (2010). The state and conservation of Southeast Asian biodiversity'. *Biodiversity Conservation,* , 19(2), 317-328.

[132] Strassburg, B. B. N, Kelly, A, Balmford, A, Davies, R. G, Gibbs, H. K, Lovett, A, Miles, L, Orme, C. D. L, Price, J, Turner, R. K, & Rodrigues, A. S. L. (2010). Global congruence of carbon storage and biodiversity in terrestrial ecosystems'. *Conservation Letters, 3* (2), 98-105.

[133] Struebig, M, Kingston, T, Zubaid, A, Mold-adnan, A, & Rossiter, S. J. (2008). Conservation value of forest fragments to Palaeotropical bats'. *Biological Conservation* , 141(8), 2112-2126.

[134] Struebig, M. J, Paoli, G, & Meijaard, E. (2010). A reality check for designer biofuel landscapes'. *Trends in Ecology and Evolution* , 25(1), 7-8.

[135] Taki, H, & Kevan, P. G. (2007). Does habitat loss affect the communities of plants and insects equally in plant-pollinator interactions? Preliminary findings'. *Biodiversity Conservation, 16* (11), 3147-3161.

[136] Tejeda-cruz, C, & Sutherland, W. J. (2004). Bird Responses to Shade Coffee Production'. *Animal Conservation* , 7(2), 169-179.

[137] Tscharntke, T, Bommarco, R, Clough, Y, Crist, T. O, Kleijn, D, Rand, T. A, Tylianakis, J. M, Van Nouhuys, S, & Vidal, S. (2007). Conservation biological control and enemy diversity on a landscape scale'. *Biological Control,* , 43(3), 294-309.

[138] Turner, I. M, & Corlett, R. T. (1996). The conservation value of small isolated fragments of lowland tropical rainforest'. *Trends in Ecology and Evolution* , 11(8), 330-333.

[139] Turner, E. C, & Foster, W. A. (2009). The impact of forest conversion to oil palm on arthropod abundance and biomass in Sabah, Malaysia'. *Journal of Tropical Ecology,* 25(1), 23-30.

[140] Waggoner, P. E. (1996). How much land can 10 billion people spare for nature?'. *Daedalus,* , 125, 73-93.

[141] Wallace, G. N, & Barborak, J. and MacFarland, C.G. ((2005). Land-use planning and regulation in and around protected areas: a study of best practices and capacity building needs in Mexico and Central America'. *Natureza y Conservacao* , 3, 147-167.

[142] Warkentin, I. G, & Greenberg, R. and Salgado Ortiz, J. ((1995). Songbird use of gallery woodlands in recently cleared and older settled landscapes of the Selva Lacandona, Chiapas, Mexico'. *Conservation Biology, 9* (5), 1095-1106.

[143] Watts, K, Griffiths, M, Quine, C, Ray, D, & Humphrey, J. W. (2005). *Towards a woodland habitat network for Wales*. Contract Science Report Countryside Council for Wales, Bangor.(686)

[144] Wells, K, Smales, L. R, Kalko, E. K. V, & Pfeiffer, M. (2007). Impact of rain-forest logging on helminth assemblages in small mammals (Muridae, Tupaiidae) from Borneo'. *Journal of Tropical Ecology, 23* (1), 35-43.

[145] Wild Asia ((2011). Bird Diversity Study in Sarawak Oil Palm Plantation'. Wild Asia Technical Support Programme. Available at: http://oilpalm.wildasia.org/1376/bird-diversity-study-in-sarawak-oil-palm-plantation/Accessed 14 August 2012)

[146] WWF ((2012). Palm oil and biodiversity loss'. WWF Global. Available at http://wwf.panda.org/what_we_do/footprint/agriculture/palm_oil/environmental_impacts/biodversity_loss/Accessed 16 August 2012)

Chapter 3

The Drivers Behind the Rapid Expansion of Genetically Modified Soya Production into the Chaco Region of Argentina

Lucía Goldfarb and Annelies Zoomers

Additional information is available at the end of the chapter

http://dx.doi.org/10.5772/53447

1. Introduction

The global demand for biofuels is increasing rapidly. Such measures as the European Biofuel Directive (2003/30/EC) promoting the use of biofuels for EU transport[1] are giving an impulse to the cultivation of soya, especially in Argentina, the source of most European biodiesel imports. The EU goal of replacing fossil fuels with biofuels, combined with the growing demand from China and the USA, is contributing to the expansion of soya for feed and fuel in the Southern Cone, turning its countries into potential energy crops providers for Northern markets. The soya frontier is rapidly expanding in Argentina and beyond its borders (e.g. in Paraguay, Bolivia, Brazil and Uruguay).

Argentina is the world's largest exporter of soya bean oil and soya meal (it supplies about 45 per cent of the world market) and the third largest exporter of soya beans, after the USA and Brazil. Its main customers are China and the EU. Soya covers about 19 million ha of the country's territory, representing about 50% of Argentina's total cultivated area. Over the last 15 years, genetically modified (GM) soya production has expanded dramatically. GM soya was first grown in the central Pampean region, initially on pasture land and in rotation with cattle wheat. Later, it was introduced to other regions, displacing other types of land use and occupying land, increasingly at great cost to native forests and *monte*.[2,3] Most of the soya

1 This Directive on the promotion of the use of biofuels and other renewable fuels for transport, which entered into force in May 2003, stipulates that EU Member States countries must take national measures to replace 5.75% of all transport fossil fuels (petrol and diesel) with biofuels by 2010.

2 Designates non-cultivated land with native vegetation that could include a range of types and density of trees, bushes and plants, but also designates a full ecosystem, which are constitutive of peasant livelihoods in northern Argentina.

produced today is GM.[4] Soya production is now moving northwards, especially into the Chaco region, and also towards other countries in South America. Conditions that made this expansion possible included the dissemination of direct tillage technologies, favourable international prices, changes in the scale and organization of agricultural companies (including new forms of land administration and the concentration of control), and the availability of new land through deforestation.[5] Soya expansion was also promoted by the government as a 'green' way to increase foreign currency reserves and to boost rural development and energy self-sufficiency.

This expansion of the soya frontier has many implications. It leads to land-use changes and deforestation, for example in the Chaco region. Soya expansion entails the 'monocultivation of the land' and the 'soyization of the pampas and forest'. Production is extremely large scale and capital intensive, which is especially a result of the introduction of zero-tillage techniques. There are, however, also indirect effects. Small-scale farmers and indigenous communities are affected, as are cattle and food producers, who are pushed away towards new zones. In addition, the expansion of the soya frontier is more than simply a change of land use: it is a basic transformation involving new technologies (zero-tillage techniques), new power relations (large companies taking over), and a shift from labour-intensive to capital-intensive production regimes, and is being accompanied by new dependencies (the reliance on a small number of firms that provide agrochemical packages). The rapid expansion of the soya frontier also puts pressure on the land, leading to speculation; it is also seen as a manifestation of land grabbing.

This chapter is based on fieldwork carried out in Argentina in between 2011 and 2012. It provides a critical analysis of the transformations that are taking place as a consequence of the rapid expansion of soya. In it, we look at the features of the current and earlier soya expansion, the actors involved and the consequences, especially for the Chaco region. More particularly, we focus on the extent to which local populations are displaced and/or enclosed, and the role of land governance and territorial planning in bringing the frontier under control.

2. The expansion of the soya frontier: How it happened

No other part of the world has devoted so much land to a single GM crop as Latin America. In 2009, the area covered with GM soya in South America was around 40 million ha [2]. In Argentina – the world's largest exporter of soya oil and meal (supplying about 45% of the world market) – more than 50 per cent of the agricultural land is covered with this crop, reaching a production level of 54 million tons[6,7] and covering an area of about 19 million ha

3 Paruelo, J.M, Oesterheld, M. Patrones espaciales y temporales de la expansión de Soja en Argentina. Relación con factores socio-económicos y ambientales. LART / FAUBA (2005) http://www.agro.uba.ar/users/lart/bancomundial/INFORME_final.pdf/ (accessed July 2012)

4 The legalization of Roundup Ready soya in Argentina in 1996 was followed by the illegal introduction of GM soya, first in Brazil and Paraguay, and lately in Bolivia.

5 Parueloet al. (idem).

http://dx.doi.org/10.5772/53447

(2009–10). After the USA and Brazil, Argentina is the world's third largest soya bean producer; it exports soya beans mainly to China and the EU.

In Argentina, the industrial production of soya started in the late 1970s in the 'nucleus zone' or pampas region, where it contributed to the intensification of land use: there was a shift from labour-intensive cattle ranching and dairy activities to capital-intensive agricultural activities.

Figure 1. South American Chaco region

An important moment for the expansion of soya (and its incursion into neighbouring countries) was 1996, when Argentina approved the production of GM soya,[8] which at the time was forbidden in all its neighbouring countries.[9,10] By doing this, Argentina became an outlet for

6 The soya yield in Argentina is almost the same as in the USA and over 20% higher than the average of the other top 5 exporters (USDA: 2010).

7 USDA. Northern Argentina Production Potential Continues to Grow. Foreign Agricultural Services. 2010 http://www.pecad.fas.usda.gov/highlights/2010/06/Argentina/2010jun24publictravelmcwill.pdf/ (acceded April 2012).

8 Today, Argentina is the world's second largest producer of GM crops after the USA. GM crops (e.g. soya, corn and cotton) cover a total area of 22 million ha.

9 'Brazilian farmers bought the illegal seeds (from Argentina) on such a scale that the official ban on GMOs became meaningless and was revoked by President Lula. Similar tactics were used to spread RR soya into Paraguay and Bolivia'. (GRAIN, 2009)

the massive and rapid expansion of 'illegal' transgenic soya beans. Soya is now the main export crop of Brazil and Paraguay and is one of the largest in weight in Bolivia and Uruguay, leading to their nickname 'Repúblicas unidas de la soja'.

2.1. The technology frontier

Following its introduction in 1996, GM soya spread rapidly throughout the continent. Large companies started to expand production by leasing land from small and medium-sized producers. Expansion began in the core Pampean region and is now moving towards the north, and even into and from Brazil, and into Paraguay, Uruguay and Bolivia in a trans-regional capital move [2], in order to reduce climate risks. Brazilian soya entrepreneurs (such as the Brazilian 'king of soya', the Andre Maggi group) are now purchasing land in Argentina, arguing that some costs are cheaper there. And soya growers in Paraguay – mostly Brazilian migrants or their descendants (the *brasiguayos*) – are expanding their soya plantations into the Paraguayan Chaco region, since they realize that the original production area in eastern Paraguay is reaching its limits.

At the same time, there are clear indications that the GM soya frontier is likely to move across the Atlantic towards Africa, where 'bridges' between countries such as Brazil and Mozambique are the starting points for connections to bring in know-how, machinery and GM technologies, and to increased interest in African land. Argentina's machinery producers and land speculators are unlikely to miss out on this business opportunity.

The production of GM soya expanded at a breakneck pace: in 1997–99, the area expanded by an average of 2 million ha per year. In 2002, Roundup Ready soya beans already accounted for 99 per cent of total soya production. The attraction of using GM soya instead of conventional soya was based not so much on yields (the productivity of conventional varieties is sometime higher), as on the compatibility with no-tilling systems, easy field operations and the simplicity of weed management.[11] The new GM seeds facilitated pest control over large areas of land and made expansion much more viable. The introduction of GM soya, however, made producers highly dependent on herbicides and fertilizers. This resulted in the rapid increase in herbicide use, raising questions about the economic and environmental sustainability of this production system.

With the massive introduction of the GM soya technological package, this production model started to expand towards areas that had not been considered suitable for agriculture. Examples are Argentina's northeast and the Chaco region, which are vulnerable in both environmental and social terms [3, 4]. Due to new technologies and soft innovations, such as new management models (introducing sowing pools and the leasing of land by extra-regional farmers), the soya frontier shifted towards semiarid areas, such as the Chaco region.[12] In the

10 GRAIN, Las consecuencias inevitables de un modelo genocida y ecocida. 2009. http://www.grain.org/article/entries/1232-las-consecuencias-inevitables-de-un-modelo-genocida-y-ecocida/ (accessed 30 January 2011).

11 Benbrook, C. Rust, resistance, run-down soils and rising costs. Problems facing soya bean producers in Argentina. Technical Paper no. 8. Benbrook Consulting Services & Ag BioTechInfoNet. 2005. http://www.greenpeace.de/fileadmin/gpd/user_upload/themen/gentechnik/Benbrook-StudieEngl.pdf/ (accessed 30 October 2011).

last 50 years, 2 million ha of forest have been cut in the centre and south of this region for the cultivation of cotton and soya, leading to environmental degradation.

After its introduction, the use of the GM soya technological package expanded very quickly. Technology has become a tool to control access to land, based on techno-fix approaches to sustainability.[13] In Argentina, the total soya bean area has expanded more than that of any other crop since 1996, and especially since the introduction of direct tilling techniques and Roundup (glyphosate) Resistant soya, combined with the intensive use of fertilizers and agrochemicals. This technological innovation made soya cultivation economically feasible, even in agro-ecological zones where soya beans would not grow before. The combination in one package of direct tillage techniques and machinery with Roundup Ready soya is fundamental for expansion to other ecological regions out of the core areas.

In addition, a new way of organizing production through planting management pools, machinery and labour contractors, and storage facilities, allowed decreasing costs through the scaling of production, which in turn generated even more incentives for territorial expansion [5]. Production is increasingly becoming large scale, because medium and smaller producers are unable to make the required investments in technological innovation (new seeds, agrochemicals, less crop rotation, tendency to monoculture and less sustainable agricultural practices). By applying techno-fix solutions, large-scale enterprises have better possibilities to 'correct' environmental and social costs, which they consider externalities of the business model.

In conclusion, technological innovation (direct tillage plus genetically modified glyphosate resistant soya) in Argentina has changed the way in and scale at which land can be devoted to a single crop, cultivated in large production units, and managed by few hands and with almost no human labour. Interested actors, such as seed and agrochemical multinationals, have actively promoted this model, with the paradigmatic case of Monsanto allowing Argentinean farmers to produce and multiply GM soya without paying royalties and for years ignoring the illegal distribution of the seed across the South American region. The commercialization of Monsanto RR soya and such inputs as glyphosate was authorized in obscure circumstances by the secretary of Agriculture in 1996: the scientific evaluations were performed by directly interested private actors and there was no public debate. The report from the secretary of Agriculture – which did not include a biosafety assessment, was not made public and lacked a peer-reviewed assessment – was approved in just 81 days, and included a large section in English written by Monsanto [6, 7]. Other regions are going to be opened for soya production through biotechnological innovation. For instance, the Obispo Colombres Experimental Station[14] is working on biotech soya development and the improvement of farmers' practices in the northwest region of the country [7]. As a result, yields have improved in this region, and storage facilities and crushing facilities are moving up north. It shows that technological

12 46% of the Chaco belongs to Argentina, 34% to Paraguay (3 departments) and 20% to Bolivia (3 departments).

13 Levidow, L., Papaioannou, T. Innovation narratives in European agricultural research,Crepe research reports. 2010. http://crepeweb.net/?page_id=339/ (accessed November 2010).

14 Estación Experimental Agroindustrial Obispo Colombres is an autonomous entity of the Ministerio de Desarrollo-Productivo del Gobierno de Tucumán.

innovation is a key factor in the advance of the agricultural frontier, allowing GM soya to be produced on various types of land and in various climates.

2.2. The land frontier

While it is widely acknowledged that international prices plus technology innovation were responsible for the expansion of soya into the 'marginal' lands of the Chaco region, other factors are also responsible for the rapid expansion of the soya frontier. It was not just the combination in a single package of the application of zero-tillage techniques and glyphosate-resistant soya seeds developed by the multinational Monsanto; it was also the fact that when the new technology appeared, large areas of underused hacienda land were available in the pampas for new and relatively quick occupation, and that extra-sector financial capital was ready to be invested.

There had for a long time (including during the period of import substitution industrialization) been a decreasing interest in farmland in Argentina; large areas of *latifundia* land (land reforms have never taken place there) were increasingly abandoned, and especially in the 1990s people lost interest in farming; capital flowed to the industrial and other sectors, especially after 1991, when in the context of the Washington Consensus, interest in agricultural production decreased. The national grain board (Junta Nacional de Granos) was closed, putting an end to state support to the agricultural sector.

It was in this context of the widespread availability of 'underused' land that technological innovation took place. Large companies and investment funds took the initiative of organizing 'sowing pools', establishing and managing the use of the technological package, controlling large tracts of land and starting to control the whole production chain, based on the previous dissemination of direct-tillage technologies among producers. It is the combination of this new form of organization (sowing pools), technological innovation (direct tillage, GM seeds) and the availability of large areas of unused land that explain the rapid expansion of soya production in the Pampas region. In the core area provinces like Santa Fe – the cradle of direct tillage – soya production advanced [rapidly, on the basis of land leases to achieve scale.]

This rapid expansion pushed annual crops and cattle ranching out of the Pampean region. In the Chaco region, deforestation also increased rapidly, due to the displacement of cattle raising to northern provinces and direct investments in land clearing for new soya plantations. This was accompanied by social conflicts, since the area was occupied by smallholders (peasants) and indigenous groups. By the end of the 1990s, high international soya bean prices and the saturation of the core area led to shifts in the investment pattern, namely from the Pampean region to the northwest of the country and into the Chaco forest, to acquire land for soya production, through either land-use change or the clearing of native forests.[15] According to Benbrook [6], between 1996 (when GM soya was legalized) and 2004, 32 per cent of the expansion of soya plantations in Argentina took place on land previously used for other crops,

15 Slutzky, D. Situaciones Problemáticas de Tenencia de la Tierra en Argentina, Prodinder, 2008.http://www.proinder.gov.ar/productos/Biblioteca/contenidos/estinv.14.situaciones% 20problematicas% 20de% 20tenencia% 20de% 20la% 20tierra.pdf/ (accessed June 2011).

http://dx.doi.org/10.5772/53447

27 per cent on former pastures and hay fields, and 41 per cent on newly cleared forest and savannah.

Figure 2. Argentina. Current soya producing provinces

Whereas soya expansion was realized in the pampas region without major transformations in property relationships, the situation was very different in the Chaco region. Here, the expansion was accompanied by large-scale land transfers, the purchase of large areas of land at 'nonsense' prices [8] and the converting new land into agricultural land. In the northwest of Argentina, soya expansion was horizontal; the cultivated area grew by 48 per cent between 1988 and 2002, especially in the provinces of Santiago del Estero and Salta, where the agricultural area doubled between 1988 and 2008 [9]. In both provinces, this process led to the destruction of native forestland (that had carob and quebracho trees) and to the end of small-scale farming, which had mainly been developed on land occupied by peasants, familiy producers and indigenous groups who lacked legal land titles. The possession and indigenous territory status compete with commercial pressure on this land, causing tensions and often even open conflicts with local land dwellers, and former land use was replaced by soya or livestock.

Because of the boom, soya bean producers in the Pampas, who already controlled the land in this area through ownership or management, made good money. Many of them decided to use their profits to purchase or lease additional land in the northern regions. It is therefore mainly the successful producers from the pampas provinces, such as Buenos Aires, Córdoba and Santa Fe, who are now buying land in the northern provinces, such as Santiago de Estero. Since land users in this area often do not hold legal land titles, land transfers are fairly easy and can depend on different mechanisms of shifting land control. For investors, in terms of opportunity costs, the purchase or occupation through fraudulent or legal eviction of relatively cheap land compensates for unstable yields, the cost of vegetation clearance and higher transport costs. The combination of ownership and other forms of access to land, such as lease contracts, continued to increase nationally by 25 per cent in the period 1988–2002, and these mixed forms, which are typical of the Pampean region, also are becoming more common in the Chaco region. The process has been described as the 'pampeanization of the Chaco', and allows new players that were already present in the nucleus region to produce in larger units. 'Accidental contracts' in particular have been instrumental as a strategy to enlarge the production units for soya bean and wheat, sunflower and livestock. Thus, the expansion of agriculture, and particularly livestock and soya, is linked not necessarily to large properties in the cadastral sense, but to large production units. Between 1988 and 2002, the agricultural area increased by 15 per cent in the whole country and but 50.3 per cent in the regions outside the pampas.

Besides the 'soyization' of the pampas and the 'agriculturization' of the Chaco region, which has brought soya into the forests and onto the land of indigenous groups and peasants, there is a third development taking place: the growth of cattle raising as a result of displacement from the pampas to other regions. In the province of Santiago del Estero, the dynamics of the land market set in train by the 'soya boom' at the end of the 1990s, soon left new investors with no 'empty' land for agriculture. Financial operations with land during the soya boom (2002–07) were among private buyers and sellers, and large-scale soya plantations were relatively easily developed since this land was the best in the province (mainly because of the availability of water); it was easily controlled and accessed through purchase and, if necessary, easily cleared, since forest protection and territorial planning legislation was not yet in place. However, once the way had been paved by soya, cattle ranching activities also increased, mainly on the remaining land, which was considered 'marginal' (or not so apt for agriculture) and was in general inhabited by holders with possession or territorial rights. In other words, soya expansion led to both a push and a pull process of cattle raising. These combined processes are reflected in a 2 per cent increase in the cattle population of the northern region. As a result, Santiago del Estero is one of the most affected provinces in terms of land-use change and the arrival of investors from other provinces. Its cattle herd increased by 40 per cent between 1988 and 2002 [10].

At the beginning of the soya era, individual entrepreneurs could afford to invest in land in the Chaco region. But prices in the increasingly dynamic land market recently rose so high that larger extra-provincial and extra-regional companies, investment pools and foreign investors have displaced local capital on the demand side. Even with rising land prices, pools are still able to invest in estates of 40–50 thousand ha. In many cases, these are only speculative investment funds, which acquire land in order to rent it out. Those who lead this process are mainly producers from the Pampean region such as Buenos Aires, Cordoba and Santa Fe. The investors find cheaper land, less regulation than in the Pampean region, and small producers with precarious tenure situations. Developing land in the Chaco region is expensive, because of deforestation, transport, irregular yields due to climate instability and the need to deal with conflicts with the holders of possession rights; nevertheless, it seems that cheap prices and short-term benefits compensate for these costs.

Although evidence shows that both rent and purchase land prices have been rising considerably in the last few years.

To sum up, the consolidation of the GM soya hegemony in Argentina's Pampean region displaced traditional cattle ranching to the north. But GM soya also made inroads into the Chaco forests, because the availability of cheap land, a lack of regulation and the underlying myth of 'empty land' [11][16] attracted new actors, such as real estate speculators, cattle growers and soya producers into the *monte*, including those of local, extra-regional and international origin. In territorial terms, soya has occupied the region's more productive land; it arrived in the early 2000s, in a violent and massive way, and in many cases at the expense of the forest, peasants' land and the *monte*. In general, extra-regional investors, with local connections especially in the field of technical support, have dominated the soyization of the Chaco. In general, the areas targeted for soya investments were easily accessible through purchase, while land conversion was carried out through the deforestation of vast areas, not necessarily empty but probably difficult to control by locals, even if they held rights to the land. The relationship with the local people was, in fact, less personalized; investors were generally represented by managers, which has both advantages and disadvantages when it comes to conflicts with the local communities and individuals with weak tenure rights.

2.3. Actors and interest groups

Thus, new frontiers of expansion are leading to new forms of land control, and to new actors and new types of alliances and relations among them. During the neoliberal period, medium- and large-scale farmers tried to alleviate the crisis by leasing land within and outside the core area to increase the size of their farms and the productivity of fixed factors using zero-tillage (*siembra directa*) techniques. Others were in a worse situation and were forced to rent out their land to those with the investment capacity. In order to minimize production costs, producers needed to use adequate technology. However, not all agricultural producers were able to incorporate new technology and upscale their level of operation. The indebted small farmer had no chance. Thus, a new type of actor appeared: the small renters, namely farmers who rent out their land to pools (*contratistas*) or to other landowners. This explains the rise of larger holdings through 'accidental' lease contracts (3 years). In this sense, there is a process of capital concentration rather than land concentration.

---

16 GAIA Foundation, Biofuelwatch, The African Biodiversity Network, Salva La Selva, Watch Indonesia, EconNexus, Agrofuel and the myth of marginal land. http://www.cbd.int/doc/biofuel/Econexus% 20Briefing% 20AgrofuelsMarginalMyth.pdf/ (accessed 19 November 2009).

Larger producers are those that can better address the costs through economies of scale, since the input sellers and traders who buy their grain have oligopolistic positions in the market. The purchasing power of larger producers allows them to get better input prices and cheaper access to transport access, as well as to store their production while awaiting better prices, and to reduce climate risks by producing in different geographical areas. Thus, small and medium-sized producers are ceding control over their production unit to other actors, such as larger landowners, contractors and sowing pools (*pools de siembra*). Actors that had different or even opposed interests in the past, now found a reason to establish an alliance, especially when political struggles against market regulation were perceived as necessary. The decrease in production units between 1988 and 2002 was mainly due to these processes of land cession (about 50,000 production units of less than 101 ha) and scaling up. The sector of small renters is now earning around 5–6 times more than they earned when they were producers. Apart from these changes in land tenure, GM soya development would not have been possible without the role of actors devoted to the dissemination and adoption of new technologies. This is the case of the direct-tillage system that was widely promoted by AAPRESID within Argentina, and that at the moment is lobbying to disseminate it globally, under the label of sustainability.

The following table presents more detail about the range of actors and interest groups in the soya sector.

Land based – traditional actors	**Description**	**Example**
Traditional large land owners	Abandon cattle growing activities and commence financial speculation Invest in modernized agriculture – technology package Keep their land and work it "/> 100,000 ha Expanding	Alzaga Unzué, Leloir, Blaquier, Fortabat, Bemberg, Duhau, Ayerza,
Medium- and large-scale producers (> 500 ha)	Increase the area of the land they rent to others (small, medium). Persisting	
Producers on 200–500 ha		
Producers on < 200 ha		
Land based – new actors		
Small renters (many of them in the group on < 200 ha)	Survived disappearance of 50,000 producers in the 1990s Now get 5–6 times their income as producers just by renting out their land Mainly in the pampas region	

Land based – traditional actors	**Description**	**Example**
New large owners coming from industrial sector	Invest in activities based on the foreignization of their industries in the agricultural sector. > 100,000 ha Expanding	Ratazzi, Terrabusi, Blanco Villegas
Absentee owners	Years later, they or their heirs claim land that is in the hands of occupants. Some falsely claim to be absentee owners	
Investment groups, trusts and sowing pools	From within and outside the agricultural sector. They work as investment funds. Rent land and manage it, and distribute benefits at the end of each agricultural cycle. Diversify production regions and contract machinery and services to avoid risks. Also work with foreign investors Expanding into Mercosur	Los Grobo, El Tejar, Cazenave, MSU-Uribelarrea, Lartirigoyen
Foreign investors in land	Comparatively cheap land, especially in the northeast and northwest Small incidence in the pampas region > 100,000 ha Expanding	Adecoagro-Soros, Lliag Argentina, Dreyfus
Real estate intermediaries & land speculators	Buy cheap land (sometimes from possessors and then get the titles) and sell it at a higher price	
Machinery contractors (*contratistas*)	Machinery and services for planting, fumigation and harvest in return for a share of production (used to be individuals, now a growing share of capitalized companies)	
Machinery producers	Increasingly concentrated and globalized with transnational and local actors.	Local companies: Deutz, Ferguson, Zanello
Storing infrastructure and Collectors (Acopiadores)	Buy the soya beans from producers and find a buyer. Deal with 80% of the production and large part of the value paid by exporters. They play an important role in ensuring the conservation and grain quality and supporting the business strategies of producers, allowing them to decide when to sell the crop Role in technical assistance and financing the producer	Aceitera General Deheza SACIC is one of the most important.

Land based – traditional actors	Description	Example
Biodiesel companies	This is a growing sector but there is already a large capacity installed	The 45% of the internal quota of biodiesel in hands of: UnitedBio, Viluco, Explora and Diaser. The remaining 55% goes to 15 companies, including Rosario Bioenergy, Biomadero and AOM
Input sellers	Few big transnationals selling seeds and some domestic companies. In the field of agrochemicals, the monopoly of Monsanto.	Seeds TNCs: Monsanto, Pioneer, Novartis, DuPont, Ciba Agrochemicas: Monsanto
Technical Assistance	Important role of the private sector Also transnationals providing inputs	AAPRESID (Argentinean Direct Tillage Producers Association) AACREA (Argentinean Association of Regional Consortia for Agricultural Experimentation)
Exporters	Buy soya beans and derivatives from producers, collectors and industry Fewer than 10 companies concentrate all exports of the chain Exporter condition industry, collection-storage, inputs selling and producers	83% soya beans exported by Cargill, Noble Argentina, ADM, Bunge, LDC-Dreyfus, AC Toepfer and Nidera. 82% soya oil exported by Bunge, LDC-Dreyfus, Cargill, ADG and Molinos Río de la Plata 90% other soya derivatives are exported by Cargill, Bunge, Dreyfus, AGD, Vicentín and Molinos Río de la Plata

*Sources: [12], [13], [14] and [15], and own elaboration.

Table 1. Actors and interest groups in the Argentinean soya sector*

2.4. The state: An actor again?

While soya expansion is promoted in the region as a 'green' way of increasing foreign currency reserves and boosting rural development and energy self-sufficiency, it is well documented that monoculture expansion is bringing about land-use change and the expansion of the agricultural frontier, which is advancing through the Chaco rainforest, small-scale farms and

http://dx.doi.org/10.5772/53447

indigenous communities, displacing food, cattle and regional crops production and leading to deforestation.

If we consider the state as an actor (which, of course, has analytical implications), in the case of Argentina – and in that of other Latin American countries – the state has returned to centre stage. The Kirchners' administrations era from 2003 to the present day have had a lot to do with re-establishing the power of the national state and its political initiative, and getting the country out of one of its deepest crises by taking progressive measures and in many cases reversing many of the neoliberal achievements. To some extent, the main success of this administration has been to give space back to politics, that is, introducing into the usual work of state institutions debates that were not widely discussed in society. Nevertheless, in the fields of rural development and the regulation of extractive activities (e.g. mining, forestry and soya monoculture), the state has failed in many ways and has acted only to create economic incentives for the further expansion of these activities. It has not properly addressed regulations for sustainability or protected rights.

First of all, the soya boom, and the income from export taxes, was used by the government to pay the external debt and provide social aid in the struggle against indigence, poverty and unemployment, in a period when more than half of the population was living below the poverty line. The main active policy measures to promote the development of the GM soya model were taken in the 1990s (when the government was openly neoliberal), namely through the Deregulation Decree in 1991, privatization, liberalization of the market and the liberalization of transgenic soya in 1996, giving impulse to its generalized expansion and the consolidation of the agribusiness model. After that, apart from the export taxes regime, state policy in relation to the expansion of the GM soya model was a complete laissez-faire, within which the agribusiness started to gain public technical and scientific knowledge arenas (such as INTA, SENASA and some areas of CONICET) as well as winning the battle of common sense in relation to the question which rural development path was desirable. Even when the government tried to be active in imposing a mobile export taxes system in 2008 that implied an increase in costs for producers, Cristina Kirchner faced the fierce opposition of the entire soya sector and a part of urban middle classes.

Nevertheless, in the last few years increasing awareness of the negative impacts of the development model – which had been denounced for years by social movements and NGOs – started to attract media interest, as well as spaces within, or in dialogue with, the state. Although with limited scope, the fact that the state is regaining the initiative in executive, legislative and judicial matters opens possibilities for regulation, whereas before there was only the market. With the conflict between the government and the soya-producing sector of 2008 over export rights, the hegemony and the problems involved in the soya model were put up for public discussion, albeit with many limitations, contradictions and omissions. While this conflict over who should benefit and how to distribute the income from soya exports was going on, another sector was ignored: *campesino* economies, the indigenous communities and the producers for the domestic food market, which are still not a target for rural development and agricultural policies.

3. Constructing a biodiesel market and sector: a new frontier of profitability?

The creation of a biofuels market through policies such as the European directive on biofuels is perceived by the governments of South American countries, as well as by actors in the production chain of soya, as an opportunity to diversify products and markets. In the Argentine case, the conditions for biodiesel production are already in place due to the existence of a highly efficient, integrated soya sector [16], which is mainly dominated by domestic and transnational industry players and traders. Biodiesel can be produced from the same biomass that produces soya oil and meal, and is highly dependent on an industrialized agriculture that is able to homogenize raw material for processing into the various by-products [17]. Argentina meets these conditions. Discourses justifying these policies refer to objectives such as diversifying the energy supply, environmental protection and rural development. In South America and in Europe, promotional discourses do not differ radically from these discourses, although they are much more centred on energy diversification and economic development – but not necessarily of rural areas.

Global drivers of GM soya expansion in South America are clearly related to the growing demand for animal protein food in Europe (and more recently also in China and other Asian countries), which demands large amounts of soya oil. In 2006, MERCOSUR member countries (Argentina, Brazil, Paraguay, Uruguay and Venezuela) signed a memorandum of understanding to encourage and promote biofuel production and consumption. The creation of a new biofuel market in the EU that will depend on Southern countries to reach the blending targets, opens a space for Argentinean biodiesel exports.[17] In this sense, the Argentinean authorities have been keen on setting policies with the aim of domestically creating value added and reduce energy imports, such as a differential export tax to promote biofuel development beyond domestic consumption, which is also being stimulated by compulsory mix quotas and presented as an alternative use for soya and thus diversifying markets. Measures to stimulate the production of biofuels have been aggressively lobbied for by the domestic soya bean sector. Therefore, on the domestic side, there is a huge demand for biofuel also in Argentina. Nevertheless, because the large companies are not interested in a domestic biofuels sector, most of the biodiesel produced by them, and the majority of total biodiesel production, continues to go to export markets, meaning that the land in Argentina continues to produce products that the population does not consume.

However, the tendency changed when the Argentine government promoted the creation of a domestic market, and thereby encouraged the soya sector to diversify, through Law 26.093 (2006), through which it established blending targets of 5 per cent for 2010, 7 per cent in 2011 and with the expectation of increasing it to 10 per cent in the next years. The exported volume started to decrease as long as the new policy was implemented. Most exports go to Europe; Spain is the main buyer. Today, 15–20 per cent of the soya bean crop in Argentina is used to

17 Friends of the Earth International. Fuelling destruction in Latin America – the real price of agrofuels 2008. http://www.foei.org/en/resources/publications/pdfs/2008/biofuels-fuelling-destruction-latinamerica/view/ (accessed January 2009).

http://dx.doi.org/10.5772/53447

produce biodiesel.[18] Investments in the sector have been growing since 2006 in the expectation of increasing exports and market diversification. Most of the investments in the biodiesel industry come from the usual big players in the international soya sector [25], such as Glencore, Nidera and Dreyfus, Bunge and Cargill, with alliances with the powerful domestic soya-processing sector and financial investors. To encourage investments in the biodiesel industry, the national government imposed a differential export tax that favours the production of biofuels over soya oil and meal (the difference is more than 17 percentage points). Local governments also encourage investments by providing tax exemptions. Prior to the legislation in Argentina, there was almost no domestic demand for products from soya beans, which made it difficult to justify a paradox for many defenders of the oilseed. However, the taxes on soya exports are estimated to account for around 12.5 per cent of the national budget [16].

The supply of seeds continues to be dominated by Monsanto and Syngenta, along with some big local companies. As mentioned, the dominant players in the biodiesel sector are the big transnational and domestic corporations that have dominated the soya sector for many years. They control about 80 per cent of the sector and in many cases integrate production and processing. As a subsector within the soya sector,[19] the biodiesel industry is a rather clustered sector at the cultivation and processing levels [16], since many of the domestic actors are located in the same geographical area, that is, around the Paraná river. But the domestic biofuels sector is also very dependent on international networks [16, 18].

	Legislation and incentives
Argentina	• Law 26.093 Compulsory biofuels blending (2006) • National Bioenergy programme • Differential export taxes vis-à-vis soya oil • Native Forests Law 26.331 (2007) • CARBIO voluntary certification scheme presented to EU • INTA. biofuels research

Table 2. Biodiesel promotion in Argentina

While additional demand for soya may lead to the further expansion of the agricultural frontier – exacerbating the social and environmental problems resulting from both past and current GM soya cultivation, and commercial pressures on land in the Chaco region – recent central government measures and an increasing conflict with EU import markets, put some question marks and bring contradiction to the biodiesel sector in Argentina. A recent announcement by

18 USDA Foreign Agricultural Services (2011) Global Agriculture Information Network Report, Argentina, Biofuels Annual 2011. http://gain.fas.usda.gov/Recent% 20GAIN% 20Publications/Biofuels% 20Annual_Buenos% 20Aires_Argentina_7-8-2011.pdf/ (accessed July 2012).

19 Actors from the soya sector particularly involved in the biodiesel promotion and production are: sowing pools, large farmers, renters, international agribusiness companies, vegetable oil refiners, service contractors, transport, storage/collection, biotechnology multinationals (seeds, fertilizer, pesticides, etc.), traders/exporters, processing industry (oil, meal, biodiesel), machinery input providers, research institutions (public and private) and business associations (AAPRESID, ACREA, CARBIO, ACSOJA, Comisión de Enlace).

Spain (the buyer of 70 per cent of all Argentinean biodiesel exports) that it would prohibit biodiesel imports from outside the EU, led Argentina to complain to the WTO. At the same time, the national government raised the taxes on biodiesel exports to the level of the other soya by-products, thereby changing the thrust of the main promotional policy. This seems to be exacerbated by the 15 per cent reduction in the price of biodiesel for the domestic market, which has negative impacts on medium and smaller non-integrated enterprises. Still the EU had been denouncing Argentinean dumping, through tax exemptions, and claiming the existence of more efficient biofuels from rapeseed and palm oil. In consequence, due to both national and international factors, there is uncertainty about the future of the biodiesel sector.

4. Main impacts of, and dilemmas posed by the expanding GM soya frontier in relation to sustainability, governance and development

The economy has recently been working quite well for the actors and interest groups linked to the soya and biodiesel industries. But the impacts of the dynamics emerging from the GM soya sector also have implications for rural actors that are not part of the production and distribution chain. In areas outside the core region, where soya has become rather hegemonic, other rural dwellers find their access to land, water and forests affected by the GM soya pressure.

4.1. Sustainable development for whom?

The rapid expansion of soya monoculture in the Argentinean Chaco region implies massive land-use changes and pressure on land whose users have weak tenure rights, or on land that is being claimed by different actors, such as companies, *campesinos*, indigenous communities and landless workers. The tensions lead to claims for the regularization of possession and indigenous rights. The problems arise because in the Argentinean Chaco provinces a larger number of peasants do not hold titles to the land they occupy and work on. For instance, around 75 per cent of peasants in Santiago del Estero are in this situation. Different mechanisms of pressure on the indigenous and peasant population as well as mechanisms of enclosure occur. These include violent evictions, unequal legal disputes, the pollution of water, soil and crops, and the clearance of forests, which are sources of livelihoods of the communities in the region.

The holders of possession rights are not so common in the Pampas, but are still a very relevant population in the northern provinces and, together with the indigenous communities, are the groups that find their access to land threatened or increasingly controlled by more powerful local, translocal and foreign actors. In particular, holders of possession rights have become vulnerable to commercial pressure on the land they occupy. As for this group, the mechanism to acquire land from them could imply the purchase of possession rights, which presents an advantage to the buyer who pays a price that is considerably under market value, since there are no property titles. A common situation that these holders face in most of the northern provinces, as well as in some areas of the core region, is the appearance of absentee owners or their descendants who claim ownership of land that have been occupied for decades by

http://dx.doi.org/10.5772/53447

families, including former workers of the owners who got paid in land rather than receiving a salary. The expansion of the frontier only speeds up this process, since the price of formerly 'unproductive' land rises, attracting not only new land users but also speculators.

Mention should also be made of swindling by dubious land claimants and of the fact that some of the holders of possession rights lose their access to land through direct displacement and enclosure. We observed in Santiago del Estero the use of lies, deceptive contracts and procedures, and intimidation and violence. In this sense, also *comodato* contracts imply only the beginning of the story or the process of gaining access, which can continue until the possessor leaves the land – perhaps still believing that he had no option but to cede the land. There are a range of situations between the extremes, such as ex-possession rights holders paying a rent to the new land user to raise cattle on the land they had lived on for decades. Incidentally, intimidation and threats occur not only in the Chaco region, but also in other areas, as Roodhuyzen (2010) showed for the islanders in the Argentinean Delta region.

The national campesino–indigenous movement (MNCI–Via Campesina) estimates that around 200,000 rural families have been forced off their land due to the advance of soya. The provinces where this process has been the most violent are Santiago del Estero and Córdoba, where heavy machinery owned by sowing pools and landowners drove over peasants' houses and parcels. REDAF documented that around 950,000 people – inhabitants of indigenous and *campesino* territories – are affected by land or environmental conflicts related to the expansion of the agricultural frontier, led by GM soya. Most of these conflicts began in 2000, when the GM soya model reached its height [22]. Many of these groups were occupants and holders of possession rights that, according to Argentinean law, imply a form of tenure security. Possessors of these rights are occupants who have been living on the land for more than 20 years and have performed 'possession acts', that is, they have worked on, cared for and improved the land. If these conditions are met, they can virtually apply for land ownership. In the northern provinces these groups are also threatened with the loss of access to the use of forests and common land for pasture. Threats usually occur explicitly and with the use of physical and verbal violence, as has been widely denounced by representatives of MOCASE.[20] In other words, the consensus that GM soya expansion had success in the core region seems to be contested at least in Chaco provinces. Here, tensions related to rights, social organization, judicial and political strategies to defend them, and claims for the institutionalization of this defence, have become a new space for land governance that might challenge dominant discourses legitimizing processes of the commoditization of land and the privatization of nature.

4.2. Environmental sustainability

Besides forest clearing for soya, there is also deforestation due to the incursion of cattle raising into the forest of the Argentinean Chaco: the equivalent of 2500 football fields is occupied every day. In Salta alone, 609,323 ha of forest disappeared between 1998 and 2006 (Greenpeace). Córdoba is the province where the process is most devastating, followed by Santiago del Estero

20 MovimientoCampesino de Santiago del Estero.

and Salta. This was particularly visible in the town of Tartagal, where violent floods occurred between 2006 and 2009, leaving behind death, destruction and homeless families, particularly indigenous families. According to local organizations there is a direct link between illegal deforestation and climate change, which led to the floods. As forests make way for monoculture, the soil cannot absorb enough water; as a consequence, there are increasing river flows and floods. Quite remarkable in this case is that expectations about the imminent implementation of a Native Forests Law (Nº 26.331) aimed at protecting forest areas, led to deforestation at an unprecedented speed, which was to blame for the violent flooding. Biodiversity is highly affected by deforestation, and by the poisoning of water and soils with herbicides, such as the massive application of glyphosate, which kills animals and microorganism, contaminates food crops and affects people's health. The excessive use of pesticides and little crop rotation turn the soil sterile through the years.

In addition to environmental harm, several (mainly poor) communities in various towns and cities in the main soya producing provinces in Argentina have been affected by serious health problems caused by indiscriminate fumigation practices in areas neighbouring the plantations, and in poor urban settlements where pesticides are stored. In 2003, Ituzaingó-Anexo reported that the incidence of cancer in children was 50 per cent higher than the national average, which was partly attributed to illegal fumigations in neighbouring GM soya fields. This diagnosis pushed a group of mothers (Madres de Ituzaingó-Anexo) to organize themselves politically and to make public their confrontation with local soya producers, by initiating legal actions against fumigation with Roundup. This was a pioneer case in the country, one that served as a precedent for many other localities where illegal fumigations had to be prohibited and legal ones regulated.[21]

4.3. Development and governance

In connection with the advance of the agricultural frontier into bio-diverse and socially vulnerable rain forests and savannahs [3, 4], we can observe the emergence of new actors and the trans-location of actors who put into practice new forms of control over land and natural resources with non-local capital. Translocal actors [19] have played and still play an important role in introducing capital, technologies and modes of production and management, which historically developed in the Pampas, into the Chaco region through the process of *pampeanization.* Some authors have put too much emphasis on the foreign character of the investments. However, the case of Argentina clearly shows that for the moment the relevance of foreignization is emblematic but not necessarily the mainstream, and that the role of the domestic and translocal private sector is more important than we thought.

As for the government, the purely economic argument of not losing an opportunity has been its leitmotiv and has led to the massive and unsustainable expansion of GM soya in the core region. The lack of policy was justified by arguing that it is best not to put limits on the market. The same type of argument was later used to justify laissez-faire policy with regard to the

21 In August 2012, the entrepreneur and pilot responsible for one of the cases of fumigation were taken to court, convicted and put on probation. This has created a precedent for future legal processes.

http://dx.doi.org/10.5772/53447

expansion into the Chaco region, allowing more powerful actors to impose their particular interests and make them seem universal. In fact, not much has been done in Argentina to define an agricultural policy for sustainable development. Associated measures were for a long time aimed at supporting the dissemination of GM soya seeds and ignored the protection of the rights of those with weak land rights (such as indigenous communities and holders of possession rights), the protection of those affected by environmental harm and the protection of the environment.

At the same time, Argentina, like some other Latin American countries, is becoming increasingly dependent on extractive activities (like GM soya cultivation), which are confronting the country with a real development paradox. The case of soya, and particularly biodiesel, raises questions regarding sustainability in a broader sense, since short-term economic development objectives and enjoyment increasingly conflict with long-term man–land relations across generations. In other words, the intrinsic logic of global capitalism makes national soya producers follow market signs without paying attention to ecological and social precautions, and the government opens the way to do that. It is therefore not surprising that the socio-historical process of the introduction, promotion, adoption and expansion of GM soya in the core region of Argentina determined to some extent the way in which the second phase of expansion is taking place in the northern provinces, where the scale of the changes has been much greater. Until 2007, tobacco and sugarcane competed with soya for land; however, soya has become dominant in recent years and is expected to continue to expand more than those other crops [3].[22] The increased demand for land is and will be accompanied by accumulation, privatization, enclosure and displacement [20, 21], redirecting resources away from local communities as commodities for distant consumers.

As stated, the expansion of the agricultural frontier in the Chaco region has been an historically ungoverned process. But if we consider governance as a synonym of control, different social groups have controlled and benefited from this process in different phases, although it was rather chaotic, with periods of land appropriation and others of land abandonment, and the taking over by other groups. In other words, the drivers of the expansion of the agricultural frontier into the Chaco were part of the process of governance. The public sector, however, remained relatively inactive for more than 15 years, that is, after the liberalization of GM soya in 1996. It was a way of acting that paved the way for development by letting it advance in an unregulated manner. This does not mean that nothing was done. Quite significant public resources were used to support scientific and technical public–private partnerships among national institutions such as INTA and CONICET and private actors in the soya chain. In addition, the taxation of crude soya exports did not address incentives to produce more, but only triggered controversies over the distribution of benefits among private and public actors.

The national government has recently taken a more proactive role in relation to policies for the agricultural sector, particularly with the launching of the new national PEA2 (Federal and Participative Agro-food and Agribusiness Strategic Plan, 2010/2016), through which Argentina seeks to achieve a privileged position as a global provider of food. The main goal is to increase

22 USDA (idem).

the amount of agricultural land in production by 6 million ha, but the plan neglects the existence of peasant production systems and excludes the Chaco region from the group defined as fragile ecosystems. An active policy of disseminating the PEA across society would show that it will seriously affect the future of rural people and the relationship between society and nature in the long run. However, it is already giving positive signals to investors in industrial agriculture and cattle raising. Moreover, some historically poor provinces that have been neglected by the state have recently been targeted for the Historic Reparation Acts, which include a range of infrastructural projects, many of them aimed at incorporating land to and converting land for agriculture production, as was the case with the Figueroa dam, which was built in 2011 to include 30,000 ha for agricultural and cattle production. Mention should also be made of the IIRSA project aimed at constructing a new infrastructural network on the South American continent, which will particularly affect the future of the South American Chaco region as a whole.

The only national and provincial legislation aimed at reducing the negative impacts of frontier expansion is the Forests Law (and the territorial planning measures linked to it), which deals with the environmental dimension of the advancing agricultural frontier through forest clearing. The results of territorial planning processes have so far been very heterogeneous in different provinces and are still open to negotiation and contestation, while the capacity of this type of regulation to protect vulnerable ecosystems has not been proven. At the same time, in many northern provinces peasant and indigenous resistance to the advance of the agricultural frontier has taken the form of a struggle organized around the recognition of possession rights, in order to gain, maintain and control access to land. However, there is no policy to defend the peasant production systems and their particular relation with the land against the social and economic impacts of frontier expansion. This means that the tensions related to possession rights, social organization, judicial and political strategies to defend peasants and claims for the institutionalization of this defence have become a new field of land governance.

5. Conclusions

The expansion of soya production in Argentina developed relatively ungoverned and without subsidies, through the initiatives of private interests in alliance with some state powers, which left its regulation to market forces. The expansion started in the 1970s in the pampas region where land and agriculture had gradually been abandoned during the period of ISI policies and where it advanced at the expense of cattle raising and diversified grain production. In the neoliberal period – when new production models (zero tillage) and new technology became available (from 1996 onwards) – the expansion of soya took place more rapidly. This was possible because large tracts of underused land were available and could easily be converted into new production units by a limited number of old and new large landowners and investors, who started to lease land. There was no major need for property change since flexible lease contracts provided the necessary scale. In other words, there was a process of capital concentration rather than of land concentration. This, together with zero-tillage practices and machinery, the introduction of GM soya (Roundup Ready (RR) resistant soya), and an innovative way of managing and organizing production by sowing pools, constituted the basis for the soya boom.

The process that started in the pampas region was accompanied by crop change and the migration of cattle farming towards the northern provinces, mainly the Chaco region. There, direct investments in soya production were generally made by soya producers originating from the pampas, since this group had succeeded in making good money from soya and decided to buy additional land in the Chaco. New land users found cheaper land, less regulation than in the Pampean region and small producers with precarious tenure situations. Investing in the Chaco implied additional costs because of the need to clear vegetation, the more expensive transport, irregular yields due to climate instability and conflicts with holders of possession rights. However, cheap land and short-term benefits compensated for these costs.

The expansion of soya production in the pampas region did not lead to major changes in landownership relations, and new alliances emerged among the actors. In the Chaco, however, forests were cleared and local peasants and indigenous groups were bought out or were legally or illegally displaced; they lost their land and were excluded from the benefits of soya growing. Nevertheless, the state largely remained inactive in the fields of rural development and the regulation of extractive activities, such as mining, forestry and soya monoculture. It acted only in the sense that is created incentives for the further expansion of these activities. Regulation for sustainability and the protection of the rights of local farmers has not been properly addressed. Therefore, the main problems are the failure to implement the existing law and to protect the rights of weaker interest groups, and the lack of a long-term vision regarding the extent to which the use of glyphosate and other pesticides – which cause land degradation, the contamination of other crops, water pollution, health problems, etc. – is compatible with the need to bring about sustainable development.

Author details

Lucía Goldfarb* and Annelies Zoomers*

*Address all correspondence to: e.b.zoomers@uu.nl, l.i.goldfarb@uu.nl

Utrecht University, Faculty of Geoscience, International Development Studies, Utrecht, The Netherlands

References

[1] Domínguez, D. & Sabatino, P. Los señores de la soja. La agricultura transgénica en América Latina. CLACSO. CICCUS. Buenos Aiers, 2010. p.31-121.

[2] Borras, S, Franco. J, Kay, C, Spoor, M. El acaparamiento de tierras en América Latina y el Caribe visto desde una perspectiva internacional más amplia. 2012 http://www.tni.org/node/70998 (accessed June 2012).

[3] Brailovsky, A. E. Memoria verde: historia ecológica de la Argentina, Sudamericana. 2004.

[4] Pengue, W. A. Bioinvasiones, Agricultura Industrial y transformaciones ambientales. Fronteras, Año 8, 2009. (accessed 1 November 2010).

[5] Gonzáles, M.C., Román, M. Expansión agrícola en áreas extrapampeanas de la Argentina. Una mirada desde los actores sociales. Cuadernos Des. Rural, Bogotá, Colombia. 6(62). pp. 99-120. 2006. http://www.google.com/url?sa=t&rct=j&q=gonzalez% 20roman % 202009% 20santiago% 20del% 20estero&source=web&cd=1&ved=0CCEQFjAA&url=http% 3A% 2F% 2Fcdr.javeriana.edu.co% 2F% 3Fidcategoria% 3D1328% 26download% 3DY&ei=2F1BULipNYSo8QTP8IG4AQ&usg=AFQjCNH4tDHTrr8BTkCBWizhBMqqSeN1OQ&sig2=wMafBNGcdOZxTT7VTUImlw/ (accessed July 2010).

[6] Verbitsky, H. Verano del '96. Página 12. 2009.

[7] Robin. M.M. The world according to Monsanto. Pollution, corruption and the control of the world's food supply. 2010.

[8] Barsky, O. & Fernández, L. Cambio técnico y transformaciones sociales en el agro extrapampeano, Teseo, 2008.

[9] Slutzky, D. Los cambios recientes en la tenencia de la tierra en el país con especial referencia a la región pampeana: nuevos y viejos actores sociales. IADE/Realidad Económica. 2010. http://www.iade.org.ar/modules/noticias/article.php?storyid=3203/ (accessed July 2011)

[10] Expansión y reconversión de la ganadería vacuna en la provincia de Santiago del Estero, Argentina. Un análisis por tipo de productor. Conference Proceedings. ALAS-RU, Porto de Galinhas, Pernambuco, Brazil, November 2010 http://www.scribd.com/doc/91883472/Alasru-2010-Ponencia/ (accessed November 2011).

[11] Borras, S, Franco, J. Global Land Grabbing and Trajectories of Agrarian Change: A Preliminary Analysis. Journal of Agrarian Change, Vol. 12 No. 1. pp: 34–59. 2012 http://www.tni.org/sites/www.tni.org/files/borras_franco_joac_2012.pdf / (accessed 13 August 2012).

[12] Arceo, E; Basualdo, E. & Arceo, N. La crisis mundial y el conflicto del agro. Universidad Nacional de Quilmes/ Página 12/ Centro Cultural de la Cooperación, Buenos Aires, 2009.

[13] Basualdo, E., Arceo, E. Ecolución y situación actual del ciclo ganadero en la Argentina. Realidad económica 221, 1º de julio/15 de agosto 2006. 2006 p: 7-29. http://www.funcex.org.br/material/REDEMERCOSUL_BIBLIOGRAFIA/biblioteca/ESTUDOS_ARGENTINA/ARG_179.pdf/ (accessed 12 March 2011).

http://dx.doi.org/10.5772/53447

[14] Giarraca N. & Teubal, M. Ayer La Rural, hoy Grobo, Página 12.

[15] Aranda, D. Argentina originaria. Genocidios, saqueos y resistencia. La Vaca, 2011.

[16] Tomei, J and Upham, P. Argentine clustering of Soy Biodiesel Production: the Role of International Networks and the Global Soy Oil and Meal Markets, in The Open Geography Journal, 2011, 4: 45-54 (accessed 5 July 2012).

[17] Transnational Institute, Econexus, CEO . Agrofuels: towards a reality check in nine key areas. Amsterdam. 2007.

[18] Mol, A. P .J. Boundless Biofuels, Sociologia Ruralis, Vol. 47, Number 4, October 2007 (Accessed May 2008).

[19] Zoomers, E.B, Westen, A.C.M. van Introduction: translocal development, development corridors and development chains. International Development Planning Review, Volume: 33, Issue: 4 (2011), pp. 377-388. 2011 (provided by authors).

[20] Zoomers, E.B. Globalisation and the current global land grab', Journal of Peasant Studies, 37: 2, 429- 447. 2010.

[21] Peluso, N., Lund, C. New frontiers of land control. Journal of Peasant Studies. 38:4. pp. 667-681. 2011 http://www.tandfonline.com/loi/fjps20/ (accessed November 2011).

Chapter 4

Integration of Farm Fossil Fuel Use with Local Scale Assessments of Biofuel Feedstock Production in Canada

J.A. Dyer, R.L. Desjardins, B.G. McConkey,
S. Kulshreshtha and X.P.C. Vergé

Additional information is available at the end of the chapter

http://dx.doi.org/10.5772/52488

1. Introduction

The viability of Canadian biofuel industries will depend on farm energy consumption rates and the CO_2 emissions from fossil fuel use for feedstock crops. The types of biofuels that are under development in Canada include biodiesel, grain ethanol, cellulosic ethanol and biomass. Each of these fuels relies on a distinct class of feedstock crops and in each case the most suitable crop is also dependent on geographic location. For example, the feedstock for biodiesel is canola in Western Canada and soybeans in Eastern Canada (Dyer et al., 2010a). For grain ethanol, the feedstock choices are corn in the east and wheat in the west (Klein and LeRoy, 2007). Cellulosic ethanol is still under development in Canada.

Technological changes in ethanol manufacturing can bring about different intensities of land use and require different land capabilities. Cellulosic ethanol and biomass can make use of land not capable of growing grains, and can exploit part of the straw from annual field crops (Dyer et al., 2011a). As a result, impacts on other land use activities with which feedstock crops compete also depend on the particular feedstock involved in the interaction and the capability of the land. Impacts on the overall sustainability of agriculture are minimal when management practices fit the local environment (Vergé et al., 2011). Therefore, to understand the different comparative advantages and impacts among regions, each landscape requires its own assessment.

Two main principles must guide biofuel industries. The first is that they must produce more energy than the fossil energy used for their production. The second is that they must displace more Greenhouse Gas (GHG) emissions than are released during their production

(Dyer and Desjardins, 2009; Klein and LeRoy, 2007). Biofuels appeal to governments for the potential to create economic opportunities in rural areas (Klein and LeRoy, 2007). Due to transport costs, feedstock crops are best grown on land that is close to facilities for processing them into biofuel. Thus, it is important to have objective criteria for determining which communities and regions are the most suitable locations for those processing plants. In addition, sustainable feedstock production requires that local suitability be established (Dyer et al., 2011a; Vergé et al., 2011). To date, a comprehensive farm energy analysis has not been done at a local scale in Canada.

The main goal of this chapter was to determine the geographic distribution of farm energy terms within each province of Canada. Due to their small sizes and limited role in Canadian agriculture, the four Atlantic Provinces were treated as one combined province. A secondary goal of this chapter was to demonstrate how much the farm energy budget contributes to the GHG emissions budget of the agricultural sector through fossil CO_2 emissions at a provincial scale. Using area based intensity, a simple demonstration was also provided of how these data could provide a baseline comparison for the fossil CO_2 emitted from growing a grain ethanol feedstock compared to current types of farms. These goals were achieved through the integration of existing models and databases, rather than by analysis of new data collected specifically for this purpose.

2. Background

The feedstock for biofuels has raised several land use questions (GAO, 2009; Malcolm and Aillery, 2009). These include: How much land will biofuel feedstock production require in order for biofuels to make an appreciable contribution to energy supply? What agricultural products would be displaced to accommodate this production? How will food supply be threatened by feedstock production? How much will meat production and livestock industries be displaced by feedstock? In large part, most of these general land use policy questions have been addressed in Canada and elsewhere. However, there have been some shortcomings of these analyses.

One of these gaps is the failure by many studies to account for carbon dioxide (CO_2) emissions caused by fossil fuel use in the feedstock production, and in agriculture, generally. One of the reasons for this gap is that under the United Nations Framework Convention on Climate Change, emissions from fossil fuels used for agriculture are reported as part of the energy sector, rather than under the agriculture sector. Although smaller in magnitude than both the methane (CH_4) and nitrous oxide (N_2O) emissions reported for agriculture, farm energy-related CO_2 emissions are an important component of the sector's GHG emissions budget, largely because it is manageable (Dyer and Desjardins, 2009). For example, reduced tillage practices which diminish fossil fuel CO_2 emissions from farm machinery (Dyer and Desjardins, 2003a), as well as conserving soil carbon, can be the difference in whether a particular feedstock or its biofuel are energy-positive or a sink for GHGs.

http://dx.doi.org/10.5772/52488

Without taking all forms of fossil energy use in agriculture into account, the GHG emissions budget for crop production is incomplete. In addition to farm field operations, the fossil fuel CO_2 emissions include agro-chemical manufacturing, equipment manufacturing, fuels for grain drying or heating farm buildings, gasoline, and electricity for lighting or cooling (Dyer and Desjardins, 2009). However, farm field operations are the most complex term and have the greatest degree of interaction with land features and crop choices. Fossil fuel consumption for farm field work has been computed using the Farm Field work and Fossil Fuel Energy and Emissions (F4E2) model (Dyer and Desjardins, 2003b; 2005). Because of their dominant role in defining regional differences in fossil fuel energy and CO_2 emissions, farm field operations have already been assessed in more detail than other farm energy terms (Dyer et al., 2010b).

3. Methodology

3.1. Selecting the spatial scale

Since decision making in the biofuel industries is limited by spatial scale, assessing the most appropriate scale was the first task undertaken in this analysis. Disaggregation of the Canadian farm energy budget to the provinces can exploit agricultural statistics available at two spatial scales. The first scale is the Census Agricultural Regions (CAR) (Statistics Canada, 2007), while the second scale is at the Soil Landscapes of Canada (SLC) (AAFC, 2011). Due to its association with agricultural census records, the geographic scale chosen for distributing farm energy use in this chapter was the CAR system which divides Canada into 55 regions (with each of the Atlantic Provinces treated as a single CAR). In spite of the soil and land variables available for SLCs, some difficult assumptions are needed to disaggregate some data to this scale. In addition to this uncertainty, the large number of spatial units in Canada at the SLC scale (nearly 4,000 units having agriculture) made presentation on the basis of SLCs impractical for this chapter.

The CARs are identified in this chapter by numbers that start from 1 in each province. In the Atlantic Provinces, with each province treated as one CAR. Hence, CAR numbers 1, 2, 3, and 4 represent New Brunswick, Prince Edward Island, Nova Scotia and Newfoundland, respectively. With the agricultural regions of Canada being spread out largely east to west, it was not practical to display the boundaries on a single page map. So, a website location, rather than a printed map, was provided in this chapter. To view the CAR sizes and locations in each province, visit: http://www.statcan.gc.ca/ca-ra2011/110006-eng.htm.

3.2. Farm energy budget

The six terms in the farm energy budget adopted for this analysis were those defined by Dyer and Desjardins (2009). All of these terms reflect operational and/or financial decisions made by farmers. For example, the energy costs of transporting products from farm gate to market that are paid for by the processer or marketer, rather than the farmer, were excluded. These terms involved several different types of fossil fuel. Based on the analytical methodologies required for spatial disaggregation, these six terms were separated into three groups. The diesel fuel used in farm field work (Dyer and Desjardins, 2003b; 2005) and the coal required to manufacture and supply farm machinery (Dyer and Desjardins, 2006a) were the first group because they were both quantified with the F4E2 model.

The fossil energy to supply chemical fertilizers and pesticide sprays was determined from a direct conversion of the weight of consumption of these chemicals (Dyer and Desjardins, 2007). Since nitrogen fertilizers are the most energy-intensive chemical inputs to manufacture, and have available sales records in Canada, this conversion was based on the natural gas to manufacture just nitrogen fertilizer. The energy conversion rate of 71.3 GJ/t{N} derived from Nagy (2001) as an average for five census years from 1981 to 2001 was used in this chapter. Although this conversion was for just nitrogen supply, it was indexed to include other farm chemicals, mainly phosphate and potash fertilizers.

The third group includes electrical power, gasoline and heating fuels. All three terms in this group had to be determined empirically since there was little basis for modeling these terms. While to some extent diesel is increasingly being used for farm owned transport vehicles, in 1996 the F4E2 model accounted for all but a small percentage (Dyer and Desjardins, 2003b; 2005) of the farm-purchased diesel fuel for farm field work. Only one percent of this diesel fuel was for household use in 1996 (Tremblay, 2000). This suggests that pick-up trucks, the sort of vehicle that would be used for both light haul farm transport and family business, were not typically diesel powered in 1996. Therefore, gasoline, rather than diesel, was likely the main fuel used for farm owned transport vehicles in 1996, the baseline year for the farm energy budget described by Dyer and Desjardins (2009). There was, therefore, no justification for including any diesel fuel in the third group of energy terms. In keeping with the conditions of the farm energy budget described above, any diesel fuel consumed by commercial trucks used for hauling grain and livestock to market or processing were not considered in this analysis.

Electrical power was a partial exception to the need for empirical determination because of a semi-empirical index of the CO_2 emissions from this term based on farm types (Dyer and Desjardins, 2006b). This index demonstrated the correlation, at least for this energy term, between energy consumption and farm types, particularly among livestock farms. Application of this index for this analysis was unnecessary because in this case livestock populations are only needed to distribute a known quantity of electrical energy among provinces and regions (CARs).

The most comprehensive source of farm energy use information in Canada is the 1996 Farm Energy Use Survey (FEUS) of Canada (Tremblay, 2000). The FEUS provided commodity-specific estimates for the three energy terms for which detailed modeling algorithms were not available. Given this empirical source, for example, it did not matter whether all gasoline was burned in farm owned transport vehicles or whether all such vehicles were powered by gasoline. What mattered was that the FEUS provided an empirical quantity of gasoline that had to be disaggregated regionally. The remaining term in the Canadian farm energy budget was a combination of three fuels, including furnace-oil, liquid propane (LPG) and natural gas, which was defined by Dyer and Desjardins (2009) as heating fuels.

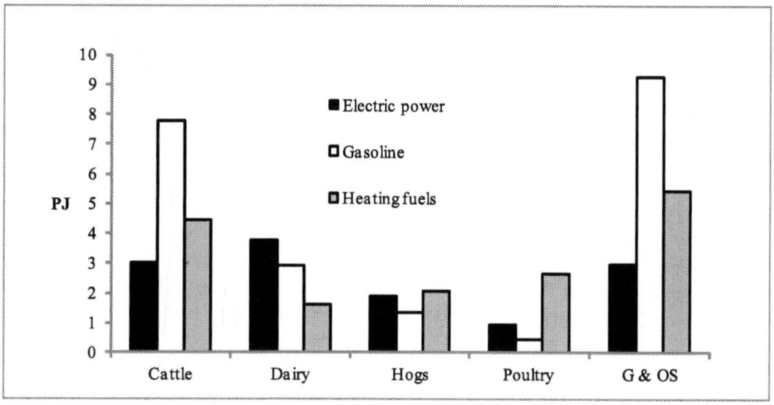

Figure 1. National consumption of three types of energy by five farm types identified in the 1996 Farm Energy Use Survey (FEUS) of Canada.

Due to confidentiality constraints, the FEUS data were not directly available at the farm level. The FEUS, however, did allow energy type data to be grouped by farm type, but only for Canada as a whole. While energy types were also grouped by provinces in the FEUS, this breakdown could not be linked to farm type uses. The FEUS also gave the consumption of diesel fuel in Canadian agriculture which was used to verify the F4E2 model (Dyer and Desjardins, 2003b). The quantities for the farm energy terms extracted from the FEUS, shown in Figure 1, illustrate the range in energy quantities that had to be disaggregated for these three energy types. These energy data were adjusted for the shares of these fuels that were used in farm households instead of farm use. These household share adjustments were only provided by fuel type, however, and not for farm type (Tremblay, 2000).

Although the purpose of the data in Figure 1 was not to compare farm types, these energy quantities still reflect both the different sizes and energy intensities of these farming systems in Canada. Grain and oilseed farms accounted for 35% of the consumption of these three energy terms. The range of total live weights in Canada for beef, dairy, hogs and poultry of 5.7, 1.1, 0.8 and 0.2 Mt, respectively, during 2001 (Vergé et al., 2012) was wider than the range in uses of these three energy types among the four livestock industries seen in Figure 1. Hence, while beef production used the largest share of this energy of any of the livestock industries, beef farms were the least intensive user on a live weight basis. Similarly, poultry, the smallest livestock industry and lowest user of these energy terms, was the most intensive user of these three types of energy.

3.3. Land use

In defining the GHG emission budgets for each of the Canada's four dominant types of livestock production, dairy, beef, pork and poultry, Vergé et al. (2007; 2008; 2009 a,b) took into consideration the land base on which the feed grains (including oilseed meal) and forage that support livestock are grown. Vergé et al. (2007) recognized that the carbon footprint for each livestock industry must include the land base that supports the crops in the livestock diet. Subsequently, the total area involved in Canadian livestock production was defined as the Livestock Crop Complex (LCC). The LCC was based on an array of crops that defined the diets of all four livestock types, including barley, grain corn, soybean meal, feed quality wheat, oats, canola meal, dry peas, seeded pasture, alfalfa, grass hay and silage corn.

The Canadian Economic and Emissions Model for Agriculture (CEEMA) was developed to estimate the spatial distribution and magnitude of GHG emissions generated by the agriculture sector (Kulshreshtha et al., 2000). Because the spatial unit of CEEMA was the CAR, this model was well suited for the analysis described in this chapter. CEEMA is composed of records of crop areas, yields, nitrogen fertilizer rates and related GHG emissions during 2001 for all field crops in each CAR. Almost 1,900 of these crop records were distributed over 55 CARs in CEEMA. While crop records identify the CAR in which they lie and define the areas of all crops within each CAR, the actual locations of crops described in the respective records within the CAR are not specified. Another limitation of the CEEMA was that these crop records were generated from analysis of optimal economic land uses for 2001 (Horner et al., 1992; Kulshreshtha et al., 2000), rather than from actual crop statistics.

The variables that determine differences among the CARs are related primarily to land use differences and farm level decisions. These variables include the selections of crops, particularly those crops that feed livestock. The CEEMA crop records do not contain soil type data. Livestock populations at the CAR scale were also not available for this analysis to preserve the confidentiality of the farmers surveyed at that scale. The variables required for assessing farm energy at the CAR scale will be discussed in more detail below.

Estimates of GHG emissions from Canada's four main livestock industries were integrated with the CEEMA. The area of each crop that was in the LCC from each CAR in each province was determined as part of a previous application of CEEMA (Dyer et al., 2011b). That study disaggregated the LCC to each crop record describing crops in the diet of Canada's four main livestock types. Some feedstock-food-livestock interactions on a national or provincial scale in Canada were analyzed in that study. It also used the CEEMA database to separate Canadian farmland into land that supported livestock and land available for other crops. However, Dyer et al. (2011b) did not separate these emissions by livestock type. Farm energy consumption and fossil fuel CO_2 emissions for farm field work have been disaggregated at a provincial scale (Dyer et al., 2010b). But no other farm energy terms have been disaggregated at a scale that allows the full farm energy budget to be quantified in the CARs.

3.4. Farm energy and livestock distributions

For the three energy terms that can only be treated empirically, electric power, gasoline and heating fuels, the FEUS provided the only link to farm types. Because of the availability of provincial livestock population data from the Canadian agricultural census, this disaggregation can be done directly at the provincial scale. Grain and oilseed production, which was defined as a farm type in the FEUS, accounted for part of each of these three energy terms.

Therefore, provincial summaries of areas in these crops were also involved in the disaggregation process.

These farm type links meant that disaggregation of these energy terms to the CAR scale could be achieved through correlation with livestock populations and crop areas. The underlying assumption was that most farm animals are located near their feed sources. This assumption was required because information on where in the provinces farm animals are actually housed was not available for this analysis (Tremblay, 2000). This limitation only affected the three empirical energy terms, including electric power needs, heating fuels and gasoline for farm transport. The farm field work and the two input supply terms can be linked directly to the CARs through CEEMA, as well as to the provinces.

Provincial estimates had to be generated for all three energy terms taken directly from the FEUS. To achieve this, the relative distribution of energy quantities across the provinces was determined for each farm type identified in the FEUS. To quantify each livestock farming system, the inter-provincial distribution was determined on the basis of the total weight of all live animals in all age-gender categories in the livestock type. The provincial live weight was calculated from the average live weight (W) of each age-gender category (*k*) of each livestock type (*a*) and the number of head (H) in each age-gender category and livestock type. The amount of energy from each energy term for each of the livestock systems from the FEUS ($E_{FEUS,a}$) was disaggregated to the provincial energy quantity (E_{prov}) by the respective shares of live weight in each province (*prov*), as follows.

$$E_{prov,a} = E_{FEUS,a} \times \left(\Sigma^{k}\ W_{k,a} \times H_{k,\ prov,a}\right) / \left(\Sigma^{Canada}\ \Sigma^{k}\ W_{k,a} \times H_{k,\ prov,a}\right) \quad (1)$$

The disaggregation of these energy terms for the farms that produce grains and oilseeds to the provinces was similar to Equation 1. The difference was that live weights (W × H) were replaced by the provincial crop areas in this farming system. The areas of each grain and oilseed crop were summed over the crop records of grain or oilseed areas in the CEEMA database. The first sum was for the crop records in each CAR to determine CAR area totals. The provincial totals for each type of grain or oilseed crop were then estimated from the sum of all areas in that crop type over all CARs in each province. This summing process was only applied to the actual grains and oilseeds crops. So rather than correlate the entire area in these crops with the energy terms, differences between these area totals and the areas of these annual crops in the LCC were used. Dyer et al. (2011b) defined these areas as the Non-Livestock Residual areas (NLR). The provincial quantities for the three energy terms and the five farm types shown in Figure 1 are given in Table 1.

A simpler computational sequence was used for the two energy terms derived from the F4E2 model and the energy term for chemical inputs. This was possible because the data for calculating these terms could be taken directly from the crop records of the CEEMA database. The main input variable from CEEMA for the F4E2 calculations was crop areas, whereas total chemical nitrogen applications were available in all CEEMA crop records for the chemical input supply energy term. Because these two energy terms were calculated on each crop record, they could be summed directly from the CEEMA database. While the calculations for grains and oilseeds used only the records for those crops designated as grains and oilseeds, calculations for these three terms used all crop records associated with the LCC or NLR. The F4E2 model took into account whether the crops were annual grains or perennial forages, along with the yields of each crop (Dyer and Desjardins, 2005).

	Beef	Dairy	Hogs	Poultry	G&OS[2]
			PJ		
			Electric power		
British Columbia	0.15	0.25	0.02	0.14	0.00
Alberta	1.71	0.26	0.32	0.10	0.75
Saskatchewan	0.59	0.10	0.22	0.04	1.60
Manitoba	0.33	0.13	0.48	0.07	0.35
Ontario	0.27	1.28	0.58	0.40	0.06
Quebec	0.12	1.20	0.70	0.26	0.02
Atlantic	0.03	0.20	0.06	0.06	0.01
Canada	3.20	3.42	2.40	1.07	2.78
			Gasoline[3]		
British Columbia	0.39	0.19	0.02	0.07	0.01
Alberta	4.43	0.20	0.23	0.05	2.33
Saskatchewan	1.52	0.08	0.15	0.02	4.95
Manitoba	0.85	0.10	0.34	0.03	1.08
Ontario	0.70	0.98	0.41	0.19	0.18
Quebec	0.32	0.92	0.49	0.12	0.05
Atlantic	0.09	0.15	0.04	0.03	0.02
Canada	8.30	2.63	1.68	0.51	8.61
			Heating fuel[4]		
British Columbia	0.22	0.11	0.02	0.40	0.01
Alberta	2.53	0.11	0.35	0.28	1.37
Saskatchewan	0.87	0.04	0.24	0.10	2.92
Manitoba	0.49	0.06	0.53	0.20	0.64
Ontario	0.40	0.54	0.64	1.12	0.10
Quebec	0.18	0.51	0.77	0.72	0.03
Atlantic	0.05	0.08	0.07	0.18	0.01
Canada	4.75	1.45	2.62	2.99	5.08

[1] 1996 Farm energy use survey for Canada

[2] Grains and oil seed farms

[3] gasoline purchased by farm operators for farm-owned vehicles.

[4] includes fumace-oil, liquid propane (LPG) and natural gas

Table 1. The provincial 2001 energy quantities for the three energy terms and the farm types identified at a national scale in the FEUS[1].

The analysis for this chapter did not disaggregate provincial livestock populations directly into the CARs. Instead, it was the LCC areas defined by these populations that were disaggregated at this scale. Like the NLR area summations, only crop records for those crops that were in each respective livestock diet were summed within the CARs, rather than the areas from all crop records in the CEEMA database. The basis for identifying these crop records was the set of provincial LCC calculations for each livestock type provided by Vergé et al. (2012).

Since the FEUS data were collected in 1996 and the CEEMA data were derived from the 2001 agricultural census, the energy quantities in Figure 1 had to be indexed from 1996 to 2001. This was done by factoring the 1996 energy terms by the ratio of the respective size of each farm system from the 2001 census records to the size of the same farm system in the 1996 census records. Updating from 1996 to 2001 was done at the same time as the farm type energy quantities from the FEUS were disaggregated to the provinces, as shown in Table 1. The different farm types required different definitions of size. For the four livestock farm types, these provincial size ratios were of total livestock weights from the two years, whereas for grain and oilseed farm areas (NLR), these provincial ratios were of total crop production (planted areas times yields) from the two census years.

3.5. Area allocation to each CAR

The allocation of LCC areas (A) to each CAR for each livestock type was determined by the aggregate share of all feed crops in the provincial LCC in that CAR. Crop areas from the crop records were converted to area totals in each CAR for each of the 12 LCC crops (listed above) that were common to both the CEEMA database and to the four LCCs (Vergé et al., 2012). The total LCC areas in the crop records (Dyer et al., 2011b) were integrated to the respective CARs for each livestock type. The allocation to livestock types was based on the share of each of the four LCCs in each province, which were derived from the diet of each livestock population (Vergé et al., 2012).

For ruminant livestock, the allocation of provincial energy quantities to the CARs required a means of equating the dietary contribution of roughages with that of feed grains. For ruminants, 1.8 kg of roughages provide the same nutrient energy as 1 kg of feed grains (IFAS, 1998; Neel, 2012; Schoenian, 2011). Using this ratio, the forages in the respective LCC areas were converted to the equivalent feed grains on the basis of crop production estimates derived from the 2001 census crop yields. This general relationship also applies to pulses and oilseed meals, but ignores the protein contributions from those feeds. This relationship is altered slightly for corn silage which provides only 42% by weight of the nutrient value of other roughages (Miller and Morrison, 1950). Rangeland was excluded because there were no available data for farm energy consumption associated with this form of land use. Very little energy would be consumed to manage rangeland because no fertilizer or chemical inputs are used and, normally, there are no farm field operations.

In addition to the different nutritional values, the bulk yield differences between grains (g) and roughages (r) also account for the importance of these two crop group areas in each LCC. For each CAR the total LCC area (A_{CAR}) was the result of the two areas ($A_{CAR,g}$ and $A_{CAR,r}$). Each area was weighted by the average total production weights for the crop group

(F) within each provincial LCC and 1.8 (the nutritional value ratio for *g* and *r*). This weighted area total was calculated for each CAR as follows.

$$A_{CAR} = \left(\left(A_{CAR,g} \times F_g\right) + \left(A_{CAR,r} \times F_r/1.8\right)\right) / \left(F_g + \left(F_r/1.8\right)\right) \tag{2}$$

These LCC area calculations at the CAR level were integrated over each province as follows.

$$A_{prov} = \Sigma^{CAR} A_{CAR} \tag{3}$$

Each energy term (E) from the FEUS for each province (Equation 1) was then disaggregated from the province to the CAR level as follows.

$$E_{CAR} = E_{prov} \times \left(A_{CAR} / A_{prov}\right) \tag{4}$$

Dyer et al. (2011b) found that occasionally the amounts of some crops were too low to meet the dietary needs of the provincial livestock populations. Because of these crop deficits, production from the surplus provinces had to be transported to the deficit provinces. Due to the reduction of $A_{CAR,r}$ by 1/1.8 and the occasional accumulation of these provincial crop deficits and surpluses, A_{CAR} was an indexed area estimate which did not equal the actual total LCC area for the CAR. Without reducing $A_{CAR,r}$ by 55%, A_{prov} would have the same difference with the provincial LCC area total (prior to these deficit corrections) as each A_{CAR} would have with the CAR total of the LCC area. Thus, using the CAR to province area ratios of these two weighted area estimates to disaggregate provincial energy terms does not result in any unnecessary distortion of the CAR energy estimates compared to the CAR-province ratios of uncorrected LCC areas.

The usefulness of disaggregating to the CAR scale depends on the sensitivity of the farm energy terms to land use parameters. Since the goal of this chapter was to determine the spatial distribution of the farm energy budget, a sensitivity analysis based on purely management-based range tests such as those described by Dyer and Desjardins (2003a) would not adequately demonstrate the sensitivity of farm energy terms to the factors that determine the spatial distribution of farm energy use at the CAR scale. This was because the only available spatial parameter at the sub-provincial CAR scale was the array of crop areas from CEEMA. Instead, the spatial sensitivity was equated to the variance of energy estimates across CARs in each province. Such sensitivity would reveal the impacts of local crop choice decisions on the consumption of different energy types.

http://dx.doi.org/10.5772/52488

4. Results and discussion

4.1. Farm energy budget at the CAR scale

The basic output from the analysis described in this chapter was the set of disaggregated farm energy terms at the CAR scale. Due to the extent of these data, they are presented in appendices, rather than as tabular results in the main body of the chapter. Some care is needed in the numbering system in these appendices since the website maps for two provinces use a different CAR numbering system than was used in CEEMA. For Manitoba, CEEMA CAR number 1 includes the online map numbers 1, 2 and 3; CEEMA number 2 includes the online map numbers 4, 5 and 6; and CEEMA number 5 includes the online map numbers 9 and 10. For CEEMA numbers 3, 4 and 6, the online map numbers are 7, 8 and 11, respectively. The online map number 12 was not used in CEEMA. To be consistent with the online CAR base map, the 10 CEEMA CARs for Ontario were combined into 5 CARs in the two Appendices.

The data presented in Appendix A are preliminary to the general (non-commodity-specific) farm energy budget in Appendix B. They resulted from the need to use farm types to disaggregate the FEUS data. The data presented in Appendix B are the intended output or primary goal of this chapter. These data represent all six terms in the energy budget described by Dyer and Desjardins (2009). The data for the three energy terms extracted from the FEUS in Appendix B were derived by integrating the data in Appendix A over the five farm types. Although it is difficult to extract any trends from these data arrays by inspection that could not otherwise be seen from provincial scale tables, these two appendices make the data at the CAR scale available for future regional investigations in farm energy use in Canada.

4.2. Provincial farm energy

Table 2 presents a re-integration of the spatially detailed data in Appendix B from the CAR to provincial scale. Even given the limited spatial detail of this table, it still puts all terms of the Canadian farm energy budget into one source, based on one integrated methodology. Not surprisingly, given its large crop area, Saskatchewan was the biggest consumer of all forms of farm energy in Canada. This was most evident in the farm machinery-related terms, which likely reflects the extensive grains and oilseeds farming system in that province. The two coastal regions (British Columbia and the Atlantic Provinces), as well as Quebec and Ontario, contribute much less to the farm energy budget than the three Prairie Provinces, simply because of the much smaller areas in agricultural use. Although fertilizers (and other farm chemicals) are the largest cause of energy consumption, the farm machinery-related terms combined are 9% higher, nationally, than the chemical inputs. The three FEUS-based terms, to which so much attention was devoted in this chapter, account for only 20% of the national farm energy budget.

	Farm field work	Machinery supply	Chemical inputs	Electric power	Gasoline[1]	Heating fuel[2]
Provinces			PJ			
British Columbia	1.1	0.6	1.0	0.6	0.7	0.8
Alberta	19.4	11.1	34.8	3.1	7.2	4.6
Saskatchewan	31.7	18.2	35.3	2.5	6.7	4.2
Manitoba	10.6	6.1	21.3	1.4	2.4	1.9
Ontario	8.6	4.9	11.1	2.6	2.5	2.8
Quebec	4.7	2.7	6.5	2.3	1.9	2.2
Atlantic	0.9	0.5	1.4	0.4	0.3	0.4
Canada	76.9	44.1	111.4	12.9	21.7	16.9

[1] Gasoline purchased by farm operators for farm-owned vehicles.

[2] Includes fumace-oil, liquid propane (LPG) and natural gas

Table 2. Provincial estimates of the six energy terms of the Canadian farm energy budget during 2001.

4.3. Assessing sensitivity through spatial variance

The spatial variance assessments of the spatial data in this chapter are shown in Tables 3 and 4. The statistic used to compare spatial variance was the coefficient of variation (CV) of the CAR energy values within each province. Being the ratio of standard deviations to their respective means, the CVs give a normalized, and thus a comparable, measure of spatial variability. In order to avoid the CVs being affected by the sizes of the CARs, the data in the two appendices were converted to energy intensities using areas of arable land extracted from the CEEMA crop records (discussed in more detail below). To illustrate, if the disaggregated energy intensities are evenly dispersed across all CARs in the province, then the crop records in the CEEMA database would have no impact on the distribution of energy consumption. Evenly dispersed energy quantities across all CARs would also result in no variance among the CARs and a provincial CV of zero.

Table 3 presents the CVs for the data presented in Appendix A, while Table 4 presents the CVs for Appendix B. In Table 3, only one set of CV estimates was needed for all three energy terms since there was no source of spatial variation associated with these energy terms prior to disaggregation to the CARs. For the pork, poultry and grains and oilseeds farm types, the two coastal provinces had the highest CVs in Table 3. Manitoba had the lowest CVs for these three farm types, which were also the lowest CVs in Table 3. For dairy, Quebec had the lowest CV, while for beef, the lowest CV was in Alberta. The poultry industry had the highest spatial variation, followed by grains and oilseeds, while dairy had the lowest overall spatial variation. Spatial variation for pork and poultry was lowest in the Manitoba. The spatial variations for pork and poultry were generally higher than for beef and dairy. On average, the Prairies had lower CVs than the other provinces. All of the CVs in Table 3 were higher than zero and there were appreciable differences among these CVs.

http://dx.doi.org/10.5772/52488

Hence, the crops that drive these five farming systems were not evenly distributed among the CARs.

	Beef	Dairy	Pork	Poultry	G&OS[3]
Provinces			CV[4]		
British Columbia	0.31	0.30	0.84	1.27	0.90
Alberta	0.25	0.18	0.34	0.33	0.39
Saskatchewan	0.35	0.25	0.30	0.26	0.27
Manitoba	0.39	0.27	0.19	0.18	0.16
Ontario	0.53	0.20	0.38	0.40	0.13
Quebec	0.30	0.13	0.65	0.65	0.38
Atlantic	0.39	0.32	0.83	1.01	0.82

[1] Farm energy use survey

[2] census agricultural regions of Canada

[3] grains and oils farms

[4] these CV estimates represent all three energy terms from the FEUS.

Table 3. Provincial Coefficients of Variation (CV) for the disaggregation of energy use by five farm types from the FEUS[1] to the CARs[2] during 2001.

Whereas there were no spatial differences among the three energy terms from the FEUS when they were separated by their farm types (Table 3), integrating over those five farm types in Table 4 created some differences among these three energy terms. Since farm field work and machinery supply were connected to each other through the F4E2 model, and had the same spatial variations, only the farm field work CVs were shown in Table 4. Farm field work, electric power and gasoline use all had similar CVs which were all lower than the CVs for heating fuels and chemical inputs. The higher CVs for heating fuels likely reflect the combining of three fuel types into one term.

Manitoba had the lowest average CV over the five energy terms in Table 4. British Columbia had the highest CVs for all energy terms except chemical inputs, which were highest in Saskatchewan. The Atlantic Provinces and then Quebec had the next highest CVs after British Columbia. The CV for electric power in Ontario was so low that it suggested almost no spatial differences for this term in Ontario. There was not as much within-province variation among the energy terms (Table 4) as among the farm types (Table 3) that determined the spatial variations for three of those terms. The CVs in Table 4 still display an appreciable amount of within-province spatial variation, however.

The more hilly and ecologically-varied terrain in the coastal provinces may account for some of the spatial variance in British Columbia and the Atlantic Provinces compared to the prairies. However, the agricultural areas in the Prairie Provinces, particularly Saskatchewan, are greater than in the other provinces, and have a greater range in latitude, and hence climate, which would result in higher spatial variation among the CARs. In spite of the relatively low CVs in some cases, Tables 3 and 4 still suggest that the data presented in the two appendices can provide some guidance on where in each province farm energy use would be the highest or the lowest for each energy term.

Provinces	Farm field work	Chemical inputs	Electrical power	Gasoline	Heating fuel
			CV		
British Columbia	0.32	0.33	0.47	0.34	0.70
Alberta	0.20	0.26	0.12	0.12	0.11
Saskatchewan	0.07	0.44	0.16	0.17	0.16
Manitoba	0.08	0.22	0.14	0.13	0.12
Ontario	0.11	0.16	0.02	0.13	0.14
Quebec	0.25	0.11	0.16	0.08	0.34
Atlantic	0.25	0.28	0.23	0.20	0.45

[1] census agricultural regions of Canada

Table 4. Provincial Coefficients of Variation (CV) for the disaggregation of the six energy terms in the Canadian farm energy budget to the CARs[1] during 2001.

4.4. Fossil CO_2 emissions from farm energy use

To satisfy the secondary goal of this chapter the farm energy budget presented in Table 2 was converted to fossil CO_2 emissions. With the variety of energy types that are used in Canadian agriculture, a different conversion was required for each of the six energy terms. For the diesel fuel for field work, coal to manufacture steel for farm machinery and gasoline, the conversion factors were 70.7, 86.2 and 68.0 Gg{CO_2}/PJ (Neitzert et al., 2005). Based on a summary of fertilizers manufacturing energy dynamics by Nagy (2001), Dyer and Desjardins (2007) used 57.9 Gg{CO_2}/PJ as the conversion factor for fossil CO_2 emissions from fertilizer supply. Even though the chemical input supply energy computations were driven by just nitrogen applications, this conversion took into account all three fertilizers, not just nitrogen, since all three fertilizers were included in this energy term. Reasoning that a very small additional share of the input energy was devoted to the supply of pesticides, which were not included in the calculations from Nagy (2001), Dyer and Desjardins (2009) defined this CO_2 emissions term as chemical inputs, rather than fertilizer supply.

Because heating fuel includes three separate fossil fuels, CO_2 emission rates had to be determined for each farm type in the same way as energy consumption rates for heating fuel were determined. This was done by converting the set of fuel and farm type estimates for this energy term and converting them to CO_2 emissions, using 59.8, 61.0 and 67.7 Gg{CO_2}/PJ, for LPG, natural-gas and furnace-oil (Neitzert et al., 2005). The conversion factor for each fuel and farm type was the ratio of these CO_2 emissions and the previously discussed energy consumption amounts. The blended factors had only minor variation among

http://dx.doi.org/10.5772/52488

the provinces, however, ranging from 61.8 Gg{CO_2}/PJ for Saskatchewan to 66.6 Gg{CO_2}/PJ for the Atlantic Provinces. Therefore, the average heating fuel conversion factor for Canada, 64.1 Gg{CO_2}/PJ, was used for all provinces in Table 5.

Since they were interested in a national farm energy budget, Dyer and Desjardins (2009) used a single average conversion factor for CO_2 emissions for the consumption of electric power. Their factor allowed for 22% of Canadian electricity generation being from coal-fired plants. However, there are great differences among provinces in the dependence of coal-based generation (NRCan, 2005), ranging from 96% in Alberta to 0% in Quebec. Because of the goal of provincial disaggregation of all farm fossil CO_2 emissions to provinces in this chapter, the conversion factor for each province was computed separately using the provincial percent of coal generation from each province. The resulting conversion factors were 41.4, 264.8, 209.6, 2.8, 44.1, 0.0 and 162.4 Gg{CO_2}/PJ, respectively, for British Columbia, Alberta, Saskatchewan, Manitoba, Ontario, Quebec and the Atlantic Provinces.

	Farm field work	Machinery supply	Chemical inputs	Electric power	Gasoline[1]	Heating fuel[2]
Provinces			Gg CO_2			
British Columbia	79	55	61	23	46	49
Alberta	1,372	961	2,014	832	492	295
Saskatchewan	2,238	1,567	2,044	533	457	258
Manitoba	749	524	1,231	4	163	122
Ontario	605	423	643	114	167	183
Quebec	332	233	378	0	130	145
Atlantic	60	42	81	59	23	26
Canada	5,435	3,805	6,451	1,566	1,478	1,078

[1] gasoline purchased by farm operators for farm-owned vehicles.

[2] Includes furnace-oil, liquid propane (LPG) and natural gas.

Table 5. Provincial fossil CO_2 emissions from the six terms of the Canadian farm energy budget during 2001.

Like Table 2, the provincial differences in Table 5 reflect the range in sizes of the agriculture sector in the provinces. Saskatchewan accounted 36% of the fossil CO_2 emissions, while the three Prairie Provinces accounted for 80%. The two coastal provinces only accounted for 4%. While fertilizer supply was the largest energy term, the two terms related to farm field work exceeded fertilizer supply as a CO_2 emitter by 50%. Heating fuels had the lowest emissions, both for Canada and for all of the provinces. The three terms from the FEUS emitted only 21% of the fossil CO_2 from Canadian agriculture. The greatest variation among provinces was from the electric power term, due to the provincial differences in the use of coal for generating power. Heating fuels showed the least variation among provinces.

With a few minor adjustments to methodology, the basic energy budget described in this chapter (prior to spatial disaggregation) was very similar to the national energy budget presented by Dyer and Desjardins (2009). Therefore, the total emissions for Canada in Table 5 can be compared to the CO_2 totals for 2001 in that paper. Dyer and Desjardins (2009) showed higher CO_2 emissions for gasoline and heating fuels than this chapter because that analysis included several horticultural farm systems that were not included in the CEEMA database. Electric power CO_2 emissions were higher in this chapter than the emissions from this term by Dyer and Desjardins (2009). This was due to the decision to use province-specific energy to CO_2 conversions for electric power generation in this chapter, which captured the greater dependence on coal in the provinces with the largest agriculture sectors. The national CO_2 emissions estimate for three energy terms that could be computed directly from the CEEMA crop records in this chapter were all equal to the 2001 estimates reported by Dyer and Desjardins (2009).

4.5. Energy use and CO_2 emission intensities

The farm fossil fuel associated with feedstock production would depend on the specific type of feedstock crop to be produced. The data in Appendix B provide a set of baseline data against which the fossil fuel required for a specific feedstock crop choice would have to be compared. These data represent the mean quantities of farm energy used either for food or livestock feed production in each CAR. These mean energy quantities, summarized by province in Table 2 and converted to CO_2 emissions in Table 5, were also converted to the area based intensities shown in Figure 2 using the crop areas presented in Table 6. These areas include annual crops and seeded perennial forages summarized from the CEEMA crop records to the CAR scale. The CARs in Table 6 are numbered in the same sequence that was used in the two appendices. Because the areas in unseeded pasture and other marginal lands account for almost no farm energy use in Canada, they were not included in Table 6. These data can be used with Appendix B to calculate the intensity of energy use in each CAR (and were used in Tables 3 and 4). Over 80% of the arable land in Canada is in the three Prairie Provinces, and almost half of Canada's farmland is in Saskatchewan.

Figure 2 integrates the six energy terms in each province. Figure 2a shows the mean energy use per ha while Figure 2b shows the mean CO_2 emissions per ha. Although the distribution of CO_2 emissions resembles the distribution of energy uses across the provinces, there are slight differences because of the different farm type mixes and fuel types associated with those farm types among the provinces. Saskatchewan had the lowest energy use and CO_2 emission intensities because that province has the lowest share of its arable land devoted to livestock feed.

The following example illustrates how to reconcile biofuel feedstock production with farm fuel use and fossil CO_2 emissions. Using their 2009 methodology, Dyer and Desjardins (2007) described theoretical CO_2 emission budgets for a wheat farm in Saskatchewan and a dairy farm in Ontario. From the perspective of carbon footprint, the simulated wheat farm would be similar to a farm growing grain as a feedstock for ethanol. Based solely on fossil CO_2 emissions, the emission intensity for the ethanol feedstock crop was only 0.26 t/ha, compared to the mean intensity of 0.49 t/ha for all farm types in Saskatchewan in Figure 2b. This

http://dx.doi.org/10.5772/52488

result suggests that diverting farmland to grow ethanol feedstock might actually lower the average on-farm fossil CO_2 emissions in Saskatchewan.

CAR #	British Columbia	Alberta	Saskatchewan	Manitoba	Ontario	Quebec	Atlantic Provinces
	Provinces						
	ha, 000						
1	11	766	1,252	1,413	822	182	113
2	22	1,036	1,351	750	1,083	94	128
3	27	940	2,198	665	373	45	90
4	17	1,993	733	781	624	73	6
5	65	1,107	2,108	397	444	102	
6	2	974	2,196	505		97	
7	57	1,513	1,342			116	
8	185		1,569			84	
9			1,849			210	
10						472	
11						219	
Total	386	8,329	14,598	4,511	3,347	1,695	337

Table 6. Areas in annual crop and seeded perennial forges distributed over the 55 Cencus Agricultural Region (CAR) of Canada during 2001.

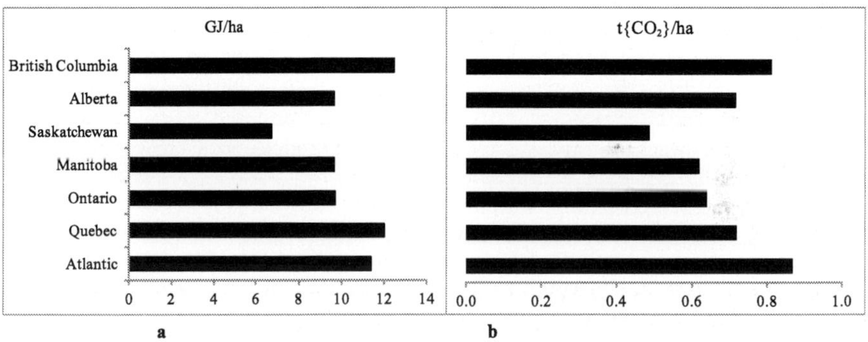

Figure 2. Area based intensity of on-farm energy use (a) and fossil CO_2 emissions (b) from all farm types in each province of Canada in 2001.

In Ontario, the simulated dairy farm emission intensity described by Dyer and Desjardins (2007) was 0.62 t/ha, compared to the 0.64 t/ha for all farm types in the province. This close agreement reflects the high share of Ontario farmland that is devoted to livestock production, much of which is dairy. These comparisons ignore CO_2 emissions from the soil, as well as the other types of GHG. A similar comparison would also be possible for the energy required to grow other biofuel feedstock crops based on data presented in this chapter. Since it is often debatable what the correct land base should be when comparing per ha intensities of different farm types, Figure 2 should be viewed with caution. Farm land has a wide range of capabilities and intensities of use. Therefore the efficiency of food or feedstock production is not necessarily determined by land use intensity.

5. Summary and conclusions

Quantifying the local impacts from land use changes driven by expanding markets for biofuel was a major focus in this chapter. The degree of spatial detail for the complete farm energy budget presented here is unprecedented in Canada. The sensitivity analysis technique for farm energy demonstrated by Dyer and Desjardins (2003a) could be used to assess scenarios for the growth of biofuel industries. While this has been done for livestock to biofuel feedstock interactions in Canada (Dyer et al., 2011c), more detailed spatial resolution for such scenario or sensitivity analysis is required. With the three Prairie Provinces accounting for 80% of both the arable land and overall farm energy use in Canada (Tables 2 and 6), the ability to assess the energy consumption patterns in this region in more spatial detail than at the provincial scale is especially important. The procedure described in this chapter disaggregated all terms to the CAR scale before re-integrating to the provincial scale. Because of this quantitative link with the CARs, and its computational flexibility, this procedure is ideally suited to this sensitivity analysis application.

5.1. Limitations of the study

The energy budget presented in this chapter does not represent all of the farm energy provided by the FEUS. This was because only those farming systems that are extensive users of farmland are relevant to the regional focus of CEEMA. The excluded energy consumers, including the horticultural enterprises such as market gardeners, fruit growers and greenhouses, are typically clustered within a few highly favourable climate zones, usually in proximity to population centres. In addition, relative to total agricultural energy use, these enterprises are very small and, consequently, small users of energy. In spite of the CEEMA data being derived from economic analysis, while the previous farm energy budget described by Dyer and Desjardins (2009) used actual crop statistics as input data, there was close agreement between these two sets of energy use estimates.

It should be cautioned that the farm energy budget described in this chapter will undergo changes, particularly since it applied to 2001. There are both uncertainties and on-going trends in several of the energy terms in this budget. The most dramatic case has been the impact of reduced tillage on farm use of diesel fuel for field operations (Dyer and Desjardins, 2005). An increasing popularity of diesel fuel for farm owned transport vehicles may mean that some use of diesel fuel for tasks other than field operations may have to be monitored and taken into account in future farm energy budget estimates. The fossil CO_2 emissions that can be attributed to farm use of electric power could also change as coal generating plants are replaced by natural gas, nuclear reactors, or renewable power sources.

For example, natural gas, with its lower CO_2 to energy ratio than coal, is becoming increasingly available for this purpose (NEB, 2006).

There are also suggestions that ammonia-based nitrogen fertilizers could consume less natural gas than other forms of this chemical input (CAP, 2008) and that allowance for increased use of ammonia-based nitrogen fertilizer is needed in the carbon footprint of farm operations. However, the estimates of CO_2 emissions associated with the supply of farm chemical inputs by Dyer and Desjardins (2009), upon which this chapter was based, is consistent with, if not lower than, other studies. For example, over the four census years prior to 2001, the average national CO_2 emissions for chemical inputs reported by Dyer and Desjardins (2009) was 9% below the same period average fossil CO_2 emissions for this term by Janzen et al. (1999). Snyder et al. (2007) reported CO_2 to N conversion rates that were the same as the 4.05 t{CO_2}/t{N} conversion used by Dyer and Desjardins (2009) for Nebraska and 10% higher for Michigan.

The assumption that most farm animals are located near their feed sources was essential to the disaggregation of the three empirical energy terms. This assumption was sound for cattle as roughage makes up an important part of their diet and, except for drought years, its long-distance transport is uneconomic. This assumption was somewhat less sound for pork and poultry as feed grains (including oilseed meal) are more easily transported. Nevertheless, for these livestock types there is an advantage to having production near the cropland that provides the feed and is available for manure disposal. The higher spatial variation for the pork and poultry compared to beef and dairy in Table 3 would support the impact of this advantage. Although pork and poultry were the smallest of the five farming systems, and the three empirical energy terms were also the smallest terms, it would be worthwhile to gather data on the distances over which livestock farmers can cost-effectively ship feed grains. Furthermore, a reliable estimate of the energy used by farmers for transport would be essential to an objective carbon footprint comparison of livestock farming with biofuel feedstock production.

5.2 Going forward: Implications for biofuels

Trends in farm energy levels will also reflect shifts in land use towards feedstock for biofuels. Providing farm type-specific energy data in Appendix A with this chapter identified the energy quantities that are most likely to shift as land resources are reallocated from livestock or food crops into feedstock if the biofuel market opportunities expand. Because of the uncertainties in the farm energy budget, such as more efficient manufacturing of farm inputs, and the land use challenges associated with the emerging biofuel industries, flexibility will be needed. The examples provided here with Figure 2 demonstrated how changes in land use can affect the area based intensity of farm energy consumption and fossil CO_2 emissions. Hence, the analytical procedures for farm energy described in this chapter are being maintained in a dynamic, integrated and repeatable computation procedure. With this flexibility it can facilitate revisions in the Canadian farm energy budget or shifts in farm management as predicted in an updated version of the CEEMA.

This chapter devoted relatively more space and effort to the electric power, gasoline and heating fuel terms than to the field work and two supply terms. Although the three terms from the FEUS were smaller energy quantities, there were two reasons for this extra attention. First, they have received almost no analysis, at least from a modeling perspective, prior this analysis. Consequently, the disaggregation of these terms was much more interpolative than process based. Second, the different levels of use by the five major farm types in Canada of these energy sources, combined with the regional differences in where these farming systems are most often found, resulted in the appreciable spatial variations at the CAR scale shown in Table 3, at least compared to Table 4.

The liquid fossil fuels burned in farm-owned vehicles (both gasoline and diesel) warrants more rigorous treatment because of its overlap with the question of the energy costs of transporting food products to processors and consumers, or feedstock to biofuel plants. Development of a predictive model for this term will depend on better understanding of how and where producers market their produce and the extent to which processers are involved in the collection of that produce, whether it is milk, wheat or canola oil. This is particularly true for biofuel feedstock where the haulage cost can grow in comparison to the production cost if the processing plants are not strategically located. Optimizing the locations of biofuel processing sites will depend on the knowledge of both energy uses and the spatial distribution of land use systems.

Much of the farm energy budget presented in this chapter was based on the 1996 FEUS. Including verification of the F4E2 model, five of the six terms in this energy budget were derived from this database. Updating the FEUS would also facilitate disaggregation of the later years in the farm fossil CO_2 emissions budget described by Dyer and Desjardins (2009) to both the provinces and the CARS. The importance of farm energy in the GHG emissions budget for both agriculture and biofuels requires a repeat of the 1996 FEUS. Since the FEUS entailed survey methodology, rather than actual measurements, an updated FEUS would be an expensive undertaking in Canada. Whereas electric power showed some promise for a predictive tool (Dyer and Desjardins, 2006b), the other FEUS-based terms, gasoline and heating fuels, offer little hope of being worked into a predictive model, although they could be indexed to changing livestock populations. Fortunately, all three of these terms contribute relatively little to Canada's farm energy budget compared to the other three terms.

Growth in biofuel industries is driving the crop selections by many Canadian farmers towards feedstock crops. But as global population expands, major land use shifts will also occur in the food industries, such as from beef or pork production, to more grains and pulses for direct human consumption. Food industries that are now minor, such as vegetable production, may see dramatic growth in response to both food demand and to a warmer climate. Canadian agriculture may well be challenged by shortages of fossil fuel to do field work and commercial fertilizer. The CEEMA database also needs to be updated to help meet these challenges. Until a repetition of the FEUS is undertaken, updated regional farm energy use, and fossil CO_2 emission estimates using more recent census years and an up to date version of CEEMA, will help to fill the information gaps caused by looming changes in the sector.

Appendix A

	CAR	Beef			Dairy			Pork			Poultry			G&OS		
		EP	Gas	HF	EP	Gas	HF	EP	Gas	HF	EP	Gas	HF	EP	Gas	HF
Provinces	#								TJ							
British Columbia	1	5	12	7	5	4	2	0	0	0	0	0	1	0	0	0
	2	9	24	13	10	7	4	0	0	0	2	1	7	0	0	0
	3	20	52	30	23	18	10	3	2	3	28	14	80	0	0	0
	4	9	23	13	12	9	5	1	1	1	4	2	11	0	0	0
	5	26	67	39	28	22	12	1	1	1	4	2	10	0	0	0
	6	1	2	1	1	1	0	0	0	0	0	0	1	0	0	0
	7	20	51	29	26	20	11	1	1	2	5	2	14	0	1	0
	8	60	156	89	145	111	62	16	11	17	99	47	277	3	10	6
Alberta	1	134	347	198	19	15	8	30	21	33	9	4	25	85	263	155
	2	178	462	264	39	30	16	66	46	72	19	9	53	161	499	294
	3	199	517	295	34	26	15	37	26	40	10	5	27	75	232	137
	4	358	928	531	60	46	26	88	62	96	28	13	78	197	611	360
	5	288	746	426	39	30	17	34	24	37	10	5	28	72	223	131
	6	298	772	441	33	25	14	22	15	24	6	3	18	48	148	87
	7	255	662	378	35	27	15	46	32	50	17	8	47	114	354	208
Saskatchewan	1	53	139	79	9	7	4	18	13	20	3	2	9	127	394	232
	2	31	79	45	6	5	3	14	10	15	2	1	7	268	437	258
	3	92	239	136	11	9	5	19	13	20	4	2	10	252	780	460
	4	39	100	57	5	4	2	6	5	7	1	1	3	127	394	232
	5	93	242	138	17	13	7	40	28	43	6	3	18	234	726	428
	6	69	180	103	14	11	6	35	25	39	6	3	17	234	725	428
	7	34	87	50	9	7	4	22	16	25	4	2	11	163	507	299
	8	51	133	76	12	9	5	30	21	33	5	2	14	179	556	328
	9	124	322	184	19	15	8	35	25	38	5	3	15	198	615	362
Manitoba	1	91	235	135	45	35	19	194	136	212	27	13	75	122	380	224
	2	74	191	109	26	20	11	78	54	85	13	6	36	58	181	107
	3	40	105	60	15	12	6	61	43	67	10	5	28	52	161	95
	4	28	71	41	14	10	6	66	47	73	12	6	33	62	191	113
	5	36	92	53	12	9	5	39	27	42	5	3	15	26	82	48
	6	61	159	91	20	15	8	44	31	48	6	3	16	27	82	48
Ontario	1	97	250	143	371	285	158	93	65	102	67	32	187	14	42	25
	2	81	211	120	397	305	169	191	134	209	131	63	367	20	61	36
	3	19	50	29	123	95	52	86	60	94	56	27	158	7	21	12
	4	17	44	25	182	140	77	164	114	179	116	55	324	11	35	21
	5	55	143	82	208	160	88	50	35	55	31	15	86	6	19	11
Quebec	1	16	40	23	134	103	57	27	19	30	10	5	22	3	9	5
	2	7	18	10	67	52	29	16	11	18	4	2	11	2	6	4
	3	4	10	6	34	26	14	13	9	14	6	3	16	1	2	1
	4	5	12	7	48	37	20	38	26	41	15	7	42	1	3	2
	5	12	30	17	89	68	38	17	12	18	6	3	18	1	2	1
	6	5	12	7	59	45	25	59	41	65	21	10	58	1	3	2
	7	12	31	18	97	75	41	31	22	34	12	6	33	1	3	2
	8	9	23	13	69	53	29	8	6	9	3	1	8	1	3	2
	9	21	54	31	168	129	71	42	30	46	16	8	44	2	6	4
	10	20	52	30	280	215	119	352	246	385	129	62	361	3	10	6
	11	15	39	22	152	117	65	101	70	110	34	17	96	2	6	4
Atlantic provinces	1	12	31	18	63	48	27	9	6	10	4	2	12	2	6	3
	2	9	24	13	59	45	25	35	24	38	35	17	97	3	10	6
	3	12	31	18	72	55	31	20	14	22	24	11	67	1	2	1
	4	1	3	2	5	4	2	0	0	0	0	0	0	0	0	0

Table 7. The 2001 energy quantities for Electrical Power (EP), Gasoline (Gas) and Heating Fuels (HF) distributed over five farm types and the 55 Census Agricultural Regions (CAR) of Canada

Appendix B

Provinces	CAR #	Farm field work	Machinery supply	Chemical inputs	Electrical power	Gasoline	Heating fuels[1]
		TJ					
British Columbia	1	28	16	34	10	17	11
	2	60	34	66	21	32	25
	3	124	71	145	74	86	122
	4	51	30	65	26	35	31
	5	137	78	178	59	92	62
	6	4	2	6	2	3	2
	7	102	58	120	52	75	56
	8	608	349	433	323	336	451
Alberta	1	1,908	1,095	1,935	277	650	419
	2	3,532	2,028	5,095	463	1,047	700
	3	2,133	1,224	3,637	355	806	514
	4	4,182	2,401	7,448	731	1,660	1,090
	5	2,594	1,489	6,622	443	1,027	639
	6	1,923	1,104	4,469	407	964	584
	7	3,139	1,802	5,582	467	1,082	699
Saskatchewan	1	2,461	1,413	2,531	211	553	345
	2	2,835	1,627	2,231	321	532	328
	3	4,926	2,828	2,978	377	1,042	632
	4	1,608	923	859	178	502	301
	5	4,925	2,827	7,523	391	1,012	635
	6	4,227	2,427	4,139	359	944	593
	7	2,958	1,698	2,556	232	618	387
	8	3,649	2,095	5,394	278	722	456
	9	4,073	2,338	7,088	382	978	608
Manitoba	1	3,288	1,887	6,916	479	798	665
	2	1,659	952	3,785	248	452	347
	3	1,650	947	2,751	179	325	256
	4	1,945	1,116	2,761	181	325	265
	5	1,005	577	2,641	118	213	163
	6	1,051	603	2,407	158	290	212
Ontario	1	1,892	1,086	3,128	642	675	615
	2	2,850	1,636	3,678	820	774	902
	3	1,106	635	1,110	292	253	345
	4	1,697	974	1,579	490	389	626
	5	1,011	580	1,605	350	371	322
Quebec	1	365	210	571	190	176	144
	2	225	129	297	96	89	71
	3	109	63	163	56	49	50
	4	216	124	266	106	85	112
	5	204	117	462	124	116	92
	6	341	196	366	144	111	156
	7	262	151	495	153	136	128
	8	151	87	306	90	86	61
	9	446	256	854	249	226	196
	10	1,749	1,004	1,871	785	586	901
	11	634	364	874	304	249	297
Atlantic provinces	1	277	159	482	90	93	69
	2	402	231	440	141	120	180
	3	162	93	439	128	113	138
	4	11	7	37	6	7	4

[1] Include furnace oil, liquid propane (LPG) and natural gas

Table 8. Energy quantities in the six terms of the Canadian farm energy balance distributed over 55 Census Agricultural Regions (CAR) of Canada during 2001.

Author details

J.A. Dyer[1], R.L. Desjardins[2], B.G. McConkey[3], S. Kulshreshtha[4] and X.P.C. Vergé[5]

1 Agro-environmental Consultant, Cambridge, Ontario, Canada

2 Agriculture & Agri-Food Canada, Ottawa, Canada

3 Agriculture & Agri-Food Canada, Swift Current, Canada

4 University of Saskatchewan, Saskatoon, Canada

5 Agro-environmental Consultant to AAFC, Ottawa, Ontario, Canada

References

[1] AAFC, 2011. Soil Landscapes of Canada (SLC), General Overview. Agriculture and Agri-food Canada (AAFC). http://sis.agr.gc.ca/cansis/nsdb/slc/intro.html. Accessed 7 May 2012.

[2] Dyer, J.A. and R.L. Desjardins. 2003a. The impact of farm machinery management on the greenhouse gas emissions from Canadian agriculture. Journal of Sustainable Agriculture. 20(3):59-74.

[3] Dyer, J.A. and R.L. Desjardins. 2003b. Simulated farm fieldwork, energy consumption and related greenhouse gas emissions in Canada. *Biosystems Engineering* 85(4): 503-513.

[4] Dyer, J.A. and R.L. Desjardins. 2005. Analysis of trends in CO2 emissions from fossil fuel use for farm fieldwork related to harvesting annual crops and hay, changing tillage practices and reduced summerfallow in Canada. *Journal of Sustainable Agriculture.* 25(3):141-156.

[5] Dyer, J.A. and R.L. Desjardins. 2005. Analysis of trends in CO2 emissions from fossil fuel use for farm fieldwork related to harvesting annual crops and hay, changing tillage practices and reduced summerfallow in Canada. *Journal of Sustainable Agriculture.* 25(3):141-156.

[6] Dyer, J.A. and R.L. Desjardins. 2006a. Carbon dioxide emissions associated with the manufacturing of tractors and farm machinery in Canada. *Biosystems Engineering.* 93(1):107-118.

[7] Dyer, J.A. and R.L. Desjardins, 2006b. An integrated index for electrical energy use in Canadian agriculture with implications for greenhouse gas emissions. *Biosystems Engineering.* 95(3):449-460.

[8] Dyer,J. A. and R. L. Desjardins. 2007. Energy-based GHG emissions from Canadian Agriculture. *Journal of the Energy Institute*. 80(2):93-95.

[9] Dyer, J.A. and R.L. Desjardins. 2009. A review and evaluation of fossil energy and carbon dioxide emissions in Canadian agriculture. *Journal of Sustainable Agriculture* 33(2):210-228.

[10] Dyer, J.A., X.P.C. Vergé, R.L. Desjardins, D.E. Worth, B.G. McConkey. 2010a. The impact of increased biodiesel production on the greenhouse gas emissions from field crops in Canada. *Energy for Sustainable Development* 14(2):73–82.

[11] Dyer, J.A., S.N. Kulshreshtha, B.G. McConkey and R.L. Desjardins. 2010b. An assessment of fossil fuel energy use and CO_2 emissions from farm field operations using a regional level crop and land use database for Canada. *Energy* 35(5):2261-2269.

[12] Dyer, J.A. O.Q. Hendrickson, R.L. Desjardins and H.L. Andrachuk. 2011a. An Environmental Impact Assessment of Biofuel Feedstock Production on Agro-Ecosystem Biodiversity in Canada. In: *Agricultural Policies: New Developments*. Chapter 3:87-115. Editor: Laura M. Contreras, ISBN 978-1-61209-630-8. Nova Science Publishers Inc. Hauppauge, NY 11788. 281 pp.

[13] Dyer, J.A., X.P.C. Vergé, S.N. Kulshreshtha, R.L. Desjardins, and B.G. McConkey. 2011b. Residual crop areas and greenhouse gas emissions from feed and fodder crops that were not used in Canadian livestock production in 2001. *Journal of Sustainable Agriculture* 35(7):780-803.

[14] Dyer, J.A., X.P.C. Vergé, R.L. Desjardins and B.G. McConkey. 2011c. Implications of biofuel feedstock crops for the livestock feed industry in Canada. In: *Environmental Impact of Biofuels*. Chapter 9, pages 161-178, Editor: Marco Aurelio Dos Santos Bernardes. InTech Open Access Publisher. Rijeka, Croatia. September 2011, ISBN 978-953-307-479-5. 270 pp.

[15] CAP, 2008. Benchmark energy efficiency and carbon dioxide emissions. Canadian Ammonia Producers (CAP).Office of Energy Efficiency, Natural Resources Canada. Ottawa, Canada. ISBN 978-0-662-46150-0, Cat. No. M144-155/2007E-PDF. 34pp.

[16] GAO, 2009. Biofuels – Potential effects and challenges of required increases in production and use. Report to Congressional Requesters. United States Government Accountability Office (GAO). August 2009. GAO-09-446. 184 pp.

[17] Horner, G.L., J. Gorman, R.E. Howett, C.A. Carter and R.J. MacGregor. 1992. The Canadian regional agricultural model: structure, operation and development. Policy Branch. Technical Report 1/92. Ottawa. Agriculture and Agri-food Canada. http:www.agr.gc.ca/pol/index_e.php?s1=pub&s2=visions&page=v_pt2.

[18] IFAS, 1998. Replacing hay with grain. Chapter 6, The Disaster Handbook - 1998 Edition. Institute of Food and Agricultural Sciences, Cooperative Extension Service, University of Florida.

[19] Janzen, H.H., R.L. Desjardins, J.M.R. Asselin and B. Grace (Editors). 1999. The Health of our Air - Towards Sustainable Agriculture in Canada. Agriculture and Agri-food Canada. Publ. No. 1981 E. ISBN 0-662-27170-X, 105 pp.

[20] Klein, K.K., LeRoy, D.G. 2007. The Biofuels Frenzy: What's in it for Canadian Agriculture? Green Paper Prepared for the Alberta Institute of Agrologists. Presented at the Annual Conference of Alberta Institute of Agrologists. Banff, Alberta, March 28, 2007. Department of Economics, University of Lethbridge. 46 pp.

[21] Kulshreshtha, S.N. .B. Jenkins, R.L. Desjardins and J.C. Giraldez. 2000. A systems approach to estimation of greenhouse gas emissions from the Canadian agriculture and agri-food sector. *World Resources Review*. 12:321-337.

[22] Malcolm, S. and M. Aillery. 2009. Growing crops for biofuels has spillover effects. *Amber Waves* 7(1):10-15. March issue.

[23] Miller, J. I. and F. B. Morrison, 1950. Comparative Value of Mixed Hay and Corn Silage for Wintering Beef Cows. Journal of Animal Science. 9(2):243-247. http://jas.fass.org/content/9/2/243.abstract

[24] Nagy, C.N. 2000. Energy and greenhouse gas coefficients inputs used in agriculture. Report to the Prairie Adaptation Research Collaborative (PARC). The Canadian Agricultural Energy End-use Data and Analysis Centre (CAEEDAC) and the Centre for Studies in Agriculture, Law and the Environment (CSALE).

[25] NEB, 2006. Natural gas for power generation: issues and implications. ISBN 0-662-43472-2. Cat No. NE23-136/2006E. National Energy Board (NEB), Calgary, Alberta, Canada. 60pp. www.neb-one.gc.ca.

[26] Neitzert, F., K. Olsen and P. Collas, 1999. Canada's greenhouse gas inventory - 1997-Emissions and removals with trends. Air Pollution Prevention Directorate, Environment Canada. Ottawa, Canada. ISBN 0-622-27783-X. Cat No. En49-8/5~9E. 160pp.

[27] Neel, James B. 2012. Rules of thumb for winter feeding. Animal Science Info Series: AS-B-260, University of Tennessee, Agricultural Extension Service. http://animalscience.ag.utk.edu/beef/pdf/RulesOfThumb.260.pdf. Accessed 18 June 2012.

[28] NRCan, 2005. Energy Efficiency Trends in Canada, 1990 and 1996 - 2003. Office of Energy Efficiency, Natural Resources Canada. Cat. No. M141-/2003. ISBN 0-662-68797-3, 64 pp.

[29] Schoenian, Susan, 2011. The truth about grain: Feeding grain to small ruminants. Small Ruminant Info Sheet, University of Maryland Extension. http://www.sheepandgoat.com/articles/graintruth.html. Accessed 6 February 2012. Accessed 18 June 2012.

[30] Snyder, C.S., T.W. Bruulsema and T.L. Jenzen, 2007. Greenhouse gas emission from cropping systems and the influence of fertilizer management. International Plant Research Institute, Norcross, Georgia, USA. 36 pp.

[31] Statistics Canada, 2007, Census Agricultural Regions Boundary Files for the 2006 Census of Agriculture - Reference Guide. Catalogue no. 92-174-GIE. http://publications.gc.ca/collections/collection_2007/statcan/92-174-G/92-174-GIE2007000.pdf. Accessed 7 May 2012.

[32] Tremblay, V. 2000. The 1997 Farm Energy Use Survey - Statistical Results. Contract Report prepared for The Office of Energy Use Efficiency, Natural Resources Canada, Ottawa, Ontario, Canada, and The Farm Financial Programs Branch, Agriculture and Agri-Food Canada, Ottawa, Ontario, Canada.

[33] Vergé, X.P.C., Dyer, J.A., Desjardins, R.L. and Worth, D. 2007. Greenhouse gas emissions from the Canadian dairy industry during 2001. *Agricultural Systems* 94(3): 683-693.

[34] Vergé, X.P.C., Dyer, J.A., Desjardins, R.L. and Worth, D. 2008. Greenhouse gas emissions from the Canadian beef industry. *Agricultural Systems* 98(2):126–134.

[35] Vergé, X.P.C., Dyer, J.A., Desjardins, R.L. and Worth, D. 2009a. Greenhouse gas emissions from the Canadian pork industry. *Livestock Science*. 121:92-101.

[36] Vergé, X.P.C., Dyer, J.A., Desjardins, R.L. and Worth, D. 2009b. Long Term trends in greenhouse gas emissions from the Canadian poultry industry. *Journal of Applied Poultry Research*. 18: 210–222.

[37] Vergé X.P.C., D.E. Worth, J.A. Dyer, R.L. Desjardins, and B.G. McConkey. 2011. LCA of animal production. In *Green Technologies in Food Production and Processing*, ed. by Joyce I. Boye and Yves Arcand, 83–114. New York: Springer.

[38] Vergé, X.P.C., J.A. Dyer, D.E. Worth, W.N. Smith, R.L. Desjardins, and B.G. McConkey, 2012. A greenhouse gas and soil carbon model for estimating the carbon footprint of livestock production in Canada. Accepted in *Animals*.2:437-454

Chapter 5

The Possibility of Future Biofuels Production Using Waste Carbon Dioxide and Solar Energy

Krzysztof Biernat, Artur Malinowski and
Malwina Gnat

Additional information is available at the end of the chapter

http://dx.doi.org/10.5772/53831

1. Introduction

The Earth's energy requirements are estimated at 14 TW/ y. Considering the economic development, and therefore high consumption and constantly increasing number of people in the world, it is estimated that energy demand in 2050 will be amount 28 - 30 TW/ y. Fuels from crude oil supply about 96% of the worldwide energy demand for transport. On the other hand, known petroleum reserves are limited and will eventually run out. According to preliminary calculations, fossil fuels will be exhausted within 150-200 years. Fuel consumption causes the emission of carbon dioxide into the atmosphere, resulting in the collapse of the balance between carbon dioxide released to environment, and gas that can be absorbed by plants. It is estimated, that in case of continued use of traditional energy sources by 2030, carbon dioxide levels will rise to 40 billion Mg per year. The correlation of carbon dioxide emissions from the world's population is shown in Figure 1:

Global emissions of CO_2 and other GHGs, despite commitments of the reduction made by developed countries, will continue to grow, because of increasing production in developing countries, for which clean technologies and investing in renewable energy sources are too expensive. Currently, such a trend can be observed, because the developed countries carbon dioxide emissions was reduced by 6.5% (according to IEA [2] data for 2009). On the other hand the developing countries increased the emissions up to 3.3% (mainly in Asia and the Middle East). In this way, the twenty-first century economy will be depend on fossil fuel resources. As a consequence of this state, GHG concentrations will increase, resulting in the continued progress of global warming. Club of Rome, in 1972, has presented a report: "The Limits to Growth" which predicted that before 2072, present industrial civilization will collapse, as a result of the lack of available energy resources, or because of polluted environment.

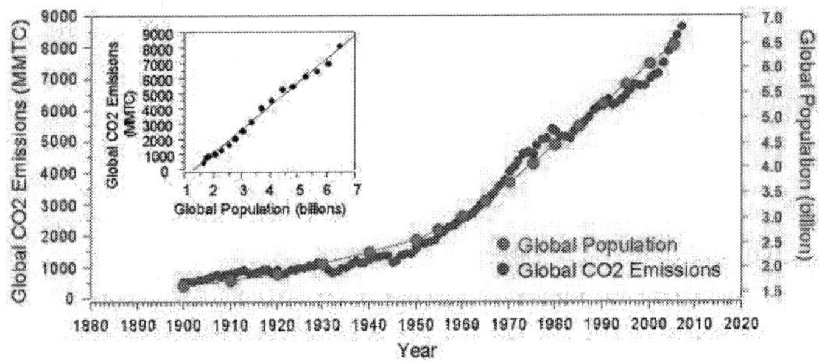

Figure 1. The correlation of CO_2 emissions with the world's population [1]

It is important to get alternative energy sources, that will be able increasingly replace the fossil fuels with reducing effect of carbon dioxide emissions at the same time. That may be renewable energy sources (RES), such as wind, water, geothermal or solar energy. The potential of the solar energy is estimated up to 100.000 TW/y. This huge amount of energy has high potential of application in thermochemical biomass conversion or artificial photosynthesis for processing carbon dioxide and water into the organic compounds.

2. Solar radiation as an energy source

Solar energy is the most important source of the energy used on the Earth. According with hypothesis of H. Bethe and C. Weizsacker made in 1938 - the energy of the Sun, has its source in the fusion reactions that occur in the interior, according to the reaction (1): [3]

$$4\,^{1}H \rightarrow {}^{4}He + 2e^{+} + 2\nu_{e} + E \quad (1)$$

Solar radiation is in the form of a wide band of the electromagnetic spectrum, which is shown in Figure 2. It covers wavelengths from about 250 nm to 1000 nm over [4].

However, the full spectral range of solar radiation (including the ultraviolet (UV)) reaches only to the edge of Earth's atmosphere. The total energy that reaches the Earth's atmosphere is marked as Fs - a stream of sunlight that the average power is 1368 W/m^2. Some of the radiation wavelength, passing through the layers of the Earth's atmosphere, are absorbed by the molecules. Even at the height of the Kármán line (about 100 km above sea level) there is the first interaction of solar radiation with occurring nitrogen and oxygen. Ultraviolet (UV), at a wavelength below 280nm, is high-energy radiation. The energy is enough to cause dissociation of molecules into atoms as is shown in following equations 2 and 3:

$$N_2 + h\nu\ (\lambda \approx 126nm) \rightarrow 2N \quad \Delta H^{\circ}=945\ kJ/mol \quad (2)$$

http://dx.doi.org/10.5772/53831

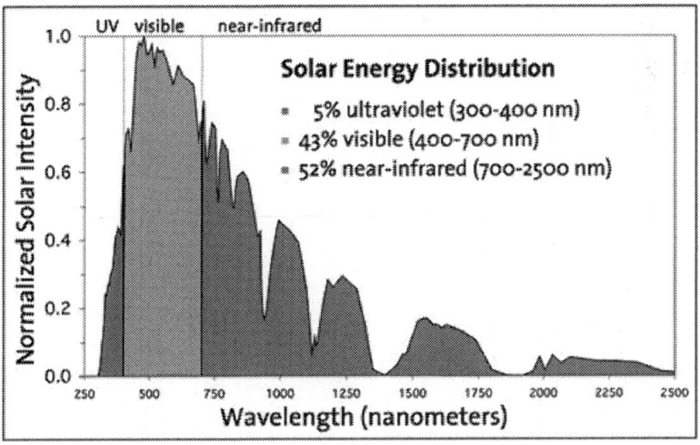

Figure 2. The spectrum of solar radiation [5]

$$O_2 + h\nu\ (\lambda \approx 240nm) \rightarrow 2O \quad \Delta H° = 498\ kJ/mol \tag{3}$$

In the lower layer of the atmosphere - called the ionosphere - solar energy is absorbed by the reactions occurring in the ionization of chemical individuals (reactions 4 and 5).

$$N_2 + h\nu\ (\lambda \approx 80nm) \rightarrow 2N^+ + e \qquad \Delta H° = 1500\ kJ/mol \tag{4}$$

$$O + h\nu\ (\lambda \approx 91nm) \rightarrow O^+ + e \qquad \Delta H° = 1310\ kJ/mol \tag{5}$$

Moving towards next layer - ozonesphere - sunlight meets molecules of ozone (O_3). In this way the next part of the radiation is absorbed by the ozone, which leads to O_3 dissociation, resulting in the formation of excited molecules and atoms of oxygen (reaction 6):

$$O_3 + h\nu\ (\lambda < 325nm) \rightarrow O_2^* + O^* \tag{6}$$

In addition, a small amount of solar radiation is absorbed during its passage through the troposphere.

As a result, after the absorption and dispersion in the atmosphere, the spectral range of solar radiation flux reaching the Earth's surface is slightly changed, mostly free of long-range radiation from ultraviolet as seen in Figure 3.

By assuming the all solar energy which reaches the earth's atmosphere amount 100 units, as many as 19 of them will be absorbed by molecules and suspensions occurring in the Earth's upper atmosphere. Other relative flows of solar radiation towards the Earth shows Figure 4.

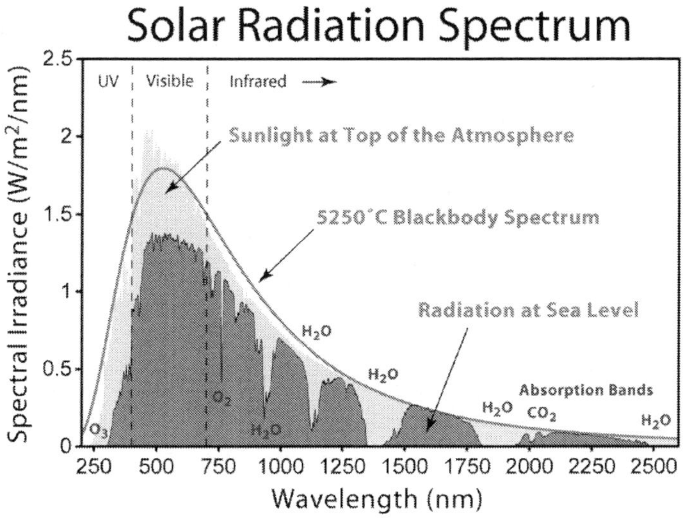

Figure 3. Sunlight spectrum before and after its passage through the atmosphere of the Earth [6]

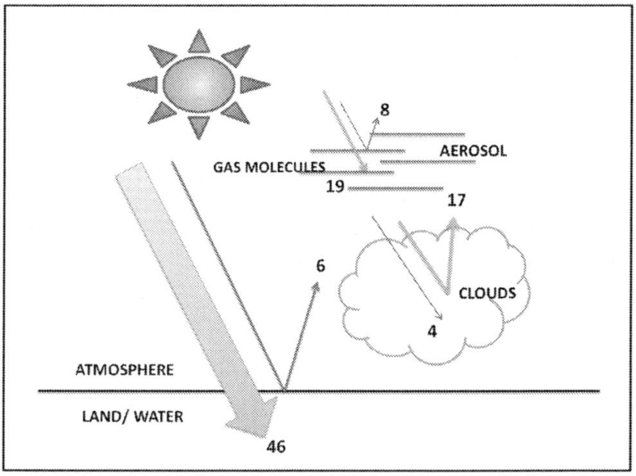

Figure 4. The flow of solar radiation reaching the Earth's surface [4 modified]

As shown in Figure 4, part of the solar radiation in general does not take part in the energy balance of the Earth - it is reflected to the space. 31 units is reflected including: 6 units directly reflected by the surface of the Earth, 8 units is reflected by aerosols, dust and other materials such as volcanic ash, and the remaining 17 units are reflected from the clouds. The value of solar radiation is expressed as a percentage by the number of *Albedo*. According to the calculations the Earth's Albedo amounts 31%. From the remaining part of the solar energy (50 units) - 4 units are still absorbed by clouds of drops suspended in the water, in consequence remains 46 units. In this way only 46% of solar radiation reaches the Earth's surface and can be absorbed by the land and water. Radiation - a relatively high-and shortwave (UV and VIS) - is partially stored and used in different processes on the Earth, and part of it is reemitted into space in the form of long-wave radiation (infrared, IR) with less energy. Therefore, the total amount of energy absorbing solar radiation reaching the earth shield can be expressed by equation 7:

$$E_s = F_s(1 - A)\,\pi r^2 \tag{7}$$

where: $\mathbf{E_s}$ - means the total amount of solar energy which is absorbed by the Earth (W)

$\mathbf{F_s}$ - average flux of solar radiation (solar constant) (1368 W/m^2)

A - (albedo) of the radiation reflected back into space (0.31)

r - the radius of the Earth ($6.37\ 10^6$)

Therefore, the absorbed solar energy is approximately $1.2\ 10^{17}$W (about 124 600 TW) [7]. This amount of absorbed solar radiation is the driving force behind all the changes that are taking place on the globe. The solar energy transformation scheme, that reaches the Earth's surface is shown in Figure 5.

Most of the solar energy stream is directly converted into the heat, which is about 67% (nearly 83 700 TW) of the radiation reaching the Earth's surface. About 32% of this energy take part in the hydrological Earth cycle (about 40 400 TW). The energy of water (oceans, seas, inland waters), tidal wave energy can be used to produce electricity in hydroelectric power plants such as the, so-called, flow power plants, which is based on a natural flow. Part of the solar energy is converted into kinetic energy of the wind. Around 400 TW is used for air movement. Wind as an energy source was used in ancient times, and today the kinetic energy of winds is used to produce electricity by using wind turbines.

Other 100 TW of solar radiation are driving forces behind the production of biomass. That organic matter is produced from the processing of solar energy through photosynthesis.

3. The processes of photosynthesis and their impact on the biomass growth

Green plants, some bacteria and protists have developed specific mechanism for synthesis of reduced carbon compounds, through which energy from the sun has been successfully transformed into a useful form of it. This process, which is one of the most important bio-chemical processes on the Earth, is called the photosynthesis. It's name comes from two ancient Greek words meaning "light" and "connect". Total formula for process of photosynthesis is a

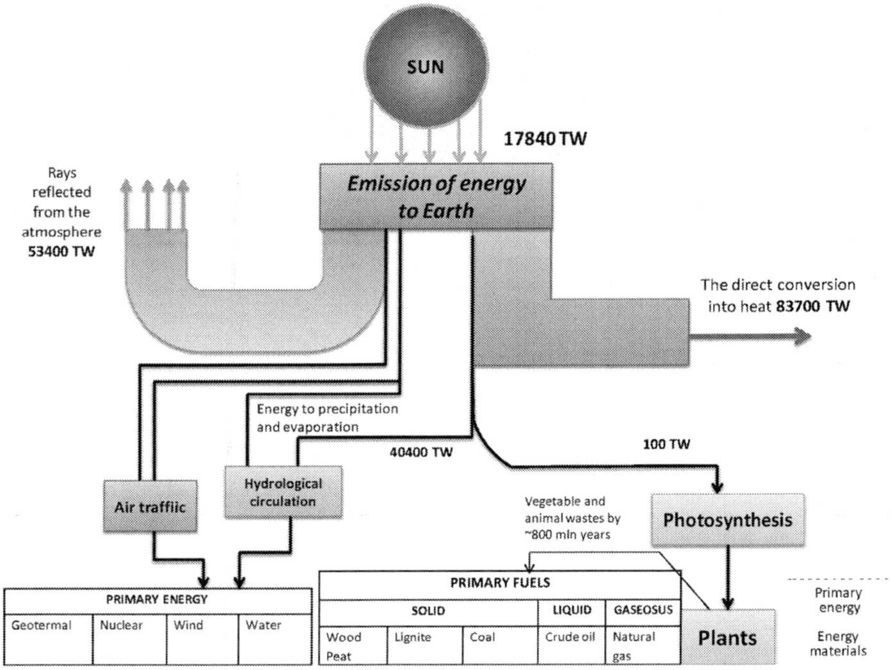

Figure 5. Diagram of solar energy transformations [7 changed]

reaction combining water molecules and carbon dioxide, in the presence of energy from the sun, to give the product as a basic sugar molecule, and oxygen as a byproduct (reaction 8).

$$nH_2O + nCO_2 \rightarrow C_nH_{2n}O_n + nO_2 \quad (8)$$

If assumed the final product of photosynthesis is glucose, simple sugar molecule belongs to a group of a hexoses. Then the total reaction formula can be written as below (Reaction 9):

$$6H_2O + 6CO_2 \rightarrow C_6H_{12}O_6 + 6O_2 \quad (9)$$

Photosynthesis can be distinguished by two sets of reactions: the light-dependent reaction and light-independent reaction (also called dark reactions of photosynthesis). The photosynthesis is initiate by solar radiation, falling on the surface of green plant leaves, is absorbed by assimilation pigments, acting as a catalyst [4]. There are two types of these pigments: chlorophyll *a* and chlorophyll *b*. Chlorophyll molecules have strong absorption properties, absorbs solar energy from the electromagnetic spectrum in the range of 400 nm to 700 nm. Radiation

http://dx.doi.org/10.5772/53831

of this wavelength range is called Photosynthetically Active Radiation (PAR). The absorption spectrum of chlorophylls differing slightly from each other. Chlorophyll *a* with total formula $C_{55}H_{72}O_5N_4Mg$ have blue-green color and absorbs light violet wavelength of 417 nm, and the red one wavelength of 657 nm. Chlorophyll *b* ($C_{55}H_{72}O_5N_4Mg$) absorbs blue light in the field of 460 nm and red one 650 nm wavelength [8]. These small shifts of the both colors absorption maximums is the result of a small difference in the construction. The methyl group present in the chlorophyll *a* has been replaced by an aldehyde group in the molecule of chlorophyll *b* (Figure 6). The Range of the spectrum, useful for photosynthesis, is enhanced by auxiliary pigments - carotenoids, which have absorption properties of solar energy, not available for green chlorophyll. The color of the carotenoids is yellow-orange, which is the result of absorption by the blue-violet wavelength of 400 nm-495 nm (Figure 7) [9].

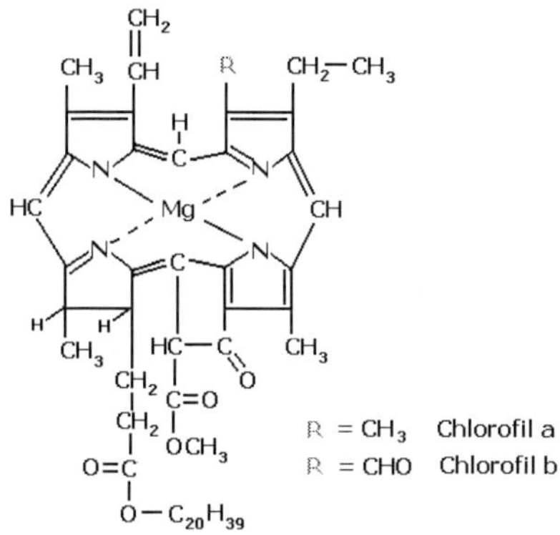

Figure 6. Differences between the structure of chlorophyll a and b [6]

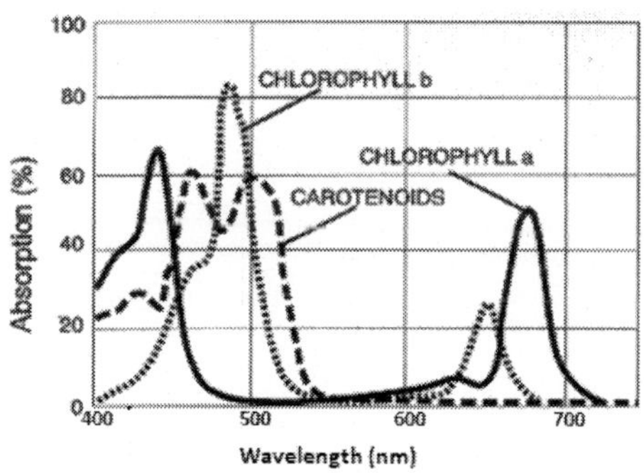

Figure 7. Absorption spectrum of carotenoids and chlorophylls a and b is a graph of light absorption by the different wavelengths of light [10]

Chlorophylls and carotenoids can be found in chloroplasts tylakoids. Dyes are arranged briefly to form units called photosystem I and photosystem II. Visible light beam falling on the chlorophyll molecule, starts the electron excitation state of the magnesium ion at the center of the porphyrin ring. Excited electron is transferred into a conjugated bonds, and then transported to the neighboring molecule dyes. Thus, to pass the initial electron acceptor in photosystem II (which is ferrodoksin molecule) to photosystem I (where particles of quinone are primary acceptor) that ends on the molecule of coenzyme NADP + and reduce it to NADPH. Chlorophyll molecule, after return the electron, has strong oxidizing properties. Thus chlorophyll molecule by a specific protein complex containing manganese ions is able to receive the missing electron from the molecule of water. As a result this process the hydrogen protons and oxygen, the product called "side" photosynthesis is obtained. The result of electron beam movement are two compounds: ATP and NADPH. Produced so-called *power of assimilation* is necessary to carry out the dark reactions of photosynthesis. A further step of photosynthesis does not require solar energy. At this stage there is a series of reactions called the Calvin cycle where carbon dioxide is assimilated from the air. Subsequently, carbon dioxide is converted and in support of enzyme Rubisco is build in natural organic molecules [9].

Formation one molecule of glucose ($C_6H_{12}O_6$) required six Calvin rotation cycle and the energy in the form of 18 ATP molecules and 12 NADPH molecules.

A series of reactions, which constitute to the process of photosynthesis, are initiated by described light beam, falling into assimilation plants dye. Falling solar rays must have sufficient energy to cause an electron excited state in chlorophyll.

For full execution of photosynthesis, and specifically to reduce 1 mole of CO_2 molecule is assumed that it takes 8 photons of red light having a wavelength (λ = 680nm), as shown in reaction 10.

$$2H_2O + CO_2 + 8\,h\nu \rightarrow CH_2O + O_2 + H_2O;\lambda{=}680nm = 1.81eV/h\nu{=}\ 174kJ/mol;\Delta G^{\circ}{=}479,1kJ/mol \tag{10}$$

One photon of this wavelength has energy = 174kJ/mol, therefore energy used in reaction is: 8 x 174kJ/mol. The free energy (ΔG°) of CO_2 reduction reaction to CH_2O, amounts 479.1 kJ/mol, than photosynthesis efficiency η_f can be calculated:

Efficiency of photosynthesis (η_f) = Energy used/Energy Supplied;

substituting: $\eta_f = 479{,}1\ kJ\ mol^{-1}/8\ 174{,}0\ kJ\ mol^{-1} = 0{,}34\ kJ\ mol^{-1}$

Thus, the ratio between the used energy to the energy put into the photosynthesis process amounts approximately 34%. However, in real conditions only this part of the radiation that can be absorbed by the Earth's surface should be considered. However only about 0.1% of this energy takes part in photosynthesis. In more precise calculations and considering such losses such as "photobreathing" of plants or microbiological decomposition, in practice the efficiency of photosynthesis does not exceed 5% [11].

Each year around the globe are synthesized billions tones of organic compounds that serve as nutrients not only for producers, but also for living organisms, which are on higher levels of ecological pyramid. Biomass as an organic matter is simply result of change-over driven by solar energy.

4. Biomass as an energy source

Solar energy, as a result of guided for over 3.5 bn years of photosynthesis, is accumulated in the form of organic matter called biomass. In this way, solar radiation is converted into a solid form, which can be used further, not only by organisms in the food chain, but also can provide a raw material for the production of effective energy.

The definition of biomass according with Directive 2009/28/WE is: „biodegradable fraction of products, waste and residues from biological origin from agriculture (including vegetal and animal substances), forestry and related industries including fisheries and aquaculture, as well as the biodegradable fraction of industrial and municipal waste"[12].

However, this definition does not fully covers the meaning of biomass, because its forms are not just a production side effect of other converting processes. There are also dedicated plantation for energetic purposes only. In other words are established the specific energy crops cultivation.

Biomass energy resources are divided into three main groups: [13,14]:

- agricultural biomass (energy crops),
- forest biomass (firewood, waste from wood industry and paper industry),
- and all organic waste from agriculture, forestry as well as gardening.
- The basic division of biomass and their byproducts are distinguished as follows due to its physical state:
- solid biomass (that is wood, straw, energy crops, briquette, pellets),
- liquid (liquid and gaseous) energy carriers from biomass processing,
- Depending on the degree of physical, chemical or biochemical processing of biomass, it can create:
- primary energy sources - wood, straw, energy crops,
- secondary energy sources - slurry, manure, sewage sludge and other organic waste,
- processed energy sources - energy carriers, that is biofuels (bioliquids) [14,15]

The potential of biomass in the world according to the German Electricity Industry Association (VDEW- German *Verband der Elektrizitaetswirtschaft*) is about 150 bn Mg per year, which equal to 120 bn Mg of coal. Resources of this biomass exceeds over ten times the current world demand for energy. However, from this potential biomass amount, only about 20-30% is suitable to be used, and in fact only 6 bn Mg of plant biomass is suitable for use [16]. Biomass resources in Poland are estimated at 30 mln Mg per year, which is the energy equivalent to 16-19 mln Mg of coal/ year and is about (110... 130) TW [16] (an average of about 11% of total consumption). For example, to supply biomass to power plant with a capacity of 1MW (= 0.000001 TW) about 5000 Mg of dry weight of the raw material is needed. This amount is equivalent to the coverage of the field area less than 500 ha, and assuming 10% of the density the crops, which is to cover an area of approximately 50 km^2 [17]. The sample calculations show, that required energy consumption for biofuels from biomass in the EU's cover area about 79 000 km^2, which can be compared with the area of the Czech Republic [18].

According to the above figures, the existing biomass resources, and thus the energy efficiency of this raw material is not enough to cover the world's growing demand for energy. It is mean the agricultural land used for food crops should be replaced to a large extent by energy crops plantations, and this way increase the food prices.

Biomass as a solid biofuel is a raw material after suitable processing is used for heat and electricity generation. Conversion of solid biomass for energy purposes can occur by two main pathways (Table 1 and 2) [14, 7]:

1. First one include processes of direct combustion of solid biomass such as in boilers and power plants or by biomass gasification and then combustion of the resulting gas (syngas).

2. The alternative path of energy accumulated in the biomass is to process biomass to liquid biofuels, which are liquid energy carriers for transport.

Process	Combustion	Gasification
Conditions	• The excess oxygen in the combustion chamber • Temperature: 800-1450°C	Phase I: drying and degassing of the material the shortage of oxygen, temperature: 450-800 ° C Phase II: combustion gases in the presence of excess oxygen, at: 1000-1200 ° C Phase III: heat transfer in the heat exchanger
Final products	CO_2, H_2O	Mixture of gases: CO_2, N_2, H_2, CO, C_xH_y, CH_4 and others.
Energy	Heat energy/ electricity	Heat energy/ electricity
Effectivity	~15-20%	~35%

Table 1. Technologies of the direct use of solid biomass (pathway 1)

Combustion of solid biofuels is the most common and easiest, but inefficient method of getting energy. Because the biomass is a mixture of organic compounds such as carbohydrates, the combustion reaction can be presented in simpler form of equation (11):

$$[CH_2O] + O_2 \rightarrow CO_2 + H_2O \qquad \Delta H = -440\ kJ \tag{11}$$

Above equation is a simple reverse of the photosynthesis reaction. It can be assumed that this form of solar energy conversion, accumulated in the form of a green plant organic matter is completely depends on carbon dioxide emitted to the atmosphere. As it has been calculated, during the combustion of fossil carbon for every 1 GJ of energy is produced 112 kg of carbon dioxide emissions. In this calculation for biomass (assuming the simplified formula - CH_2O) the amount of CO_2 emitted per each 1 GJ of energy is around 100 kg, which is slightly advantage over the coal. However, it is suggested, the final energy balance of biomass combustion process is zero, because of CO_2 emitted during combustion process is equal to the amount of gas which is absorbed by the organism during the plant growing season [4]. However, the rate of biomass growing in time is not enough, to compensate the CO_2 emitted into the atmosphere. The plants during the growing season, green organisms need about 10 years [4] in ideal conditions for example with proper lighting, temperature and humidity. In order to support the performance of biomass energy crops increase are also used different mixes, fertilizers and pesticides, which are produced from natural gas. It is estimated, to get a 25% share of biomass fuels will lead to increase use of NPK fertilizers by 40% [4, 19]. Plant protection products and supporting substances causes additional emissions of greenhouse gases into the atmosphere such as N_2O. Because of combustion of the solid biomass is not only CO_2 or N_2O emission, but also carbon oxide (CO), sulfur compounds, or other combustion residues such as fly ash and dust.

<table>
<tr><td colspan="10">BIOMASS</td></tr>
<tr><td colspan="3">Edible parts of plants (seeds, grains, fruits)</td><td colspan="7" rowspan="2">Non-edible parts of plants which are agricultural waste or derived from energy crops</td></tr>
<tr><td colspan="2">Containing starch</td><td>Containing vegetable oils: rapeseed, palm oil, sunflower oil</td></tr>
<tr><td colspan="2" rowspan="2">Fermentation</td><td rowspan="2">Transesterification</td><td colspan="2">Fermentation processes</td><td colspan="3" rowspan="2">Gasification</td><td colspan="2" rowspan="2">Thermolysis</td></tr>
<tr><td>Fermentation: anaerobic ethanol butanol</td><td>Fermentation</td></tr>
<tr><td>Ethanol</td><td>Butanol</td><td>Biodiesel</td><td>Methane</td><td>Short fatty acids</td><td colspan="3">Syngas</td><td>Light gases</td><td>Oil fractions</td></tr>
<tr><td colspan="5"></td><td colspan="3">Fischer-Tropsch Synthesis</td><td colspan="2"></td></tr>
<tr><td colspan="5"></td><td>DME</td><td>Methanol</td><td>Diesel</td><td colspan="2"></td></tr>
</table>

Table 2. Examples of processes to convert biomass to liquid biofuels [21]

The efficient combustion of such material needs advanced techniques for boilers, allowing the complete combustion of volatile products of biomass pyrolysis. The incomplete combustion of biomass significantly reduces the efficiency of the process. In addition, low-density of biomass causes difficulties in its storage and transport, which creates extra costs. A large range of moisture content, within the ranges 50... 60% gives the energy value (6... 9) MJ/ kg. On the other side biomass dried has calorific value up to 19MJ/kg [20]) which is also difficult and costly for preparation to use in the heat and electricity production. Consequently, because of the above reasons, considering the whole process, namely from the operation of farm machinery, through the transport, processing of biomass (drying, degassing of material, seal or change of boilers). Finally the combustion of biomass is not efficient to cover the still increasing demand for energy. A good solution of the shortage problem of biomass used for energy purposes could be manufacturing biomass by microorganisms such as algae, which will be discussed later in this chapter.

Mentioned alternative process for the recovery of solar energy collected in photosynthetic organisms is subjected to physical, chemical or biochemical conversion of biomass (pyrolysis, gasification, fermentation, distillation) [4]. In this way the converted form of biomass is called liquid biofuels, which sample preparation processes are shown in Table 2.

5. Biofuels

In 1900 at the World Exhibition in Paris, Rudolf Diesel introduced the high-pressure engine driven with clean oil from arachidic peanuts. Twenty years later Henry Ford introduced the possibility to power the internal combustion engines with spark ignition by ethanol. Mixtures of alcohol and gasoline as a fuel (30:70 v/v) was reported for using in 1929 in Poland. One of the first scientific publications regarding the possibility of using these compounds to power the engines of tractors has been published in Polish journal "Chemical Industry" [22].

In the expected terms of production and use of biofuels (alternative fuels), it is assumed that those fuels should:

- to be available in sufficiently large quantities;
- demonstrate a technical and energetic properties of determining their suitability to supply the engines or heating devices;
- to be cheap in production and has attractive price for costumers;
- has a low risk for environment than the fossil fuels, by less emission of toxic compounds and greenhouse gases in the combustion process;
- provide an acceptable economic indicators of engines or boilers and safety of their use, and enable the lower operating costs of the equipment;
- increase the energy independence.

Until now, there are several complementary definition of biomass. Previously cited European definition mentioned in the European Directive 2009/28/EC and the biomass as the main source of raw materials for the biofuels production. The European definition distinguish two basic raw material pathways and corresponding them technologies, namely BtL processes ("biomass to liquid"), alternatively BtG ("biomass to gas") and WtL ("waste to liquid"), alternatively WtG ("waste to gas").

The Directive also introduced bioliquids as liquid biofuels for other than transport purposes, like production of electricity, heat and cooling. The introduction of "xtL processes" identifies general processes for conversion of biomass instead of distinguishing of second, third and fourth generation biofuels. Apart from this there was defined the term "synthetic biofuels" describing them as synthetic hydrocarbons or their mixtures gained from biomass, such as SynGas produced in the gasification process of forest biomass or SynDiesel.

This division resulted from the conditions described above, and first from the assess the suitability of fuels in modern engine technology and availability of raw materials and their impact on the environment. The formal division of biofuels on the appropriate generations has been published in the report "Biofuels in the European Vision, the Vision 2030 and Beyond". This report divided biofuels into the first generation biofuels, so-called "conventional biofuels" and the second generation biofuels, so-called "future biofuels".

The first generation biofuels („conventional biofuels") includes:

- bioethanol (BioEtOH, BioEt), understood as a conventional ethanol gained from hydrolysis and fermentation processes from raw materials such as grain, sugar beets and so on.
- pure vegetable oils (PVO-pure vegetable oils), got from cold pressing processes and the extraction of oilseeds;
- biodiesel, which is rapeseed oil methyl ester (RME) or methyl and ethyl esters (FAME and FAEE) of higher fatty acids or from the other oleaginous plants, obtained by the process of cold pressing, extraction and transesterification;
- biodiesel, which is the methyl and ethyl esters gained by transesterification of waste oils;
- biogas got by purification processes of damp waste or agriculture raw biogas;
- bio-ETBE, got from the chemical transformation of bioethanol.
- To the biofuels of the second category ("future biofuels") have been classified as follows:
- bioethanol, biobutanol and mixtures of higher alcohols and their derivatives got by the advanced hydrolysis and fermentation process of lignocellulose biomass (except of raw materials for food purposes);
- synthetic biofuels, which are products of biomass change-over by gasification and follow by catalytic synthesis for hydrocarbon fuel components in BtL processes.
- fuel for diesel engines gained from lignocellulosic biomass processing through the Fischer-Tropsch processes;
- biomethanol obtained by the processes of lignocellulose transformation, including the Fischer-Tropsch synthesis and also with use of waste carbon dioxide;

- biodimethylether (bioDME) obtained in thermochemical biomass conversion processes, including methanol, biogas and synthesis gases which are derivatives of biomass conversion processes;
- biodiesel as a biofuel or as a biocomponent for diesel engines derived bý hydrogen refining (hydrogenation) of vegetable oils and animal fats;
- biodimethylfuran (bioDMF) derived from the processing of sugars, including cellulose, in biochemical and thermochemical processes;
- biogas as a synthetically derived natural gas - biomethane (SNG), obtained through the lignocellulose gasification processes and the appropriate synthesis and by purification processes of biogas from agriculture, landfills and sewage sludge;
- biohydrogen obtained by gasification of lignocellulose and synthesis of gasification products or through the biochemical processes.

The concept of the second generation biofuels is based on the assumption, the raw material for its production should be biomass as well as waste vegetable oils and animal fats, and any residual organic materials, unsuitable for the food industry or forestry.

Department of Transport and Energy of the European Commission proposed to separate the third generation biofuels, as those for which the technological development and their implementation into operation can be estimated for the years 2030 and beyond. To those fuels the biohydrogen and biomethanol were included as well.

The fourth generation biofuels has different definitions but one of the simplest says about crops that are genetically engineered to consume more CO_2 from the atmosphere than they'll produce during combustion later as a fuel. Both of these biofuels groups are included in the future biofuels group ("*advanced biofuels*"). Thus, the third generation biofuels can be got through the similar method as the second generation biofuels, but from the modified (at the cultivation stage) raw material (biomass) by means of molecular biological techniques. The purpose of these modifications is to improve the process of converting biomass to biofuels (biohydrogen, biomethanol, biobutanol), for example by growing trees with low lignin content, development of crops with respectively built enzymes, etc.

The proposal of split the new, fourth generation of biofuels was established due to the need to close the balance of carbon dioxide or eliminate its impact on the environment. Road Map 2050 prepared by the European Commission talks about CCS (Carbon Capture and Storage) [23], but repeatedly claimed that it is a commercial unimplemented technology. Many companies and organizations have social and environmental objections to implement it. Something that is more reasonable and what should be lean is CCU (*Carbon Capture and Utilization*). It would be profitable to capture CO_2 from the atmosphere or the exhaust of power stations and convert it to fuel by using a sustainable source of energy such as sunlight. Photocatalytic or termochemical conversion of CO_2 to fuels by using semiconductors and metal oxides are two main routes. There is also combination of these routes by using water or hydrogen co-fed with CO_2 for fuels generation.

http://dx.doi.org/10.5772/53831

There are developed new, forward-looking technologies in the U.S. and Europe characterized by high reduction of CO_2 by means of LCA (*Life Cycle Assessment*) parameter of specific biofuel like:

- biofuel production technologies, including Jet type, by culturing algae without sunlight from agricultural sludge, grass and waste substances, with the use of carbon dioxide (technology "SOLAZYME");
- technology of plasma gasification of waste biomass and municipal and industrial waste (BtG and WtG processes), followed by processing upgraded synthetic gas towards liquid biofuels like GTL diesel and Jet type fuel ("SOLENA" technology, carried out in the UK and Italy);
- technology of carbon dioxide use in the carrier energy production processes;
- complex biorefinery technologies.

European strategy mentioned in the "European Strategic Research Agenda, Update 2010" defines the following biofuels as a prospective and showing the technological pathways of its getting:

- synthetic fuels/hydrocarbons from biomass gasification (application: transport fuels from RES for jet and diesel engines);
- biomethane and other gas fuels from biomass gasification (substitutes of natural gas and other gas fuels), (application: engine fuels and high efficient energy production);
- biofuels (bioliquids) from biomass got through the other thermochemical processes such as pyrolysis (application: heating fuels, electricity production or indirectly through the **xTL** processes for transport fuels);
- ethanol and higher alcohols from sugars containing biomass (application: transport fuels from RES or as a petrol biocomponents, E85);
- hydrocarbons from biomass, got from sugars, created in biological or chemical processes (application: renewable transport fuels for jet and diesel engines);
- biofuels gained from the use of carbon dioxide for the microorganisms production or from the direct synthesis of carbon dioxide in thermal and biochemical processes (applications: transport fuels and for aviation).

International Energy Agency, have been identified also "solarfuels" ("solar fuel").

It is necessary to determine the best and universal technologies of their production. The technologies of biofuels production have to provide the possibility of safe operation of engines while reducing exhaust emissions of toxic ingredients under the needs of the relevant class "EURO" in Europe and the U.S./ California ULEV in the USA. At the same time it is important that biofuels with a similar composition and properties should be available in every country, because of the motor requirements.

That's why the developed by the International Energy Agency, road map for biofuels in transport ("Technology Road map. Biofuels for Transport") determines the perspective biofuel technologies in the world till 2050. Therefore it was proposed to divide the conventional

("conventional") and future ("advanced") biofuels. Distribution and severity of various biofuels in both groups, according to the MAE, is shown in Figure 8.

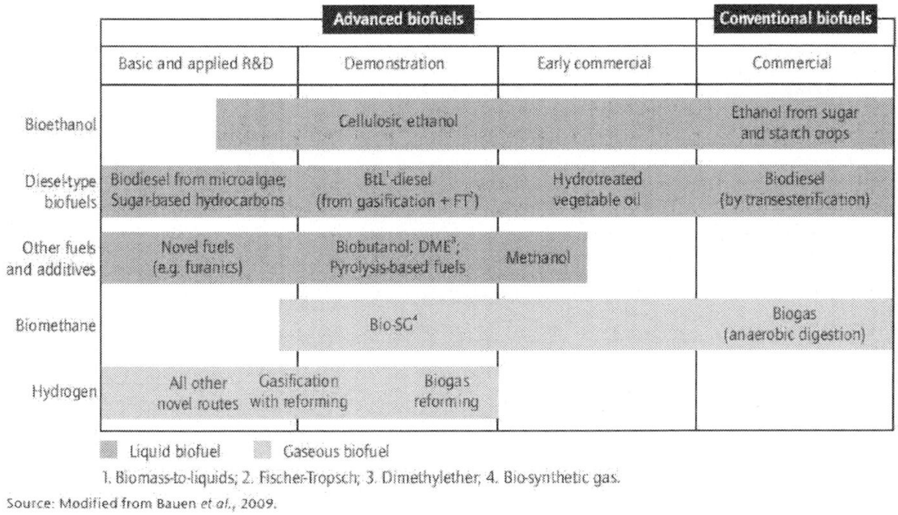

Figure 8. *Biofuels division and stage of its production* (Technology Roadmap Biofuels for Transport © OECD/IEA, 2011) [24] with the consent of: *OECD/IEA.*

The target biofuels production in 2050 required a biomass material with a total energy potential estimated at 65 EJ, which is equivalent to 100 mln ha of cultivations in 2050. Assuming that 50% of the feedstock for the advanced biofuels production will come from waste substances (XtL processes). This means that it is necessary to increase the area devoted for cultivation of biomass for energy purposes. In an optimistic scenario, it is assumed the possibility to get 145 EJ per year of energy from biomass and all kinds of waste substances intended for liquid energy carriers for transport and for heat and electricity in the polygeneration. Estimated reducing of CO_2 emissions is determined at 2.1 Pg per year, with the share of biofuels of 27% (v/v) of the total amount of transport fuels.

Considering the demand for biofuels that meet the future needs for transport, including air transport, and their ability for CO_2 reduction, it was established by MAE the following biofuels and their technology paths:

- **fuels from BtL processes** (synthetic hydrocarbons compositions), get by fast pyrolysis, by heating biomass at temperature between (400... 600) ° C, and then applied fast cooling. The volatile compounds may be converted to a bioliquid or further catalytic deoxygenation, distillation and refining toward fuel component. The solid residue is called "Bio-char"

http://dx.doi.org/10.5772/53831

("charcoal") and as a by-product can be used as a solid fuel, or used as a sequestration agent and for soil fertility;

- **diesel oil from BtL processes**, so-called FT-diesel obtained by conversion to the synthesis gas and catalytic Fischer-Tropsch synthesis (FT) in a wide range of liquid hydrocarbons, including synthetic diesel and JET type biofuels;
- **hydrorafinated vegetable oil (HVO)** as a fuel for diesel engines or heating oil produced by the hydrogenation of vegetable oils or animal fats (non-food and waste). The first commercial plants started in Finland and Singapore;
- **cellulosic bioethanol** from lignocellulosic raw materials produced by the biochemical conversion of cellulose and hemicelluloses followed by fermentation of sugars (IEA, 2008a);
- **biogas** obtained by anaerobic digestion of raw materials, such as organic waste, animal manure and sewage sludge, than purified to the biomethane form (SNG!!) by removing CO_2 and hydrogen sulfide (H_2S). Can be applied in transport as methane fuels (bioCNG, bioLNG) or as a source of hydrogen, also for the fuel cell;
- **dimethyl ether (bioDME)** as gas fuel for diesel engines, derived from methanol catalytic dehydration process, from synthesis gas by gasification of lignocellulose and other biomass. BioDME production from biomass gasification is at the demo phase (September 2010 in Sweden (Chemrec));
- **biobutanol** have a higher energy density than ethanol in petrol. It can be shared through the existing network for gasoline, Biobutanol can be produced by sugars fermentation using bacteria, *Clostridium acetobutylicum*. Demonstration plants are working in Germany and in the USA, the others are under construction.
- **furan fuels** in the latest "Technology road map, biofuels for transport", compiled by the International Energy Agency, were classified as prospective biofuels. Compounds such as furfural and 5-hydroxymethylfurfural (HMF) can be obtained with good yield through dehydration of monosaccharides, such as hexoses (e.g. fructose) or pentoses (e.g. xylose) in the presence of various catalysts. Fuel is suitable for spark-ignition engines with the advantages when compering with ethanol.
- **solarfuels**, obtained by gasification of biomass towards synthesis gas by using heat produced by concentrating solar energy. They can also be obtained by decompose of water (water vapor) and the use of carbon dioxide to form synthesis gas catalytically transformed into fuel fractions. In terms of the production of these fuels can be included technology of so-called "Artificial leaf";
- **biorefinery** for liquid fuels production and chemical by-products, which discussion exceeds the scope of this chapter.

In terms of the most promising biomass for the future biofuels production, considering so-called "land hunger" ("ground competition") and the needs in terms of CO_2 emissions, can be distinguished following cultivation: algae, camelina, jatropha, and halophytes. Under development are new technologies without sunlight ("dark"), marine photosynthetic membrane systems for the algae production, as well the technologies for biomethanol production.

6. Technologies of liquid energy carriers production based on waste substances

6.1. WtL (*Waste to Liquid*) and WtE (Waste to Energy) processes

Dynamically observed population increasing in the world stimulates fast economic growth. This is related with growing demand for all kinds of products and energy. In result of progressive consumption, waste generation is increasing. According to the data total amount of waste, which is generated in the European Union, maintains an upward trend. Municipal solid waste (MSW) amounted 150 million Mg in 1980. In 2005, increased up to 250 million Mg, and forecasts for 2015 growth is estimated up to 300 million Mg [25]. The constantly increasing amount of waste contributed in overfilled existed landfills, forces society for development of new landfills. The overfilled landfills are big the social and environmental problem. Therefore, novel energy technologies have preferences for waste treatment WTE (Waste to Energy), or for receiving new products from waste in the process of WTP (Waste to Product.). Waste treatment processes for liquid energy carriers are processes WTL (Waste to Liquid). The trends for conventional energetic waste utilization are based on direct combustion and more advanced thermochemical methods such as pyrolysis, gasification, and plasma technologies under development since 1970. Figure 9 shows the main alternative methods of converting waste to energy in WTL and WTE processes [25].

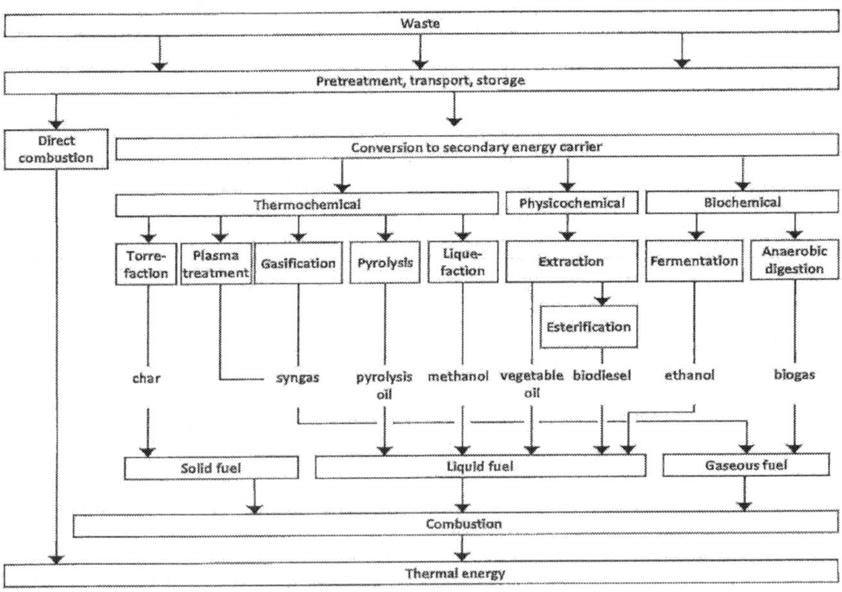

Figure 9. Waste to Liquid and Waste to Energy conversion technologies [25], with the consent of: Anouk Bosmans

In the WTL and WTE processes, pre-treated waste is converted into secondary energy carriers using thermochemical methods: physico-chemical or biochemical. As a result of these changes is obtained biofuels solid, liquid or gas. These fuels are burned and converted into thermal energy. WTL processes path is shown in Fig.9..

A HTU is processing waste into liquid biocrude oil (Hydro-Thermal Upgrading), at high temperature and pressure near supercritical condition of water, which is usually the solvent in this process, the basic conditions are shown in Table 3.

Conditions	Temperature: (300 ...350) °C Pressure: (120...180) bar Reaction time : (5... 20) minutes.
Feedstock	All types of biomass, domestic, agricultural and industrial residues, wood. Also wet feedstocks, no drying required.
Chemistry	Oxygen removed as Carbon Dioxide
Products	45 Biocrude (%w on feedstock, dry basis) 25 Gas ("/> 90% CO_2) 20 H_2O 10 dissolved organics (e.g., acetic acid, methanol)
Thermal efficiency	70 - 90 %

Table 3. Basic physical and chemical conditions of the process HTU [26]

HTU process developed by Shell in 80's occurs at a temperature up to 350°C and a pressure about 180 bars. HTU can be used to convert liquid fuel from a wide range of biomass feedstock, without the need of drying raw material. It has been designed to carry out the reaction in the excess of water, under supercritical condition of water. The final product is a "bio-oil" ("bio-crude") with properties similar to crude oil, so it can be used after upgrading as fuel in boilers, turbines and so on. The calorific value (LHV) of biocrude obtained this way amount 30... 35 MJ/ kg [27].

Another an example of a technology for waste conversion is based on the catalytic thermal depolymerization patented by Alphakat GmbH, which principle of operation is shown in Figure 10.

KDV process takes place in a special industrial installation known as KDV Unit. As can be see from Figure 10, the whole system is a closed with strictly defined conditions: the process is carried out at ~350°C, under low vacuum 0.9 bar. The vacuum is maintained in the system by special vacuum pumps. As shown in the figure, the pump mix the raw material provided with a catalyst causing the circulation of the reaction mixture in a closed system. After heating the batch by thermal oil up to ~350 °C, volatile organic compounds (VOC) are formed, which at the main distillation column undergo separation for diesel and gasoline fraction [29, 30].

Using KDV technology a wide range of raw materials derived from biomass or waste materials, either organic or mineral, can be process. Table 3 summarizes the basic substances that can be used as a feedstock in KDV plant. We can see all kinds of biomass waste, agricultural, municipal and industrial sewage, as well as synthetic materials.

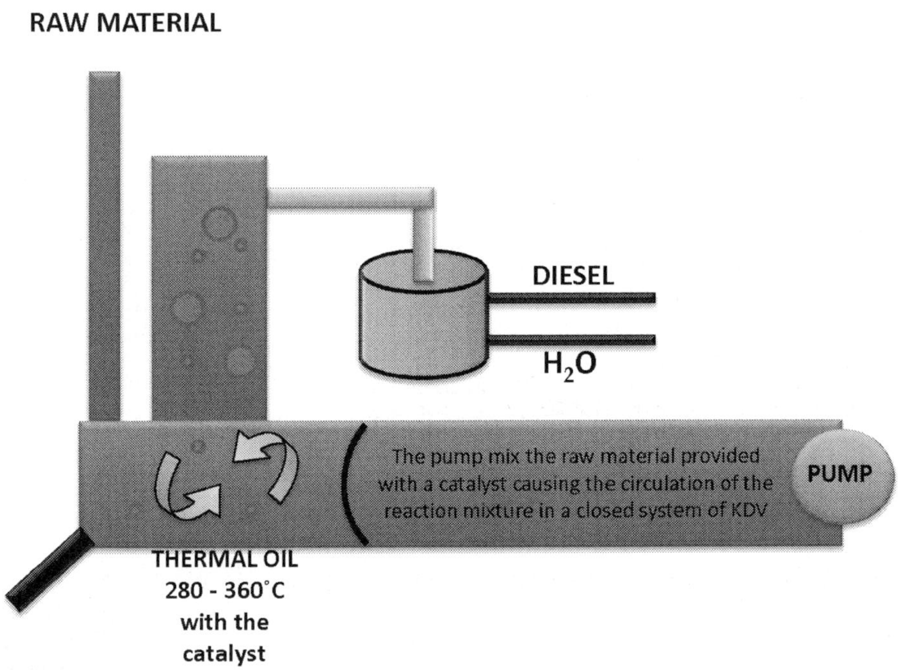

Figure 10. The principle operation of KDV technology [28 modified]

Lp.	Type of material	Sort of material
1	Biomass (C-3 and C-4 plants)	• grown energy sources like Jatropha, • wood, biogenous residues like leaves, straw, etc.
2	Waste	• Industrial Waste (IW), • Municipal Solid Waste (MSW), • agricultural waste, • waste oil (also contaminated oil), • refinery residues, bitumen. etc, • dried sludge from sewage treatment plants, • rubber and tires.
3	Synthetic materials	• All kinds of plastics and synthetic materials (PVC, PP, PET, etc.)

Table 4. Basic substances which can be used as raw material for the KDV installation [29, 30]

http://dx.doi.org/10.5772/53831

The main products from KDV plant is synthetic diesel fuel or kerosene depends on process parameters [30]. However these fuels are available after catalytic hydrotreatment process (like hydrodesulphorization in refinery), because the product after distillation has still properties of biocrude oil. KDV were started as a demo units in Germany, Canada, Spain and Mexico and recently in Poland [31].

6.2. Overview of selected technologies for liquid biofuels from waste carbon dioxide

6.2.1. Conversion processes

Scientists all over the world are trying to learn the essence of the photosynthesis process. Already in 1912, the Italian chemist Luigi Giacomo Ciamician in the Science magazine, said the society should be transfered from a civilization based on oil and coal and start using clean energy from the sun. He thought that by using appropriate photochemical reactions and using the new compounds is possible to discover the mystery of the photosynthesis guarded plants.

In July 2010, in the United States was established Joint Center for Artificial Photosynthesis - JCAP, funded by the U.S. Department of Energy (DOE). The aim of this project is to develop a method able to copy the photosynthesis process, available to establish the path for the production of liquid fuels using waste carbon dioxide.

According to information provided by JCAP [32], major challenges facing scientists dealing with the artificial photosynthesis is a system that consists of key elements of the process: the visible light absorber, catalyst and membrane switches.Visible light absorber is a molecule that is designed to capture and convert the sunlight with the high efficiency, like the chlorophyll in plants. Absorbed energy puts absorber molecules into excited state, in which all the light energy is submitted to electron. Properly designed absorber pushes the electron from the shell, directing him to an adjacent molecule (in the same way as it does in natural photosynthesis, where electrons jump from photosystem I to photosystem II). In this way formed electron beam passes through a further conveyor releasing the energy required for further electrochemical reaction. At the point of detachment of the electron take place creation so-called, "Hole," or positive charge. At this point is necessary a catalyst that efficiently breaks the water molecule, this way is freed the missing electron in place of created "hole". In natural photosynthesis complex of the protein, which contains manganese ions act as the catalyst. This compound is still a great mystery because it is not yet possible to copy of its complex structure in the laboratory. Therefore, research is continuing to find a suitable replacement, which will perform a similar sequance cleavage involving water molecules processed by the solar energy absorber. Catalyst must not only be efficient and sustainable, but also cheap, and widely available on the Earth, preferably non-Nobel metal or his complex.

Copying plants, we can create something like artificial leaf. The function of the matrix is not only to act as a micro bioreactor, but also play role of separation reactants and products (fuel and oxygen). Oxygen is not a desirable product in artificial photosynthesis. Instead of molecular oxygen (O_2), it can happen that forms oxygen radicals (O), very reactive with strong oxidizing properties.

With a listed part of the process of artificial photosynthesis, there is only lack of one element - the connector, through which various components interact with each other and create a functional whole.

In order to build a system composed of the described elements, there are carried out intensive research in the USA, Switzerland, UK, Japan and Poland.

The project regarding artificial photosynthesis in Poland started in 2006 at the University of Maria Curie Skłodowska (UMCS) University in Lublin and in collaboration with EKOBENZ company.

The concept of methanol synthesis based on the carbon dioxide formed as a result of different technological processes has been developed by Professor Nazimek from UMCS. The invention is based on the oxidation of water using titanium dioxide (TiO_2) as a photocatalyst and at the same time the reduction of carbon dioxide (dissolved in $NaOH_{aq}$). Reaction is carrying out in assistance of fixed ultraviolet radiation (UV) as is shown by reaction 12.

$$\left(2CO_2 + 4H_2O\right)^{hv} \rightarrow 2CH_3\text{-}OH + 3O_2 \tag{12}$$

This technology will enable the production of methanol from the two cheap, universal components: carbon dioxide and water vapour. The proposed technology is one of the most promising methods of utilization of CO_2, since this compound can be converted to a valuable product, under natural conditions by exposure to UV light.

Implemented technology is characterized by the synthesis reaction of carbon dioxide and water which takes place in aquatic environment at the presence of a photocatalyst at a temperature of 20 ° C and atmospheric pressure. Specifications referred this reaction on a laboratory are shown below:

- Energy consumption 0.75 kWh/dm^3 of methanol
- the solution flow rate 8 dm^3 / h;
- carbon dioxide flow 370 dm^3/ h;
- the concentration of methanol in the aqueous solution was 15%.

Based on these parameters has been developed preliminary schematic diagram of the planned production process, as shown in Figure 11.

Under the proposed scheme carbon dioxide is stored in the reservoir of raw material from which goes to the mixer where is absorbed in demineralized water. Then, in the reactor under the influence of UV radiation takes place photoreduction reaction. When reaction is completed, the distillation take place and methanol (CH_3OH) and water (H_2O) are separated. The methanol is sent to the product tank, and the water is reused in the production process – is transfered to the mixer. The active and stable catalyst is the key for obtaining methanol by CO_2 photoreduction. For photocatalytic tests were applied TiO_2 in the form of anatase supported by alumina as a carrier of the active phase.

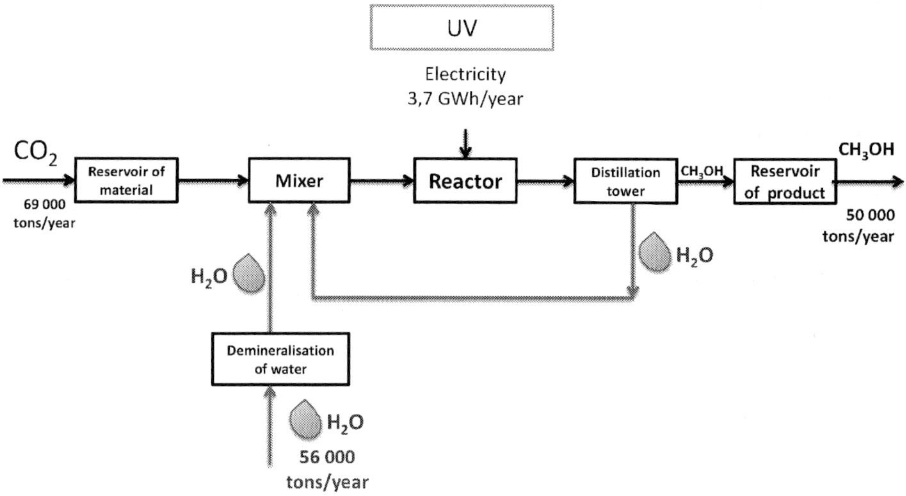

Figure 11. Block diagram showing a manufacturing process of methanol carbon dioxide and water, with the consent of: Ekobenz Ltd

The study at laboratory scale was carried out in flowing system, at atmospheric pressure and ambient temperature, equipped in quartz pipe reactor. In typical experiment 100cm^3 of distilled saturated by CO_2, and air or nitrogen flowed through reactor. Bed catalyst was exposed on UV lamp radiation at fixed wavelength. The effluent was analyzed for content of organic compounds by using gas chromatograph. The conversion of CO_2 to methanol was determined by the formula (13).

$$X(\%) = \frac{C_0 - C_1}{C_0} \tag{13}$$

where: C_o – initial concentration CO_2 [mol/dm^3]

C_1 – final concentration CO_2 [mol/dm^3]

The conversion degree of carbon dioxide amounted only 6% [33]. It was found the titanium dioxide catalyst had a defect in the absorption spectrum, which was located only in the UV light range. Visible radiation even during long-term irradiation was not able to induce the active phase of the pure TiO_2. The suspension of titanium dioxide could be the solving the problem, but it is related with separatation of the catalyst from the product liquid phase after reaction completion.

Therefore, the above measurements were carrying out involving a new nano-structured wall catalyst. This catalyst is protected by Polish national patent No. 208030. The catalyst is characterized as the active sites in the form of clusters TiO_3 with aluminum III ions and sodium I ions (TiO_3 content in the range 4.36 to 5.34% (m/m) of the total weight of the catalyst) supported by alumina. These clusters are deposited on the inner wall of the aluminum tube of the reactor.

The catalyst described above as the wall form allows to implement an artificial photosynthesis process in the new technology. Developed a new efficient photocatalyst created the conditions for process with performance sufficient for commercial applications, for example by design 0.5 m length of processing pipe reactor, photocatalysts (with $\varphi = 5$ cm) allowing to convert the 370 dm^3 of CO_2/h into methanol with the degree of conversion at 97%.

Figure 12 shows the schematic diagram of the installation for methanol production applying artificial photosyntesis.

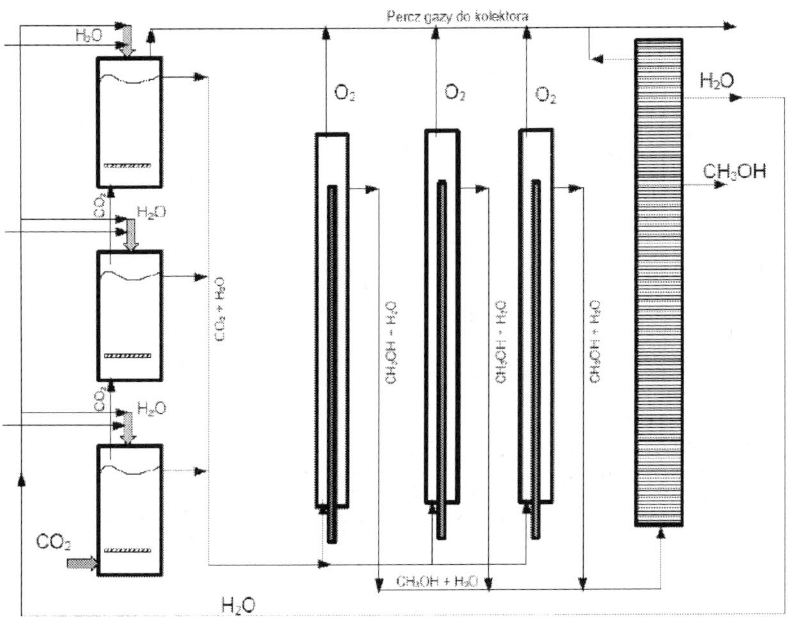

Figure 12. A schematic diagram of the measuring system for the production of methanol in the artificial photosynthesis, with the consent of: Ekobenz Ltd

As shown in Figure 12 carbon dioxide with water are transfered into mixers arranged in series, which supply the reactors system. The catalyst is located on the reactor walls, as a result of reaction are products like oxygen molecule, water and methanol. Molecular oxygen is discharged to the gas collector, and the mixture of methanol and water is further transferred into the distillation tower, where separation these compounds take place. Water is returned to the mixer, where it is used in the subsequent cycles of the process, and desired the final product is separated as a pure methanol.

The methanol obtained through the described technology can be directly applied as a fuel in dedicated engines and adjusted fueling systems with seals, proper polymer pipes because methanol is known as a good solving agent for many materials used in cars now. On the other hand methanol can be catalytically converted towards gasoline fraction by using very well known from 70's Mobile process- MtG (Methanol to Gasoline). MtG method involves the synthesis of a higher hydrocarbons mixture (which are synthetic petrol) from methanol, the intermediate product according to the general reaction (14) are:

$$nCH_3OH(\text{zeolite catalyst eg. ZSM-5}) \rightarrow C_nH_{n+x} + H_2O^* \tag{14}$$

*The **n** and **x** values are variable depending on the temperature and pressure in the system, as well as catalyst used in the process.

Reaction (14) occurs at a temperature T = 723K and under the pressure of p = 200 bar. Water is a by-product of the reaction is, which is a limiting factor in MtG synthesis because it affect the catalyst performance. The catalysts used in the synthesis of synthetic gasoline from methanol are based on aluminosilicate matrix. These type of catalysts in some excess of water vapour, in the reaction mixture, lose their activity because of extraction for example Al ions from the matrix of catalyst. A new catalyst has been developed, which is water-resistant in the reaction mixture used in MtG method as well as related technology EtG (ethanol to gasoline, which will be discussed later in this chapter). Nano-structure catalyst developed for the synthesis of gasoline from the ethanol according to equation (14) is characterized by the active centers in the form of a copper ions coordinated in octahedral structure. Copper is supported by nano-structured aluminosilicate matrix with metal content in the range of 0.2-0.5% w/w. This catalyst has a high water resistance with content even more than 10% water by volume of the reaction mixture.

The final product of the described method MtG is appropriate mix of higher hydrocarbons, which are synthetic gasoline with an octane rating of up to 108, characterized by the same properties as gasoline derived from crude oil processing. Summary of the final products of described MtG method is shown in Table 5 [34].

Gasoline until 108 LOB (Q = 44,8 MJ/kg)
ON - 56 cetane number
SNG (~ 95% CH_4)
aromatic fractions

Table 5. Summary of the final products of the MtG technology

According to preliminary estimates, this method is able for application within three years, and what is most important can cause the reduction of CO_2 in the atmosphere by 25%.

Some inaccuracies in the calculation of the energy balance above process should be noted. The synthesis of methanol from CO_2 and H_2O is endothermic reaction, which means that for the occuring reaction is required determined amount of energy. It was found out, the total power consumption in order to obtain a 1 dm^3 methanol is 0.75 kWh. The balance of energy showed the following conclusions:

if: 1 kilomole of CH_3OH = 586 MJ

1 kilomole of CH_3OH = 32 kg

therefore: 32 kg CH_3OH = 586 MJ

*Then 1 kg CH_3OH is:*586 MJ / 32 kg = 18.31 = 5.08 kWh

specific weight CH_3OH is: 0.48 kg/l.

therefore: 1 kg = 5.08 kWh, *which gives* 0.48 kg (=1liter) = 2.44 kWh

Conclusion: To make 1 liter of methanol from CO_2 and H_2O is required 2.44 kWh, it is three times higher than reported by the inventor of the method - 0.75 kWh.

Therefore, the cost obtaining of 1 liter methanol will be about 0.31 USD, and not as referred value 0.03 USD [35]. Regardless, a drawback of this method is the fact the developing technology is based on the pure components. In the synthesis reaction of the methanol with carbon dioxide and water, is used 96.5% CO_2 from the pressure cylinder, which occurs very rarely, only in some chemical processes (including fermentation). The effects of using carbon dioxide mixed with air in the reaction, such as is found in the nature, wasn't fully investigated yet.

In summary, the method of converting carbon dioxide into methanol at the moment is not applicable, because there is no beneficial economic and energy effects. For these reasons, the project in methanol synthesis in the process of photoreduction of carbon dioxide was suspended. The chance for this interesting project is replacing artificial sources of UV radiation by the natural sunlight. Then the project possibly would be cost and energy-effective. It depends on the active catalyst that would work efficiently in the visible light range.

The part of the project under investigation is still EtG process (*Etanol to Gasoline*). The coupling process of ethyl alcohol towards of higher hydrocarbons also take place with a suitable zeolite catalyst, at higher temperature and pressure. Replacing ethanol by its homologue methanol causes a significant reduction of pressure required for a reaction, from about 200 bar to 30 bar. Also, the temperature is much lower and amounts 653 K [36]. Reaction of ethanol conversion into higher hydrocarbons is exothermic, and therefore there is a problem with removal of the heat. if combined with endothermic process of methanol synthesis could improve the energy balance of carbon dioxide conversion into methanol. The tubular reactor is cooled at the moment by medium, which at the beginning of the research was air. Now it was changed to a more efficient oil. This method of catalyst cooling is not very effective, because it reduce the temperature only near the outer wall of the reactor. The gradient temperature within radius can reach the temperature up to 300 °C. Then, the catalyst located in the center of the reactor can be affected by sintering process, and then lose activity. The current research aim is to optimize the temperature of the entire length of the reactor, to avoid deactivation of the catalyst.

At the present stage of research, the experiments are carried out in one reactor, and the target is to operate six reactors cooperating with each other, forming so-called hexagonal reactor. The overall diagram of the EtG is shown on Figure 13.

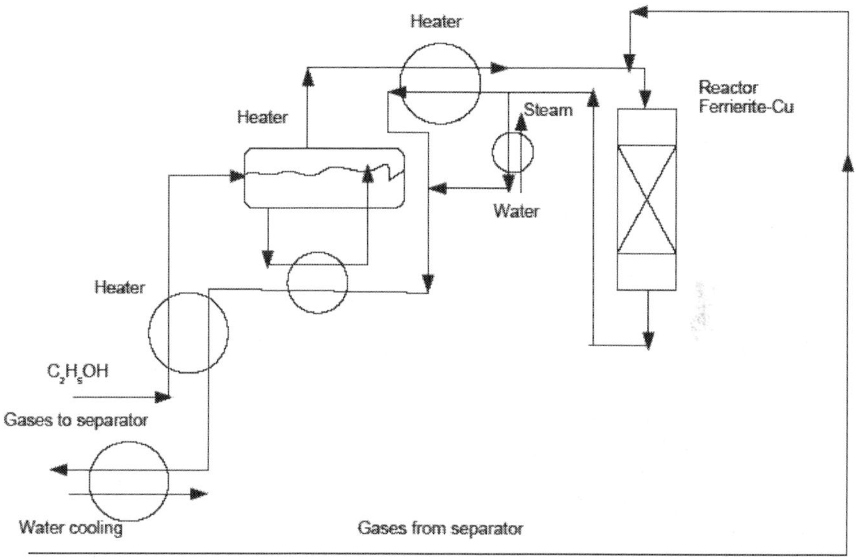

Figure 13. Diagram of the research installation for the process EtG, with the consent of: Ekobenz Ltd

Under the pictured diagram, ethyl alcohol is sampled into the heater, where in the electric heater spiral is heated to about 270 °C, and through the isolated pipes reaches the reactor. Outlet reaction mixture flows to the condenser, where the liquid products are out dropped. However the ethanol is converted not only towards the liquid fraction, but as a result of this reaction the gaseous products are also formed. Gases are recycled to the reactor, mixed with a new part of ethanol steam and are converted into gasoline fraction. The liquid fuel is further analyzed from point of view requirements of gasoline and diesel. The typical content of such products are benzene (as aromatization compound of ethanol on zeolite), olefins, paraffins and other aromatics. Research in this area is continued in Poland on the frame of "Operational Programe of Innovative Economy". The process is called ETG (ethanol to gasoline) is similar to the MTG (methanol to gasoline) process which was successfully used by Methanex in New Zealand.

6.3. Photoconversion technology using "*artificial leaf*"

On 241 Congress of *National Meeting of the American Chemical Society* in Anaheim, Califfornia, it was announced the invention of the first functional artificial leaf. *"A practical artificial leaf has been for decades the one of the Holy Grails of the science"* - said the leader of the team, Dr. Daniel Nocera - "We believe that we did this" [37].

A breakthrough moment in the study of artificial leaf is development of a suitable catalyst, nickel (Ni) and cobalt (Co). These elements are widely available comparing with noble metals, what is important they act at ambient conditions and have a high stability during the reaction. Inventor reported very effective work of artificial leave for production hydrogen and oxygen.

The artificial leaf is a matrix composed of material used in most of the solar cells, semiconductor silicon. Surface is covered with a matrix of cobalt catalyst on the one side, and on the other side with the molybdenum-zinc alloy of nickel. Solar radiations falling on the surface are transformed into electricity which allow to decompose bond of water molecules in the presence of catalyst. On the first side of the matrix, the electrons are knocked out from the water molecules using cobalt catalyst. The water molecule is decomposed into hydrogen ions H^+ and oxygen. Oxygen remains on first this side of the membrane, while the hydrogen ions are transported to the other side of the matrix, where the nickel is catalyzing the recombination reaction of hydrogen ions with the previously made free electrons. The result of the process is H_2 molecules formation [37-40]. Diagram of the act "artificial leaf" is shown in Figure 14.

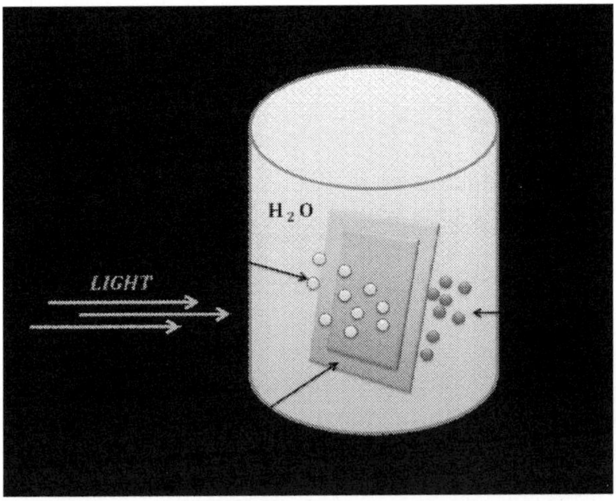

Figure 14. Diagram of the artificial leaf [38 modified]

In this way, as it is shown in Figure 14, one surface of an artificial leaf produces oxygen and the other - hydrogen.

According to data the authors team, with one gallon of water, exposed to solar radiation using artificial leaf; it is possible to produce enough electricity required for the daily demand for household energy in developing countries.

6.4. Solar reactor in technology of synthesis gas getting („*solarfuels*")

The research group of CalTech from United States published in December 2010 the paper regarding their latest achievement in artificial photosynthesis. Prototype solar device was constructed, where the heart of device is cerium dioxide (CeO_2) catalyst. Diagram of the device is shown in Figure 15.

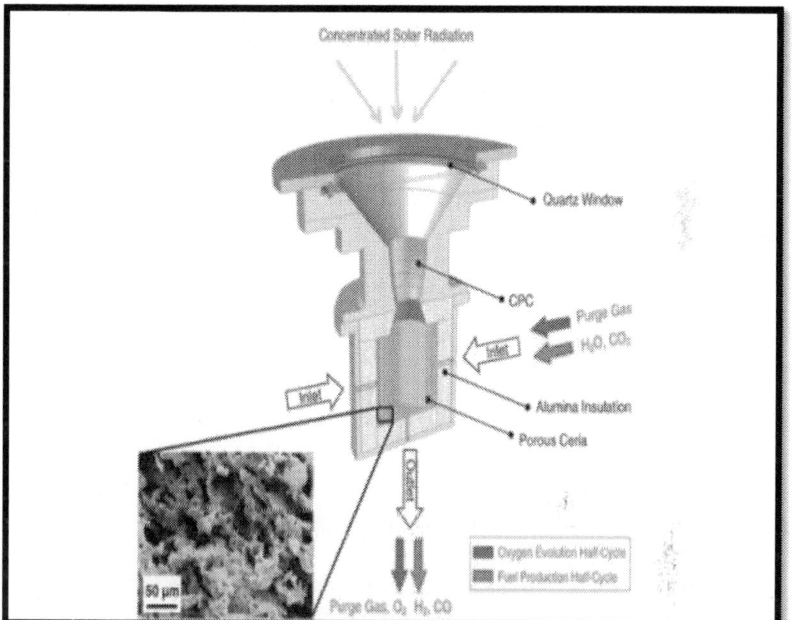

Figure 15. Prototype reactor uses the sun to produce storable hydrogen fuel [41 modified]

Reactor operation is based on the absorption of concentrated solar radiation, which falls on a quartz apparatus window, then passes into the insulated chamber of the receiver, which is filled with a porous cerium oxide. This compound has a specific property of the oxygen binding at low temperatures, and desorption it at high temperatures, without damage of its crystalline structure, according to the reactions (15), (16), and (17).

$$1/n\ MO_2 \rightarrow 1/n\ MO_{2\text{-}n} + 1/2\ O_{2\,(g)} \quad T_H \tag{15}$$

$$H_2O_{(g)} + 1/n\, MO_{2\text{-}n} \rightarrow 1/n\, MO_2 + H_{2(g)}\;\; T_L \tag{16}$$

$$CO_{2(g)} + 1/n\, MO_{2\text{-}n} \rightarrow 1/n\, MO_2 + CO_{(g)}\;\; T_L \tag{17}$$

where: M - pure cerium or cerium with mixtures; T_H - at high temperatures; T_L - at low temperatures [42]

In this way, it absorbs oxygen gained from supplied to the reactor carbon dioxide (CO_2) and water vapor (H_2O). Thus, inside the reactor are formed the carbon monoxide (CO) and hydrogen (H_2), which is an ideal source of energy. Hydrogen is considered to be so-called "clean fuel" because in the result of its combustion the water is just produced. However, the reaction of hydrogen with carbon monoxide if presented by stechiometric reaction can be directed towards methane formation, popular heat, transport fuel and source of chemicals production (reaction 18). The synthesis gas in proper molar ratio of CO/H_2 and using FT catalyst (Co or bimetallic with Fe) can be converted towards higher hydrocarbons.

$$CO + 3H_2 \leftrightarrow CH_4 + H_2O \tag{18}$$

In the CalTech process the reactor was exposed to solar radiation at 1.9 kW of power. Such power is widely used in various types of solar antennas or solar towers; any of these devices has not reached the efficiency of the presented prototype. This can be explained by the fact, the reactor uses the entire spectrum of light, not only the selected wavelength. Disadvantages of the presented solar reactor is the energy losses as a result of energy radiation through the walls of the reactor (about 50% of the energy input) and the release of radiation through the quartz window (about 41% of supplied energy).

It can be assumed that further improvement of the process will be able to reduce the temperature of the reactor (today reactor operates at 1648 °C), and less energy losses through more efficient use of solar radiation. An initial analysis of the process effectiveness showed that improvements structure of cerium oxide will allow to increase activity of catalyst in the range of 16-19%.

This concept can be used for conversion carbon dioxide emitted from coal power plant to liquid energy carriers. The pure carbon dioxide is not required to carry out this reaction. This allows at least twice using of the same coal compund [42].

6.5. Biological processes using carbon dioxide as a raw material

As stated previously, the world's biomass resources are not enough for producing renewable energy with at the same time growing needs of modern civilization. For example in UK, replacement of diesel by biodiesel produced from rapeseed would required plantation more than half area of the country [43]. The areas for the production of liquid biofuels is competitive with the same areas for the cultivation of food crops. There is a conflict between energy and food use of agricultural land. In addition, oilseeds cultivation leads to monocultures over large areas, which is inconsistent with the Directive 2009/30/EC. As stated in mentioned Directive production of biomass for energy purposes should go hand in hand with biodiversity. Nowadays more and more we hear about a new raw material for the production of liquid fuels which can be tiny microorganisms such as algae. Algae growing is a good option for limited areas of cultivation of biomass for energy purposes. Algae can be grown in areas not suitable for the cultivation of the soil, such as deserts or oceans. In addition, they are a kind of "factory", able to convert waste carbon dioxide into valuable energy products.

The idea of using algae as a feedstock for bioenergy production has already appeared in the mid-twentieth century. In 1940, it was discovered that many species of microalgae growing in strictly definite conditions can produce large quantities of lipids. The concept of using the stock of lipids as an energy source has been caused by the oil crisis which arose from the imposition of the embargo in 1973 by the OPEC countries. The rapid increase in oil prices and the reduction of the energy carrier from Middle East to the United States contributed to the search for new sources of energy such as micro-algae. In 1978, *Program of the Department of Energy for Aquatic organisms* have been created *(DOE's Aquatic Species Program)*, which focused on the acquisition of the fuel in the form of pure hydrogen supply by aquatic organisms. Already in the early 80's, this was changed and focused on trying to manufacture liquid fuels, mainly biodiesel. Over 300 000 existing algae strains were collected from different extreme environments. The next seven years were carried out research regarding tolerance of different salinity on temperature, pH, and the ability for the production of neutral lipids. After these tests, the study was limited up to 300 promising strains, in which the main role was played: diatoms algae (*Bacillariophyceae)*, and green algae (*Chlorophyceae*) [44].

Algae are a valuable source of raw material for energy transformation because of fast growth of the biomass. Autotrophic algae contain chlorophyll molecule in their cells so they have the ability to carry out the process of photosynthesis. They are able to use about 10% [45] of the sunlight falling on it, in consequence able to double their weight during the day, and in the experimental conditions this time was reduced up to 3.5 hours [46]. The study shows that wit1hin one year cultivation of algae on the area of one hectare, it is possible to get 8,200 liters of biodiesel from extracted algae lipids, while for other oleic plants like *jatropha* only 2700 liters, 1560 liters from canola, and from soybean only 544 l [43]. These data are summarized in Table 5.

Crop	Biodiesel yield (L/ ha/ year)
Oilseed rape	1560
Soya	544
Jatropha	2700
Chlorella vulgaris	8200

Table 6. Estimation of oil productivity from different crops [43]

According to the literature, the production algae biomass is 40... 60 times higher than previously cultivated energy crops. The yield estimated with different cultures growing algae in closed photo bioreactors is about (400... 500) Mg/ ha/ y, which coresponds (400... 500) thousand m3 of bioethanol per year [47].

Another advantage is that algae cultivation is possible in wide variety of water sources: seawater, postproduction water, as well as in wastewater.

One of the key reasons to use algal biomass for energy purposes is these microorganisms absorb and convert a significant amount of residual carbon dioxide. To provide 100 g of biomass, algae consume about 183 g of carbon dioxide, which is associated with a 50% share of CO_2 in dry weight of these microorganisms [48]. With the ability to bind CO_2 algae opens the possibility of using waste carbon dioxide from the exhausted gase, for example coal-fired plants. In this way, they close the circulation of carbon in nature. They also show the ability to absorb other waste components, such as nitrogen or phosphorus for example from of chemical fertilizers. With this ability algae can lead the process of bioremediation of contaminated environments, thus contributing to reduce contamination of soil or surface waters (eutrophication) and groundwater.

Energy sources that can be produced from algae depend on the substrate used and the method used to process the obtained algae biomass. The main path of bioenergy production using algae is shown in Fig.16.

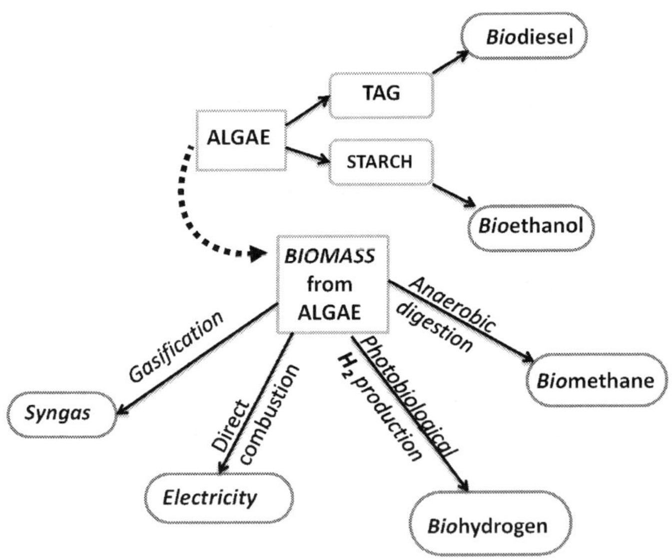

Figure 16. Path of bioenergy produced by using algae.

As shown in Figure 16, we can get bioenergy directly from algal biomass include electricity, syngas, biohydrogen and biomethane. However, two main liquid energy carriers gained from the cultivation of algae are biodiesel and bioethanol.

For the production of biodiesel from algae, the most important is choosing the right kind of algae for triacylglycerols (TAG) obtaining as well as made the ideal cultivation system of these microorganisms. It is necessary to use not only the knowledge of the bioenginering of the various strains of algae, but also integrate this knowledge with the best matching further process technologies [44].

It was observed that in the optimum conditions for the development of algae, which is a sufficient amount of nutrients (such as nitrogen and phosphorus) biomass increase, but the percentage of the dry weight of TAG is rather small. For example for the species Chlorella Vulgaris it is about 14-20% of TAG. However, in the lack of nutrients, algae forms a reserve of energy in the form of so-called TAG. The percentage of TAG per dry mass for Chlorella Vulgaris reaches up to 70%, when nutrition substances are limited. One proposed solution for this phenomenon is a two-step process of algae cultivation. It is based on the fact that in the first stage of breeding algae grown in optimal conditions, with a sufficient amount of nutrients, which results in a rapid increase of their dry biomass. In the second stage the algae are transferred to the place with significantly reduced amount of nutrients. So these micro-organisms, as they are stored in a nutrient deficient conditions lead to increased production of TAG cells.

Another solution is to use microtrophic algae. They are characterized that under conditions of sufficient amounts of nutrients, mainly easily assimilable carbon such as glucose, are transferred to the heterotrophic mode during which produces significantly greater amounts of lipids. Unfortunately, by introducing additional nutrients such as glucose, this solution entails the risk of contamination algae culture by heterotrophic fungi and bacteria [49]. Moreover, one of the main advantages of using algae for biodiesel production is the use of carbon dioxide from the combustion of fossil fuels.

Properly grown biomass is collected and processed in such a way to maximize the production of triacylglycerols cells for the production of biodiesel (Fig. 17).

Algae farm is usually isolated as was shown in Figure 17. In order to reduce the water content, of the harvested biomass, is added the flocculant. Lipid fraction extraction take place with help of hot diesel fuel, than the mixture is separated by using centrifudge.The oil fraction is separated from the watery phase and the biomass waste. Oil fraction can then be used in transesterification reaction to produce FAME or synthetic hydrocarbons using HDO process. As a result, from showed processes already are produced fuels by Sapphire Energy, Petroalgae and Solix Biofuels.

For example, an American company *Sapphire Energy* in 2008 has successfully produced 91-octane gasoline from algae, which complies with U.S. standards. They made a flight test with the twin-engine Boeing using algae-based jet fuel in 2009. They started the construction of an integrated minibiorefinery based on biomass from algae in the southern part of the state of New Mexico in 2010 (called *Intergrated Algal Bio-Refinery*) [50, 51]

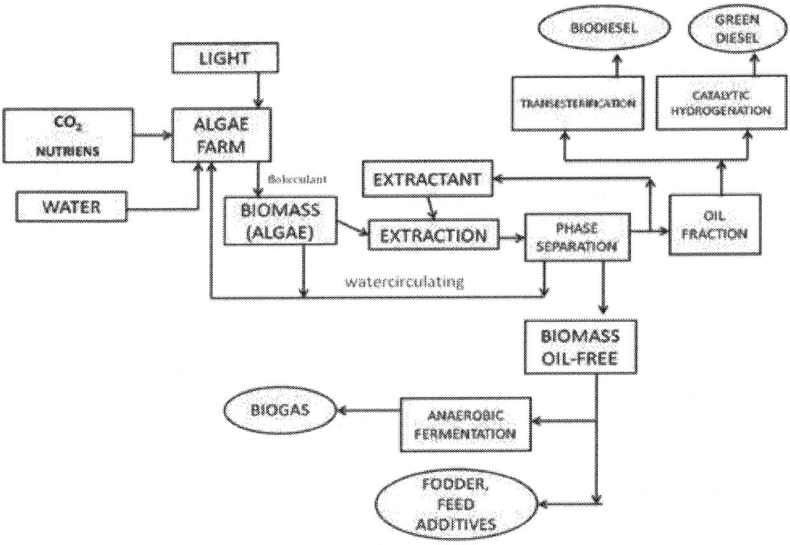

Figure 17. Schematic diagram of the fuel receiving a "biodiesel" from algae [45]

In the process of producing ethanol from starch contained in algae cells, the most plentiful in this hydrocarbon algal classes are: *Chlorophyceae* (green algae) of the types: *Chlamydomonas* and *Chlorella* and *Cyanophyceae* of species: *Arthrospira, Oscylatoria* and *Microcystis*. Bioethanol from the sugar gained from algal biomass can be obtained by:

- Breeding of certain species of algae in anaerobic and without the sun conditions (then the algae produce ethanol within the extracellular space);
- Extraction starch from algal cells and then, through the enzymatic hydrolysis and fermentation ethanol is produced [45].

For example, patented by the company *Algenol Biofuels* process is called *Direct to Ethanol* ® and is based on breeding of algae *Cyanophyta* in specially designed flexible plastic photobioreactors. Using seawater and waste carbon dioxide (fed directly from industrial plants) and subjected to direct sunlight, the algae perform photosynthesis. Sugar molecule (starch) get by photosynthesis process is fermented to ethanol, which diffuses through the cell walls of algae to the solution of breeding. Because of raised temperature during the day, ethanol is evaporated from the solution. On the inner wall occurs condensation of ethanol vapors when the temperature decrease at night outside of photobioreactor,. The condensate drops down of the reactor, where on special internal basins is collected and transferred to distillation unit.

Except lipid molecules in the form of triacylglycerols, or hydrocarbons in the form of starch, which are precursors for the production of biofuels, from algae biomass may be obtained also other products such as proteins. Therefore, in the production of biofuels using algae is formed a wide range of products such as animal feed, fertilizers, pigments, bioplastics, detergents, cosmetics and even food [52].

6.5.1. *Offshore membrane enclosures for growing algae – OMEGA system*

OMEGA technology, originally developed by NASA, is derived from the space program aimed to closing the loop (called „*close the loop*") between the waste stream and the resources necessary for astronauts during long flights [53]. The modified system is to grow cultures of algae in specially designed, floating on the water surface, photobioreactors (PBR) composed of polymers. On the figure 18 is presented the idea of this system.

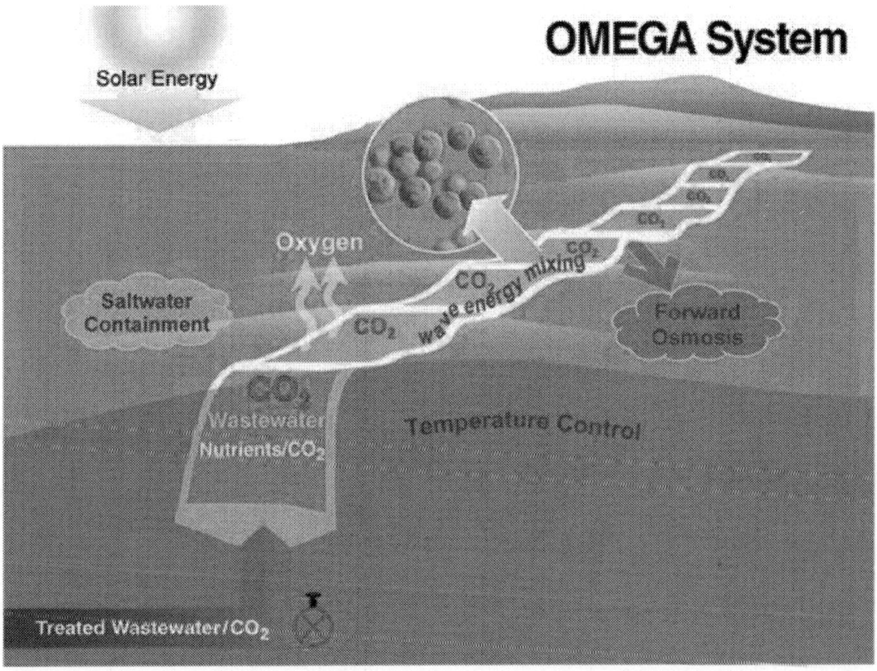

Figure 18. *The OMEGA System* [54], with the consent of: *Jonathan Trent*

The system consists of a number of connected, floating photobioreactors, which are pumped by waste water from the mainland (the actual appearance of the plastic photobioreactors - "bags" floating on the water are shown in Figure 19).

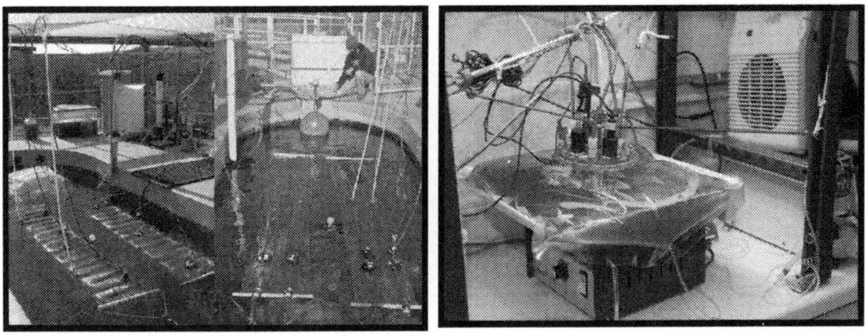

Figure 19. Examples of prototype photobioreactors for technology OMEGA System [54], with the consent of: Jonathan Trent

Algae consume nutrients contained in the sludge, and they associate the carbon dioxide from the air, or waste CO_2 emitted directly from industrial gas plants. By using solar energy, they embed CO_2 into their cells and give off the oxygen into the atmosphere, at the same time producing biomass and oils in their cells. The temperature inside the photobioreactor is controlled by the heat capacity of the surrounding ocean waters, and the gradient between the wastewater and seawater was used for the drainage of photobioreactors. In case of leak the photobioreactor content into the ocean, freshwater algae are breaking up, because they are not able to survive in saltwater. These biochemical species of algae are decomposed into simpler chemical compounds. Thus sea water where are mounted plantations algae are not contaminated. The algae after the growing phase are transfered into the osmotic chamber to make them more dense. Then biomass material is transported into a collection chamber, from where are send to biorefineries. Waste water is returned to the cultivation unit, to maintain the adequate stock and optimum concentration of nutrients for carry out the photosynthesis process by algae. The OMEGA technology final products are biodiesel, jet fuel and by-products such as cosmetics, fertilizers, and animal feed.

The method of producing biofuels using microorganisms is an agree with the nature. There is not used genetic modifications organism and the risk of GMO species invasion into the natural sea environment. Strains of algae are biodegradable, which means it does not involve any risk of pollution of the seas, and oceans, and may even provide food for fish living there. The only added cost is the pumping wastewater into the photobioreactor. However, it was showed the idea of driving the process using wind turbines or solar panels [54, 55].

6.6. Processes of sunless photosynthesis

Term of sunless photosynthesis may be used to define the process carried out by heterotrophic algae without the solar energy, but using the previously synthesized organic compounds as a source of carbon and energy.

http://dx.doi.org/10.5772/53831

Heterotrophic algae growth is dependent on many factors. A significant role in the stimulation or inhibition of heterotrophic microorganism's growth will play the access of sources organic compounds. Depending on the concentration as well as the amounts of organic compounds in a given nourishment, the algae biomass growth runs with different efficiencies. It was also found the positive effect on the growth of microorganisms in environment with poor carbon source mixture. Another factor that has an impact on the growth of algae biomass is the concentration of nutrients such as nitrogen. For example, the studies show the most effective source of nitrogen for the growing of microorganisms on the glucose nourishment is urea [56]. One of the companies dealing with the issue of sunless photosynthesis of algae on an industrial scale is a private company Solazyme Inc., which was found out in 2003. Method for waste biomass change-over to liquid fuels, developed in the company, uses various strains of heterotrophic genetically changed algae to produce different types of fuels. Figure 20 shows the main production pathway in Solazyme technology:

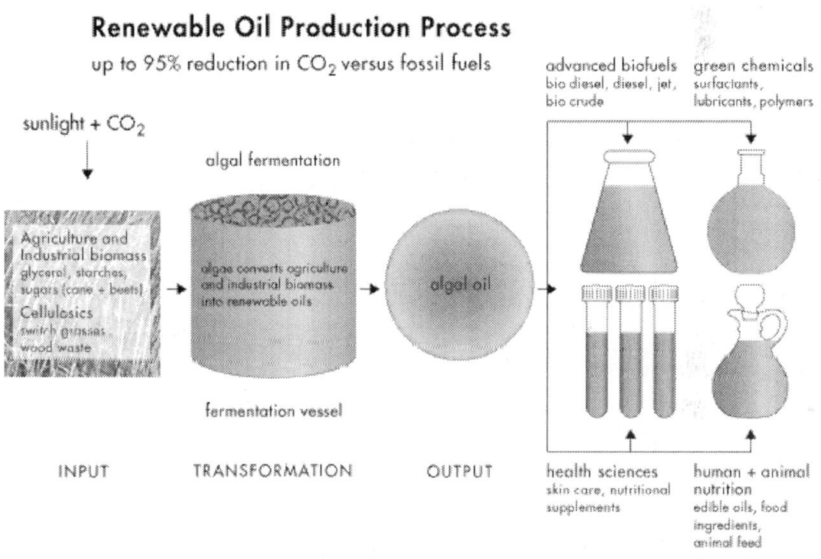

Figure 20. Schematic diagram drawing the production process of different industrial products by Solazyme technology [57]

Alpha Algae, used by Solazyme, are heterotrophic organisms, which means they are developing without solar energy. Identified strains of algae, as it was shown in Figure 20, are transferred into the stainless steel containers. There, in darkness algae are fed by nourishment in the form of various types of raw material. This is mostly lignocellulosic biomass, so all kinds of grass, wood chips, agricultural sludge, as well as other waste substances. Depending on the type of nourishment, and the breeding strain, the algae produce different products. Some of them produces lipids, others provide a rich mixture of hydrocarbons, similar to those included in the light oil [58].

The biggest achievement of California's company happened on 13 March 2012. The ship LM 2500 fleet of U.S. Navy frigate, sailed successfully from the port in Everett (WA), to San Diego (CA), powered by fuel: Soladiesel HRD-76 ®. The ship was driven using 25,000 liters of 50/50 mixture of Soladiesel ® and pertoleum F-76. That was the first ever demonstration of alternative fuel mixture in the ship of naval fleet [59].

7. Biohydrogen and Liquid Organic Hydrides (LOH) application

In the context discussed above solar fuels effective production of hydrogen could be the chance for application of LOH as chemical hydrogen storage medium.. The use of liquid organic hydrides in hydrogen storage provides high gravimetric and volumetric hydrogen density, low potential risk, and low capital investment because it is largely compatible with the current transport infrastructure, see Fig.21. Despite its technical, economical, and environmental advantages, the idea of hydrogen storage in liquid organic carriers has not been commercially proved yet. This is because of technical limitations related to the amount of energy required to extract the hydrogen from liquid organic hydride and dehydrogenation catalyst is not stable enough. Renewable hydrogen as well aromatic compounds can be obtained by direct catalytic conversion of methane or biomethane in dehydrocondensation (DHC) reaction. DHC processwas intensively was studied for the last decade by several groups of researchers, mostly from Japan [60-63], China [64], USA [55-67], and Hungary [68]. Following endothermic reaction represents methane direct conversion to benzene and hydrogen without participation of oxygen:

$$6\,CH_4 \Leftrightarrow C_6H_6 + 9H_2, \quad DH_{298} = 88{,}4\ kJ/mol$$

The most effective catalyst for this process is Mo/H-ZSM-5 and Re/H-ZSM5. Oxygen-free conditions used for this reaction result in high benzene selectivity (up to 80%). Nowdays, methane is mostly used for heating purposes, as a transport fuel (CNG) and for chemical synthesis. DHC is promising process from point of view petrochemical feed stocks synthesis, hydrogen production for fuel cells and possible conversion of the waste and difficult accessible resources of natural gas. It can be clathrate (methane hydrate), coal bed methane, post fermentation biogas, land fill and recently shale gas able to be converted into liquid fuels, chemicals easily transportable liquid products.

http://dx.doi.org/10.5772/53831

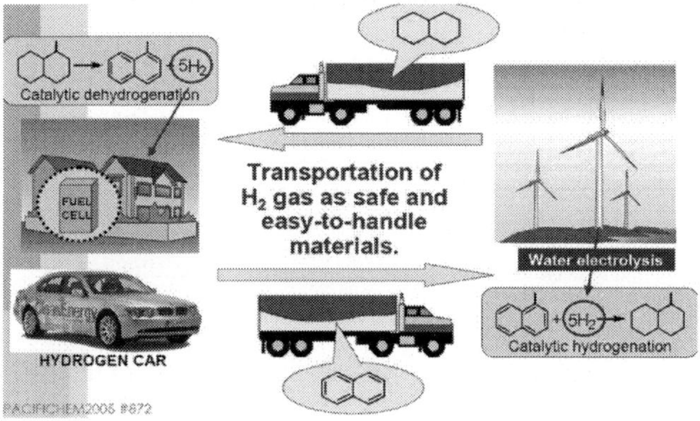

Figure 21. The idea of application LOH compounds as a hydrogen carrier (PACIFICHEM Conference 2005, #872) [69]

8. Environmental conditions for xtL processes implementation

"XtL" processes include an indirect path of energy generation from different types of biomass by converting it to liquid energy carriers [70]. XtL processes means "x to liquid fuels", that is the conversion of various types of raw materials as feedstock for liquid biofuels. The previously presented innovative technologies of waste carbon dioxide conversion and other forms of organic matter into liquid energy carriers are part of a convention of XtL processes.

The environmental aspects of XtL technologies are related with:

1. Emission to the atmosphere the additional amounts of GHG.
2. Generating, recycling and removal of waste.
3. The use of natural resources, raw materials and energy.
4. Sludge production.
5. Space planning (development of the new areas on the Earth).
6. Impact on biodiversity.
7. The destruction of animal species and their habitats.

These methods are directed to the reuse, recycling of existing CO_2 for its processing into useful forms of energy. Algae used for production of liquid fuels are characterized by an effective reduction of significant amount of wastewater, through their purification from toxic substances such as heavy metals. The result of this microorganisms activity are the extra amounts of clean drinking water. It also shows that they have no negative impacts for ecosystems biodiversity, there are also not the reason of the disappearance of certain group of animals or plants.

As it is clear from the review of literature, cultivation of algae for fuels production, seems to be a forward-looking technology for producing liquid energy carriers. The algae have many important advantages from the environmental point of view:

- they are able to sequester waste carbon dioxide, which is used for production of oil and sugars, that are a substrate for the biofuels production,
- they have the ability to bioremediation of contaminated environments,
- they create a "biofertilizers" by absorbing elements such as nitrogen and phosphorus,
- they do not compete with the food market,
- they have a greater oil production capacity than terrestrial energy plants (more than 10 times greater than cultivated oilseed crops),
- they are a group of organisms, with possibility of genetic changes for improvement of their production ability.

Concluding, it is clear from the above considerations, the discussed projects may be an excellent alternative to conventional resources of crude oil, gas and coal. Finally by reducing the concentration of carbon dioxide in the atmosphere, without any negative impact on the environment, above projects have large potential for limitation so-called Global Warming.

9. SWOT analysis of selected technologies

9.1. Technology of „Artificial leaf"

STRENGTHS	WEAKNESSES
• The artificial leaf is able to carry out the water hydrolysis efficiently comparing with green plants. • Artificial leaf is able to carry out the water hydrolysis process according to the reaction: $H_2O \rightarrow O_2+2H_2$ producing hydrogen from noncarbon materials. Its combustion release pure drinking water and oxygen as a byproduct. • Technology does not generate extra amounts of waste. • This method uses the solar energy. • It uses an innovative catalyst to accelerate the reaction, using common metals - nickel and cobalt. • Developed artificial leaf does not need large areas to be used - it has the size of a playing card, in addition, it is very thin and light.	• Artificial leaf is still not able to bind residual carbon dioxide. Research work on copying the entire process of photosynthesis, which takes place in green plants are still under development. The effective method of CO_2 sequestration has not been invented yet. • To effectively trigger the energy, artificial leaf requires integration with existing fuel cells and solar panels, which must first be improved earlier.

OPPORTUNITIES	THREATS
• Developed process of artificial photosynthesis does not require cleaning of water. This technology would be able to use water from rivers, polluted postproduction water, pretreated wastewater, or even seawater to produce hydrogen as a fuel. Conversion of salty seawater resources to fresh drinking water can be one of most desired applicationt of this process. • This technology will be used as a cheap source of electricity for households. An opportunity is a vision to use this method to produce the electricity also in the villages of India, or Africa.	

Table 7.

The strength of "artificial leaf" is ability to carry out the hydrolysis process of water molecule into oxygen and hydrogen, which is the cleanest known fuel. It cannot be possibly limited in future only for use pure water but also polluted one. In this way, perhaps in the future could be used even salty water. Another strong point of this method is the efficiency of the artificial leaf, which now comes to (20... 30)%. If Artificial Leaf would be able to bind waste carbon dioxide from the exhausted gases then it can be an effective method of producing so-called solarfuels.

9.2. Methanol synthesis in carbon dioxide photoreduction process

STRENGTHS	WEAKNESSES
• High quality of biofuel gained in EtL process, meets the requirements for liquid fuels. • Long lifetime of the catalyst estimated at 4-5 days.	• Lack of direct CO_2 sequestration. • Risk of catalyst sintering - no temperature control over the entire height of the reactor. • Low efficiency of the process
OPPORTUNITIES	**THREATS**
• The opportunity will be to further research funding and a return to the study of *artificial photosynthesis*. • Methods for producing second-generation biofuels are more likely for industrial implementation, because they are more technologically advanced than third generation biofuels. • These fuels are compatible with existing infrastructure for conventional fuels. • Start of cooperation with institutions leading the projects/ installations for the production of ethanol from waste biomass.	• Lack of research funding. • Lack of abolition of excise duty on alcohol as a raw material for the biofuels production in the ETL process.

Table 8.

The opportunity for the development of EtL processes is a partnership with institutions leading the projects of ethanol production of all kinds of waste material.

9.3. The process of photosynthesis without sunlight

STRENGTHS	WEAKNESSES
• The use of genetic engineering to increase the productivity of breeding strains of algae. • Production of a wide range of products from oils derived from algae. Besides the basic scope products such as fuel, among others biodiesel, aviation biofuel, biofuel for ships, *Solazyme* also produces various types of coproducts such as cosmetics, bioplastics, and animal feed. • The biggest advantage of technology developed by *Solazyme* is its industrial implementation. There are already commercial biofuel production using this method and are known early success such as an American United aircraft passenger flight from Houston to Chicago powered by blend of 40% biofuel jet. This fuel has been certified by ASTM.	• Lack of direct CO_2sequestration. • Nutrition for algae is lignocellulosic biomass, which means dependence of algae biofuel production from local crops availability.
OPPORTUNITIES	**THREATS**
• The production is not dependent on the availability of solar energy. • Further genetic adjustment of algae cultures can contribute to an even better and more efficient production of oil and algae biomass. • Postproduction waste biomass can be used anaerobic digestion to produce CH_4 and gain of electricity that can power the algae biomass production plant instead of conventional energy consumption.	• Lignocellulosic biomass is not able to satisfy the energy needs of even highly developed countries such as the U.S.. According to preliminary estimate, energy from biomass can cover only a 1/3 of demand for energy. • Some of the substances in the algae nutrition may contaminate the culture by heterotrophic fungi or bacteria. These processes involve additional costs. • The high competitiveness of biofuels third and fourth generation, those involving both algal biomass and GMOs. Preparation of autotrophic algae biomass appears to be more effective and efficient process than *Solazyme* technology based on heterotrophs.

Table 9.

The advantage of sunless photosynthesis is the genetic modification of the strains cultivated algae. Genetic engineering allows increasing of algae productivity, and efficient development of microbial species without the solar energy. This in turn results in the elimination of risks in the form of competition from autotrophic algae.

9.4. OMEGA system – Marine membrane algae cultivation farms

STRENGTHS	WEAKNESSES
• Autotrophic algae are cultivated which are even thirty times more efficient in the oil production than oleaginous plants. • The fed raw material for algae is waste water from which algae receive essential nutrients, such as nitrogen or phosphorus. • Used in the method of OMEGA algae are able to bind waste CO_2 coming directly from the flue gas include plants and directly from the air. • The duration of the production cycle last only 10-14days. • Algae cultured in membrane systems are freshwater species that under conditions of salt are not able to survive. This means that in the event of a system failure will not spread microorganisms and does not attack the existing ecosystems. What's more, algae, which escapes from "farm" will be able to provide an extra source of food for fish, and the rest will be biodegradable, does not cause any pollution of the sea, or ocean. So it is a safe technology for natural ecosystems. • Do not compete cultivation land area, because they can be grown in the desert, wasteland, or on the seas and oceans.	• Localization issues – society is not prepared for existence of algae farm in vicinity for example bays. • OMEGA method involves activity on a large-scale, and not developed any alternative for smaller projects.
OPPORTUNITIES	**THREATS**
• Opportunity to set up algae farms by OMEGA method in places of highest concentration of waste water entering the oceans, such as between the San Diego and San Francisco, where is produced over 7 billion liters of wastewater per day. • The development of biorefineries for the integrated production of biodiesel from algae would reduce production costs. Biorefinery, by the way, could also utilize the way post-production waste biomass such as anaerobic digestion process to produce CH_4 and acquisition the electricity to power the plant of biomass production from algae, rather than consuming conventional energy.	• The bay area of about 1 ha is necessary for continuation of development OMEGA system. Unfortunately, still no pilot plant on a larger sea area has not been completed. • The effects of inclusion of this idea in the sea, on a larger scale is unknown. There is a risk not review the practice of light osmosis membranes of biological pollution and reduction of light penetration.

Table 10.

Strong side of OMEGA system is the raw material for fed of algae. This allows to use the opportunity that comes from the cultivation of algae for example, between San Diego and San Francisco, where is discharged from industrial plants about 7 billion gallons per day of sewage into the sea. This would be a very effective method of disposing of waste, and even that would be a form of recycling, as a result of microbial activity would arise alternative fuels.

10. Conclusions

Cultivation of oil plants for fuels production creates strong competition with crops for consumption. It's hard to talk about the promotion of biofuels, which production is based on energy crops and in many cases is a major cause of the food crops shortage. In many regions of the World are starving societies of so-called *third world countries*. Therefore as the most desirable features of technology is to use waste CO_2 to produce liquid fuels related with:

- reality;
- affordability;
- efficiency;
- the environmental aspect.

In addition, all products and waste that may be generated during the manufacturing process for liquid energy carriers, should use the strategy **3R** – *Recycling Reduce Reuse* in order to reduce energy costs and protect the environment. It is believed that the process of hydrolysis of water by solar energy, in the artificial leaf technology, occurs and is energy efficient.

Harry B. Gray, winner of many awards, including in chemistry, in an article on "*Solar Fuel*" says:

„We need to stop burning hydrocarbon as soon as it is possible, because they are wonderful resources that we desperately need for the production of dyes, pharmaceuticals, T-shirts, chairs and cars - is irresponsible to burn them!"[71].

Waste Carbon dioxide is appropriate material. It is a resource that exists in plenty and causing a negative impact on the environment by including contributing to the greenhouse effect. Artificial photosynthesis is in the research phase and need more decades to make it economically profitable. Today we can see many successful trials to transform CO_2 into various liquid fuel.

The presented analysis shows, that the most important goals for the energy sector should be utilization of carbon dioxide emissions by mentioned CCU idea. This target can be realized by using modern technology of CO_2 reuse to produce liquid energy carriers.

http://dx.doi.org/10.5772/53831

Acknowledgements

The authors gratefully acknowledge financial support from the EKOBENZ Ltd

Author details

Krzysztof Biernat[1], Artur Malinowski[1] and Malwina Gnat[2]

1 Automotive Industry Institute, Department for Fuels and Renewable Energy, Warsaw, Poland

2 Institute for Ecology and Bioethics, University Cardinal Stefan Wyszynski, Warsaw, Poland

References

[1] What the Future Holds in Store, in: World Climate Report, The Web's Longest-Running Climate Change Blog, 30 January, 2008, available from: http://www.worldclimatereport.com/index.php/2008/01/30/what-the-future-holds-in-store/ (accessed 20 August 2012)

[2] CO_2 emissions from fuel combustion IEA Statistics, available from: http://www.iea.org/co2highlights/CO2highlights.pdf (accessed 12 October 2011)

[3] W jaki sposób świeci Słońce?, available from: http://urania.pta.edu.pl/bahcall1.html dn. 12.12.2011 (accessed 25 October 2011) ("How does the Sun shines?")

[4] Van Loon G., Duffy S. J., Gazy cieplarniane związane ze stosowaniem paliw opartych na węglu; Bilans energetyczny; Chemia klimatu globalnego, in Chemia środowiska, Pub. PWN SA, Warsaw, 2007. (Greenhouse gases associated with the use of carbon based fuels; The energy balance; Chemistry of the global climate, in Environmental chemistry)

[5] Solar energy distribution graph showing that infrared radiation makes up a large portion of the solar spectrum, available from: http://continuingeducation.construction.com/article.php?L=68&C=488&P=6 (accessed 24 October 2012)

[6] Solar Spectrum, from Wikipedia, the free encyclopedia, available from: http://en.wikipedia.org/wiki/File:Solar_Spectrum.png (accessed 24 April 2012)

[7] Płaska J., Sałek M., Surma T., Wykład pierwszy. Charakterystyka odnawialnych źródeł energii, „Energetyka" – March 2005. („The first lecture. Characteristics of renewable sources of energy")

[8] Strebeyko P., Fotosynteza, Pub. PWN, Warsaw, 1964. ("Photosynthesis")

[9] Staroń K., (ed.) Przetwarzanie energii i materii podczas fotosyntezy, in Biologia, Pub. WSiP, Warsaw, 2003, ISBN 83-02-08628-2) pp. 106-120. ("The processing of energy and matter during photosynthesis in Biology")

[10] Luminos output of grow lights is not the right parameter, available from: http://www.myledlightingguide.com/Article.aspx?ArticleID=39 (accessed 24 April 2012)

[11] Burda K., Dlaczego warto zajmować się fotosyntezą?, FOTON 93, Summer 2006, available from: http://www.if.uj.edu.pl/Foton/93/pdf/04%20fotosynteza.pdf (accessed 20 September 2012) („Why is photosynthesis?")

[12] Directive 2009/29/EC, available from: http://eur-lex.europa.eu/LexUriServ/LexUriServ.do?uri=OJ:L:2009:140:0016:0062:pl:PDF (accessed 12 October 2011)

[13] Beret-Kowalska G., Kacprowska J., et al., Energia ze źródeł odnawialnych w 2007 roku, Information and Statistical Papers, Central Statistical Office, Department of Industry, Ministry of Economy, Department of Energy, Warsaw, 2008, pp. 5-14. (ISSN: 1898-4347) ("Energy from renewable sources in 2007")

[14] Grzybek A., Analiza możliwości wykorzystania surowców rolnych na potrzeby produkcji biopaliw, „Problemy inżynierii Rolniczej", No. 1/2007. („Analysis of agricultural raw materials for the production of biofuels")

[15] Grzybek A., Zapotrzebowanie na biomasę i strategie energetycznego jej wykorzystania, „Studies and reports IUNG – PIB", Issue II, 2008, Warsaw.) („The demand for biomass energy and its utilization strategies")

[16] Lipski R., Orliński S., Tokarski M., Energetyczne wykorzystanie biomasy na przykładzie kotłowni opalanej słomą we Fromborku, MOTROL, 2006, 8A, pp. 202-209. ("Energy use of biomass on an example of straw-fired boiler in Frombork")

[17] Roszkowski A., Efektywność energetyczna różnych sposobów produkcji i wykorzystania biomasy, the Institute for Building, Mechanization and Electrification of Agriculture, Warsaw, Studies and Reports IUN, issue 11/2008, available from: http://217.113.158.23/wydawnictwa/Pliki/pdfPIB/zesz11.pdf#page=100 (accessed 12 April 2012) („The effectiveness of different methods of biomass production and use")

[18] Biopaliwa – ratunek dla klimatu czy zagrożenie dla przyrody, Warsaw, 5 April 2011, OTOP, available from: http://otop.org.pl/uploads/media/informacja_prasowa_biopaliwa.pdf (accessed 23 March 2012) („Biofuels - the rescue of the climate or the threat to wildlife?")

[19] Wójcicki Z., Energia odnawialna, biopaliwa i ekologia, IBMiER in Warsaw, „Problemy Inżynierii Rolniczej" No. 2/2007. ("Renewable energy, biofuels and ecology")

[20] Niedziółka I., Zuchniarz A., Analiza energetyczna wybranych rodzajów biomasy pochodzenia roślinnego, Department of Agricultural Machinery, Agricultural University of Lublin, „MOTROL", 2006, 8A, pp.233, available form: http://www.pan-ol.lublin.pl/wydawnictwa/Motrol8a/Niedziolka.pdf (accessed 23 April 2012) ("Energy analysis of selected types of plant biomass")

[21] Kijeńska M., Kijeński J., Biopaliwa czy wyczerpaliśmy już wszystkie możliwości?, „CHEMIK", 2011, available from: http://chemia.wnp.pl/biopaliwa-czy-wyczerpalismy-juz-wszystkie-mozliwosci,7511_2_0_2.html (accessed 22 October 2011) („Biofuels – did we exhauste all the possibilities?")

[22] Taylor K., Ivanovsky W., Spirits mixture of diesel, "Chemical Industry" No. 11-12, Lvov, Poland, 1926

[23] http://ec.europa.eu/clima/policies/roadmap/index_en.htm (accessed 09 October 2012)

[24] Technology Roadmap Biofuels for Transport © OECD/IEA, 2011, International Energy Agency 9 rue de la Fédération 75739 Paris Cedex 15, France

[25] Bosmans A., Vanderreydt I., Geysen D., Helsen L., The crucial role of Waste-to-Energy technologies in enhanced landfill mining: a technology review, Journal of Cleaner Production. In press. Available from: http://www.sciencedirect.com/science/article/pii/S0959652612002557(accessed 19 September 2012)

[26] Naber J. E., Goudriaan F., (BIOFUEL BV), Successfully using biomass to Harness renewable energy in an efficient and cost-effective way, available from: https://docs.google.com/viewer?a=v&q=cache:_SpnMHqVSawJ:www.cpi.umist.ac.uk/eminent/Confidential/meeting/RigaMeeting/Riga%2520Workshop/PresenatieHTU-Biofuel.ppt+&hl=en&gl=pl&pid=bl&srcid=ADGEESgMHqR-XurQxLkzf8s6voe2T2AsmrxLMOGwHrqmqRYlcUMMekL57T8pTe-LtMDgxMN6dLSYz30U37W5gpDoAKfAISwapjspGTUrVB9XwICS5SkcnnJ0uka5wrwNfz6AD27qedYLa&sig=AHIEtbTXFACuXwfDtR9YJho3R2xu1V8VTw (accessed 9 September 2012)

[27] Li G., Kong L., Wang H., Huang J., Xu J., Application of hydrothermal reaction in resource recovery of organic wastes, State Key Lab of Pollution Control and Resource Reuse, School of Environmental Science and Engineering, Tongji University, No. 1239 Siping Road, Shanghai 200092, China

[28] Malinowski A., Perspektywiczne technologie biopaliw drugiej generacji na rynku europejskim, conference: „Rynek biopaliw w unijnym pakiecie dot. energii i klimatu", Warsaw, 27 September 2010 („Promising second-generation biofuels technologies in the European market")

[29] Official website of Alphakat Diesel: http://www.alphakatdiesel.pl/technologiakdv.html (accessed 19 April 2012)

[30] Technology modules of KDV, available from: http://www.kdv-projektmanagement.com/technology (accessed 19 April 2012)

[31] Official website of AlphakatPoland: http://alphakatpolska.pl/char_pro.html (accessed 19 April 2012)

[32] Official website of JCAP: http://solarfuelshub.org/hub-overview

[33] PL Patent 208030 B1, „Katalizator do syntezy metanolu i jego pochodnych" by Nazimek D., Czech B., Jabłoński B., Zaniuk W., Wasińska Z. ("The catalyst for the synthesis of methanol and its derivatives").

[34] Nazimek D., (2009) Sztuczna Fotosynteza CO_2 - pierwszy etap produkcji biopaliw, available from: http://www.gmina.zgierz.pl/cms/files/File/KONFERENCJA%20Rogozno%20-%20wczoraj,%20dzis,%20jutro/SFCO2-2009-Krakow.pdf ("Artificial Photosynthesis CO2 - the first stage in the production of biofuels").

[35] Report: Polish Power Plant and University to Cooperate on CO2 to Methanol Trial, available from: http://www.greencarcongress.com/2009/07/nazimek-20090707.html (accessed 21 April 2012)

[36] Nazimek D., Niećko J., (2010) Coupling Ethanol with Synthetic Fuel, Polish J. of Environ. Stud. Vol. 19, No. 3 (2010), 507-514, available from: http://www.pjoes.com/pdf/19.3/507-514.pdf (accessed 20 September 2012)

[37] MIT's Daniel Nocera Announces Artificial Leaf With Goal To Make Every Home a Power Station, Signs with Tata, available from: http://www.freeenergytimes.com/2011/03/28/mit-chemist-daniel-nocera-announces-artificial-leaf-goal-to-make-every-home-a-power-station/ (accessed 10 May 2012)

[38] Van Noorden R., Secrets of artificial leaf revealed, Nature, doi:10.1038/news.2011.564, published online 29 September 2011, available from: http://www.nature.com/news/2011/110929/full/news.2011.564.html (accessed 10 May 2012)

[39] Chandler D. L., Taking a Leaf from Nature's Book, Technology Review published by MIT, January/February 2012, available from: http://www.technologyreview.com/printer_friendly_article.aspx?id=39259 (accessed 24 April 2012)

[40] Regalado A., „Reinventing the Leaf. The ultimate fuel may come not from corn or algae but directly from sun itself.", „Scientific American", October 2010, pp.68-71.

[41] Prototype reactor uses the sun to generate storable hydrogen fuel , available from: http://www.mobilemag.com/2010/12/28/prototype-reactor-uses-the-sun-to-generate-storable-hydrogen-fuel/ (accessed 24 April 2012)

[42] Chueh W. C., Falter C., Abbott M., Scipio D., Furler P., Haile S. M., Steinfeld A., High-Flux Solar-Driven Thermochemical Dissociation of CO2 and H2O Using Nonstoichiometric Ceria, „Science", 24.12.2010.: Vol. 330 No. 6012 pp.1797-1801.

[43] Scott S. A., Biodiesel from alga: challenges and prospects, „ScienceDirect", Current Opinion in Biotechnology 2010,21:277-286.

[44] Fishman D., Majumdar R., Morello J., Pate R., National Algal Biofuels Technology Roadmap, U.S. Department of Energy, Office of Energy Efficiency and Renewable Energy, Biomass Program, U.S. DOE 2010.

[45] Burczyk B., Biomasa. Surowiec do syntez chemicznych i produkcji paliw, OWPW, Wrocław 2011, pp.180,244. („Biomass. Raw material for the chemical synthesis and the production of fuels")

[46] Frąc M., et al, Algi – energia jutra (biomasa, biodiesel), Acta Agrophysica 2009, 13(3), 627-638, Instytut Agrofizyki im. Bohdana Dobrzańskiego PAN, Lublin, pp.634. („Algae - tomorrow's energy (biomass, biodiesel")

[47] Zakrzewski T., Biomasa mikroalg-obiecujące paliwo przyszłości, „Czysta Energia", 2/2011(114). ("Biomass microalgae-promising fuel of the future", "Clean Energy")

[48] Kozieł W., Włodarczyk T., Glony-produkcja biomasy, Acta Agrophysica 2011, 17(1), 105-116, Instytut Agrofizyki im. Bohdana Dobrzańskiego PAN, Lublin, pp. 109. („Algae-biomass production")

[49] Cieśliński H., Biodiesel z alg – paliwo przyszłości?, available from: http://www.silawiedzy.pl/images/stories/artykuly_pdf/837.pdf (accessed 12 April 2012) ("Biodiesel from algae - fuel of the future?")

[50] Sapphire Energy technology, available from: http://www.sapphireenergy.com (accessed 11 April 2012)

[51] Malinowski A., Biopaliwa dla Lotnictwa (Aviation Biofuels), Czysta Energia (Clean Energy), Nr: 121, 9/2011; http://e-czytelnia.abrys.pl/index.php?mod=tekst&id=13518 (accessed 08 October 2012)

[52] Algenol's carbon platform modular technology, available from: http://www.algenolbiofuels.com/direct-to-ethanol/direct-to-ethanol (accessed 23 April 2012)

[53] OMEGA, in AlgaeSystems, available from: http://algaesystems.com/technology/omega/ (accessed 24 April 2012)

[54] Schwartz D., AIM Interview: NASA's OMEGA Scientist, Dr. Jonathan Trent, August 21, 2011in Algae Industry Magazine website, available from: http://www.algaeindustrymagazine.com/nasas-omega-scientist-dr-jonathan-trent/ (accessed 24 April 2012)

[55] Trent J., Wiley P., Tozzi S., McKuin B., Reinsch S., Biofuels, Volume 3, Issue 5, The future of biofuels: is it in the bag?, pp. 521-524, 01 September, 2012

[56] Sikora Z., Heterotroficzny wzrost glonów, „Wiadomości Botaniczne", Volume XIII – Issue 2, 1969. ("Heterotrophic growth of algae")

[57] Gunther M., Solazyme's amazing algae, available from: http://www.marcgunther.com/2010/03/18/solazymes-amazing-algae/ (accessed 11 April 2012)

[58] Fuel from algae, available from: www.technologyrewiev (accessed 24 August 2011)

[59] U.S. Navy Successfully Tests SoladieselHRD-76® During Operational Transit Voyage in USS Ford Frigate, Solazyme Inc., March 13, 2012, available from: http://investors.solazyme.com/releasedetail.cfm?ReleaseID=656801 (accessed 13 March 2012)

[60] Ohnishi R., Liu S., Dong Q., Wang L., Ichikawa M., Catalytic Dehydrocondensation of Methane with CO and CO_2 toward Benzene and Naphthalene on Mo/HZSM-5 and Fe/Co-Modified Mo/HZSM-5, J. Catal. 182, 92 (1999).

[61] Shu Y., Ma H., Ohnishi R. and Ichikawa M., Highly stable performance of catalytic methane dehydrocondensation towards benzene on Mo/HZSM-5 by a periodic switching treatment with H_2 and CO_2, Chem. Commun., 86, (2003).

[62] Malinowski A., Ohnishi R., Ichikawa M., CVD Synthesis in Static Mode of Mo/H-ZSM5 Catalyst for the Methane Dehydroaromatization Reaction to Benzene, Catalysis Letters Vol.96, No 3-4, pp 141-146, (2004)

[63] Patent JP2002336704 Japan Steel Works LTD, Aromatizing reaction catalyst of methane and method for preparing the same, 26 Nov. 2002, Ichikawa M., Ohnishi R., Malinowski A.,. Ikeda K., Yagi K.; The European Patent Office: http://ep.espacenet.com

[64] Shu Y., Xu Y., Wong S., Wang L., Guo X., Promotional Effect of Ru on the Dehydrogenation and Aromatization of Methane in the Absence of Oxygen over Mo/HZSM-5 Catalysts, J. Catal. 170 11 (1997).

[65] Weckhuysen B.M., Wang D., Rosynek M.P., and Lunsford J.H., Conversion of Methane to Benzene over Transition Metal Ion ZSM-5 Zeolites: II. Catalyst Characterization by X-Ray Photoelectron Spectroscopy ,J. Catal. 175, 347 (1998).

[66] Wang D., Lunsford J.H., Rosynek M.P., Characterization of a Mo/ZSM-5 Catalyst for the Conversion of Methane to Benzene, J. Catal. 169 347 (1997).

[67] Solymosi F., Cserényi J., Szöke A., Bánsagi T., and Oszkó A. J., Aromatization of Methane over Supported and Unsupported Mo-Based Catalysts ,J. Catal. 165, 150 (1997).

[68] Liu Z., Nutt M.A. and Iglesia E., The Effects of CO_2, CO and H_2 Co-Reactants on Methane Reactions Catalyzed by Mo/H-ZSM-5, Catal. Lett., 2002, 81, 271-279.

[69] PACIFICHEM Conference 2005, #872

[70] Biernat K., Biokomponenty i biopaliwa – możliwości rozwoju i zastosowania, „Forum Czystej Energii" X edycja, 23-25 listopada 2010 r., Poznań, „Technology Roadmap. Biofuels for Transport" Pub. OECE/IEA, Paris 2011. („Biocomponents and biofuels - opportunities for development and use")

[71] Gray, H. B., *Solar Fuel*, „Engineering and Science" 60 (3). pp. 28-33, 1997, ISSN 0013-7812, available from: http://calteches.library.caltech.edu/3924/1/Fuel.pdf (accessed 20 September 2012)

Chapter 6

Oil Palm Plantations in Indonesia: The Implications for Migration, Settlement/Resettlement and Local Economic Development

Suseno Budidarsono, Ari Susanti and
Annelies Zoomers

Additional information is available at the end of the chapter

http://dx.doi.org/10.5772/53586

1. Introduction

Palm oil is the world's most traded vegetable oil: in August 2012, the share of palm oil (including kernel oil) in world supply was 37.6% [1]. Palm oil is extracted from the fruit of the oil palm tree (*Elaeisguineensis*); the main products are crude palm oil (CPO) and palm kernel oil (PKO). In terms of land use, the oil palm tree is more efficient than any other oil crop [2], and in economic terms palm oil is highly competitive. The value chain of palm oil and its derivatives has a strong degree of vertical integration [3], and its production costs are relatively low compared to other vegetable oils. It is therefore seen as one of the cheapest and most attractive vegetable oils traded on the world market [4, 5].

The palm oil sector provides income and employment for a significant number of individuals in developing countries [6]. A study of the Indonesian palm oil industry carried out as part of a global study under the coordination of the Australian National University, concluded that palm oil developments have had a positive impact on the incomes and living standards of all involved [7]. According to an assessment carried out in Sumatra, oil palm plantations have high labour requirements and show high return to labour [8].

Palm oil, being a multi-purpose vegetable oil, offers good prospects for further expansion. There is a growing demand from the commercial food and oleo-chemical industries that use oil palm in processed foods, cosmetics, soaps, pharmaceuticals, industrial and agro-chemical products, and as a feedstock for bio-diesel. The growing worldwide interest in bio-diesel as an alternative to fossil fuel is expected to lead to the further expansion oil palm plantation [9,10,11]. Tilman and colleagues [12] assert that this might lead to a food–energy–environment 'trilemma'. Even though oil palm trees are not a problem (they are 'green'), the rapid expansion of oil palm plantations across Southeast Asia, and particularly Indonesia and Malaysia, could cause the destruction of rainforests, as well as a lot of social problems, including food security challenges [13]. There are concerns that oil palm expansion will lead to the loss of biodiversity and the conversion of forest area [14 - 16] and the aggravation of social conflicts. According to Colchester and colleagues [17], who analysed the situation on and around six oil palm plantations, local communities face serious problems with the companies and there are many land conflicts. There is a widespread feeling in the communities of being cheated by the companies, and of being pushed into agreements through false promises without having a voice in decision-making. To the extent that people are employed, labour conditions are often not favourable. According to Marti [18], who conducted research on the labour conditions of plantation workers, much will depend on 'whether Indonesia's policy-makers intend to keep a large labouring class in low-paid, low-skill jobs as the rest of the country develops, or whether the country anticipates inviting millions of workers from even less fortunate countries to work on their plantations in future.'

With these controversies in mind, this chapter provides an overview of the pros and cons of oil palm development in Indonesia, paying attention to changing government policies and focusing on the implications of increasing oil palm investments for migration, settlement/ resettlement and local economic development.

Since the 1970s, the Indonesian government has stimulated oil palm expansion in various ways, initially in the form of plantations. For a considerable time, the government played a direct role in stimulating investments in oil palm plantations through state agencies (with direct interventions in service provision, institutional support, agricultural extension, access to land and capital, etc.). Plantation development policies were carried out in close relation with other policy objectives, namely population redistribution through resettlement schemes or transmigration to stimulate the development of the outer islands (Sumatra, Kalimantan and Papua) and revitalizing the by then huge transmigration settlements that had often failed to produce more than rice and subsistence crops [19] and enhancing regional development (i.e., the increase in agricultural production, employment generation etc.); and the political objective of promoting national integrity and increasing national security [20].

This chapter is based on empirical data collected in Riau province (one of the booming oil palm producing areas that has undergone rapid expansion in the last 10 years) as well as on desk research. The field data collection focused on the economic activities related to oil palm development. The desk study included an analysis of legal and technical documents related to historical records of oil palm development in Indonesia in general and in Riau province in particular. These documents were collected in libraries and government offices, and from companies and online sources.

After providing some background information on palm oil investments, we give a historical overview of how oil palm expansion took place in Indonesia in close relation to changing policies. This is followed by an assessment of the implications of rapid oil palm expansion for migration, settlement/resettlement and local economic development.

http://dx.doi.org/10.5772/53586

2. Investment in palm oil production, settlement development and rural economy

Investments in palm oil production require large-scale plantations to be economical and the establishment of palm oil processing units (mills) close to the plantations, as the oil palm fruits have to be processed within 48 hours after harvesting. Once investments in palm oil production have been made, the plantations are made to last as long as possible, or at least for one production cycle (25 years). A large-scale oil palm plantation requires a significant amount of labour to establish the plantation, maintain the palm trees and harvest the fruits. A detailed study on palm oil production in Indonesia noted that the labour requirements of a plantation during one production cycle (25 years) vary between 59 and 144 person-days (pds) per ha per year, which is an average of 91 pds per year[21]. Although a mill requires less labour, managers, technical staff and labourers have to be on hand. However, in many (if not all) cases, large-scale plantations are established in scarcely populated areas, and even in 'empty' frontier areas. Large-scale plantations are therefore always accompanied by settlement/resettlement, for which staff and labour facilities must be established [22]. For example, a block of 1000 ha of oil palm trees needs 91,000 pds/year. Assuming that the optimum working time for one person is 250 pds per year, 364 employees and labourers will be needed. As most of them will live with their families, significant provisions have to be made.

The establishment of large-scale oil palm plantations and processing units is an important economic activity that strongly influences the rate of land development in a region. It speeds up the development of infrastructure (e.g. the construction of roads to open up less accessible areas, and the provision of health and educational facilities) and stimulates the growth of the local economy through income spending by the plantation workers and the expenses of the company. Economic diversification will take place in the surrounding area, for example from subsistence agriculture to market-oriented cash production. The most fundamental effect of cash-cropping is that it increases the separation between the consumption and the production activities of the household. This in turn tends to change the pattern of family and communal dependencies and to strengthen the role of the nucleus family as a productive enterprise [23]. Another effect is the emergence of non-agricultural activities (trade, home industry and services). Urban employment opportunities subsequently increase.

Empirical research carried out in Riau province assessing the development impact of large-scale oil palm plantations on the local economy [24] revealed that the investment in palm oil production strongly induced local economic growth. The study found that marginal propensity to consume locally (MPCL) of the community living from oil palm plantation was 0.8415. This means that about 84% of additional income was spent locally. The percentage of money spent locally that becomes local income (PSY) was 71%; nearly three quarters of the necessary inputs needed could be provided locally. Based on these figures, the calculated income multiplier effect was 2.48 (1/(1 – (MPCL x PSY)). This means that an autonomous IDR 1 million change in income (or investment expenditures) in the area results in a change of IDR 2.48 million. The multiplier effect refers to the increase in final income arising from any new injection of spending.

3. Oil palm plantation development in Indonesia: A brief review

3.1. The early stage of palm oil development

The oil palm (*Elaeis guineensis*) originates from the tropical rainforest of West Africa, and was originally cultivated mostly by independent small farmers with landholdings of up to 7.5 ha. Palm oil has been commercially traded to Europe since 1811 [25, 26]. In Indonesia, oil palms have been cultivated commercially since 1911, when they were first developed in the east coast area of Sumatra under Dutch administration [2]. While the tree was cultivated successfully in this area in large plantations, the native population did not replace their coconut palms with this new palm species. They planted it only for decorative purposes.

In the east coast region of Sumatra, palm oil production (CPO) grew dramatically, from 181 tons in 1919 to 190,627 tons of CPO and 39,630 tons of kernel oil in 1937 [27]. This was in line with the oil palm plantation development in the area. The size of the first estate, which was established in 1910–14, was 2,620 ha. The plantation area increased to 6,920 ha in 1919, and steadily increased until 1936, when the total area planted amounted to 75,000 ha, of which 63,234 ha were in a productive stage. In this regard, Deasy [28] wrote that:

> … the oil palm industry in Sumatra during the last two decades has bordered on the phenomenal. Production of palm oil in that island has increased from a few thousand metric tons in the early twenties of this century to almost two hundred thousand metric tons in 1937. In 1920, Sumatra accounted for less than one-half of one per cent of the world's annual export of palm oil; by 1937 over 40 per cent of the total export of that commodity was originating in the island. From an insignificant crop with an output far below that of most other Sumatran estate crops in 1920, the oil palm by 1937 had risen to a position of joint supremacy with rubber in the agricultural economy of the island.

The east coast of Sumatra became the site of one of the most intensive and successful pursuits of foreign agricultural enterprise or plantation [29]. Bio-physically, the area is quite suitable for oil palm growing, with high rainfall (minimum 1600 mm/year) and a tropical climate within 10° of the equator. Land and labour, the most important inputs, were available. The land was made available by getting long-term land leases from the local indigenous rulers through the colonial government [29, 30]. In the 1870s, the Dutch government started to withdraw from direct involvement in economic production, and increasingly concentrated on creating incentives that would stimulate the private initiative. As Breman [30] wrote:

> The 1870 Agrarian Act, which officially put an end to the forced cultivation system in Java and started transition to an era of unbridled liberalism, indicated the orientation of the new policy: the archipelago's natural resources were henceforth to be made accessible to capitalist interests in the mother country.

---

http://dx.doi.org/10.5772/53586

The establishment of the palm oil industry in Sumatra started at the right time. The demand for palm oil was growing enormously worldwide, due to the increasing uses of palm oil products. Palm oil had already been traded to Europe and the United States since the early nineteenth century. Industrial development in Great Britain stimulated the demand for oil, a demand that was partly satisfied through trade with the West African coast. However, the increased mechanization of industry created a much larger need for lubricants, and palm oil proved useful in the manufacture of grease. Of equal importance, though, was the value of palm oil in the making of soap, candles and medicinal ointments, the increased use of which was also a result of the Industrial Revolution [26]. Scientific research supported the production of oil palm by modern, well-equipped estates and mills [25]. As a result, oil palm plantations in the Sumatra region became five times more productive than in Africa. The plantation development culture acquired from the cultivation and processing of latex rubber was a good foundation on which to introduce the large-scale cultivation of palm oil. The emergence of corporate capitalism since the 1850s, especially during the 'Tobacco Deli era' in the east coast of Sumatra, provided an ideal environment for large-scale plantation investments [29]. The capital market, labour market and remuneration systems had already developed. There was a banking system created by the tobacco planters association. Plantations that found themselves short of labour could easily acquire more through labour suppliers in Malaysia and Singapore.

Export-oriented plantation culture in Sumatra East Coast provided local employment that also attracted immigrants, which brought about more economic activity in the region. However, there are some criticisms regarding the development of large-scale plantations during the colonial era. The plantations did not lead to balanced economic growth and regional development [31]. The large-scale monocultures of capital-intensive enterprises created little wealth for unskilled workers (migrants) or local farmers, since wage levels were low. Many local farmers (especially shifting cultivators) suffered from land grabbing by the estate companies and the local economy suffered from the siphoning off of the locally produced surpluses (profits, dividends, staff salaries) to the head offices in urban centres and abroad. Moreover, the highly specialized plantations often suffered from instabilities on the world market, while having high fixed costs (for machinery, buildings, staff and maintenance). Together with the low purchasing power of the local population, this caused a low demand for urban goods and services and, as a result, a weak development of local market centres. There were consequently few opportunities for the local diversification of production and employment. Historical studies by Stoler [29] and Breman [30] to some degree provide a similar nuance from the labour perspectives.

However, those observations do not negate the fact that during the early stage of oil palm plantation, the east coast of Sumatra underwent a high population growth. As shown in Table 1, the annual population growth rate in the area in 1900–30 was 10.4%. Chinese and Javanese workers were the largest groups among the plantation workers (54% and 34%, respectively in 1902; [30]). It seems that the region was quite attractive for people to move to. The share of local people was assumed to be 56%, with an immigrant population of 44% in 1930.

According to Stoler [29] and Breman [30] investments in the plantation agriculture on the east coast of Sumatra (and the emergence of a plantation society) induced local development. The large number of people coming to and residing in the region created a market for consumer goods. To some degree, agglomeration economies evolved in the region. It is also noted that plantation support services, such as banking systems, capital and labour markets, were introduced, established and worked well. Within about five decades, the east coast of Sumatra region had been transformed from a frontier area into a well-developed and integrated plantation area. However, no study has looked at how the investments in such large-scale and modern plantation activities led to settlement development in the surrounding area, how the rural service centres developed and grew into urban centres, or, more in general, what kind of transformation took place in the region because of the new technology and the increase in population caused by the inflow of different ethnic groups.

Year	European 1)	Inlanders 2)	Chinese	Other 3)	Total
Population					
1900	2,079	306,035	103,768	9,208	421,090
1905	2,667	450,941	99,236	15,573	568,417
1915	5,200	681,000	132,000	14,320	832,520
1920	7,882	1,042,930	134,750	11,992	1,197,554
1930	11,079	1,470,395	192,822	18,904	1,693,200
			Proportion (%)		
1900	0.5%	72.7%	24.6%	2.2%	
1905	17.1%	79.3%	17.5%	2.7%	
1920	0.7%	81.8%	15.9%	1.7%	
1930	0.7%	87.1%	11.3%	1.0%	
Annual growth rate 1900–30	14.9%	13.1%	3.0%	3.6%	10.4%
Annual growth rate 1900–20	14.7%	12.7%	1.6%	1.6%	9.7%
Annual growth rate 1920–30	4.5%	4.6%	4.8%	6.4%	4.6%

Sources: Adapted from Het Oostkust van Sumatra Instituut, 1938.

Note: referring to Breman [30]

1. Includes Dutch, Belgian, German, Swiss, French, British and Austrian

2. Consists of local community and immigrants: Javanese, Banjarese, Baweanese, Bataks, Malay, Gayo, Mandailing

3. Consists of Thai, Indians

Table 1. Population of the east coast of Sumatra, 1900–30: population, proportion by ethnic group and annual growth rate

http://dx.doi.org/10.5772/53586

3.2. Palm oil production after independence and recent development

Not much has been written about plantation development - or more specifically oil palm plantations – in the period from World War II until the independence of Indonesia, or during the transitional phase after independence. In general terms, Benjamin Higgins, a financial advisor for the Indonesian government in the early 1950s under the UN Technical Assistance Programme, wrote that the plantation industry was facing serious problems, which was shown by the fact that the area under plantation crops was still only two thirds of the pre-war level [32]. Figures showed that palm oil exports amounted to only 109,000 tons in 1960 [33], compared to 240,000 tons of CPO and kernel oil in 1937 [27]. However, since the earliest national reconstruction plan the Indonesian government has included plantation agriculture in its development policy. In 1955, the Ministry of Transmigration submitted a 5-year plan to the National Planning Bureau to resettle 400,000 families from overpopulated Java to unsettled areas in the outer islands (mainly associated with traditional export commodities from plantation agriculture, such as rubber, coffee, tea, etc.). According to Higgins, the objectives of the transmigration programme were not achieved, mainly due to a lack of assured financing, inadequate manpower and a lack of technical education [32].

Looking at time series data of oil palm plantation in Indonesia (Figure 1), it appears that Indonesian oil palm plantation increased significantly only after the 1970s. This is related to the New Order government policy for the agricultural sectors, which included plantation development.

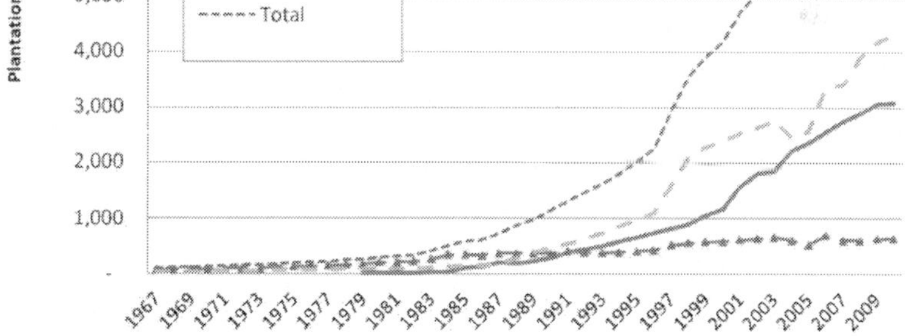

Figure 1. The development of oil palm plantation area in Indonesia, by actors. Source: Series of Statistik Perkebunan Indonesia (estate statistics)

Much has been published about the post-1970s development of the palm oil industry in Indonesia, focusing on topics such as germ palm development [34], competitiveness [7], socio-economic improvement [19], agrarian issues and social transformation [5, 17], and even human rights [18].

As mentioned, the Indonesian government began to put sustained effort into promoting tree plantation crops in the late 1970s. It established a scheme called Nucleus Estate Scheme (NES) (*Perkebunan Inti Rakyat*; PIR), whereby state-owned plantation companies (the 'nucleus') helped farmers (namely plasma farmers) to grow oil palm. The plantation companies provided seedlings, technical assistance and financing to small holders. Their crops would be purchased by companies' mills [35]. The policy was not implemented for solely plantation development; it was linked and integrated with other policy objectives: population redistribution through resettlement schemes or transmigration (i.e. by moving people from densely populated regions to scarcely populated areas), socio-economic progress (regional development, increased agricultural production, employment generation) and political consolidation (promotion of national integrity and security) [36]. Thus, the initial programme consisted basically of direct state investments through state-owned companies (PTPN) and was integrated with government-sponsored transmigration programmes to provide a labour force for the new plantations [22]. This integration was embryonic for smallholder engagement in state-led agribusiness; the emergence of smallholder oil palm planters constituted a spread effect of plantation development led by the government, and, most importantly, it was the time when settlement development was started in the surrounding of large-scale oil palm plantation.

Analysing the evolution of oil palm plantation development since the 1970s, McCarthy [5] differentiates into three phases of development: (1) the New Order state developmental period (the late 1970s to 1994), which was characterized by direct state intervention; (2) the transitional period towards private initiative through the KKPA model (1994–98); and (3) the 'laissez-faire' period (the term used by McCarthy to name the *Reformasi* era) since 1998.

First phase: During the New Order period, the government pursued a developmental agenda that combined the aim of ensuring political and macro-economic stability by financing infra structure and providing subsidies derived from oil revenues [5]. There were direct interventions by the state that enabled state-owned companies to have more access to land and capital; institutional support was also provided [37]. The introduction of the Nucleus Estate Scheme (the PIR) in combination with the transmigration programme (the PIR-Trans) took place in this period. In general, in terms of area, the ratio between the nucleus estate and the smallholder plasma was 20:80; the nucleus estate held 20% of the total area and the remaining 80% was owned by smallholders – plasma farmers with technical assistance from the nucleus estate. Under the PIR-Trans scheme, trans-migrants played significant roles as labour for the plantation (crop care and harvesting) and constitute an important component during the establishment years. The government provided financial support to smallholders to establish their plantations and to finance their living and housing expenses, while the nucleus estate (the company or investors) was responsible for extension services and for collecting and processing the fruit bunches. The government also facilitated access to land (mostly

http://dx.doi.org/10.5772/53586

state forest land and village lands), developed some infrastructure and granted credit at concessionary rates for plantation development. The PIR-Trans schemes, which existed between 1986 and 1994, benefited smallholders, as 'plasma' farmers. The state provided access to 'free land' and concessionary credit in exchange for submission to a particular agribusiness model, namely the inclusion of smallholders in peripheral areas. The farmers would obtain fully private rights over their holdings upon settlement of the oil palm development loan. By incorporating trans-migrants, new settlements were established in the area surrounding the plantation, besides those of the local people.

Second phase: In the transitional period towards private–community initiatives (the KKPA model), the government changed its policy by seeking to encourage private sector initiatives, facilitate foreign direct investment and accelerate estate crop development. The new orientation was in response to the World Bank's criticism of the on-going state support of smallholder oil palm schemes. It advised the government to officially abandon its direct subsidizing role and leave oil palm development to the market. The government ignored this advice for some time, before changing its policy as a result of increasing pressure on the state budget, as well as donor advocacy for a more direct social–private partnership model. Accordingly, the next phase in the development of oil palm schemes marked a new milestone in an on-going trend towards state withdrawal.

The new scheme was known as the KKPA (*Koperasi Kredit Primer untuk Anggota*; Primary Cooperative Credit for Members) and covered the period 1995–98. It involved a more direct private–community 'partnership' model, with the plantation firm being responsible for nearly all of the project, working directly with the participating farmers to resolve land problems, and providing training and extension. During this period the door for foreign direct investment in large-scale plantations was opened. Local communities, including settlements of trans-migrants, which had often been unsuccessful to move beyond the production of rice and subsistence crops, could be transformed into oil palm plantations, depending on the agreement between both parties. More independent smallholder oil palm farmers emerged in that time, and led to more spontaneous migration into the oil palm area.

Third phase: McCarthy [5] asserts that in the 'laissez-faire' period (1998 onwards), the shift to decentralization, public–private partnerships between market actors and the government, and social–private partnerships between market actors and communities have affected Indonesian policy. The year 1998 was, in fact, a key moment in the transition from a developmental approach to a more neoliberal, market-driven model [5]. A series of policy changes provided for the development of 'community plantations' (*perkebunan rakyat*) under various partnership (*kemitraan*) models (see the two ministerial decrees: the Decree of Forestry and Estate Ministry No. 107/Kpts-II/1999 and the Decree of Agricultural Ministry No 26/2007). Existing estates entered into partnerships with large, capital-intensive companies willing to invest in labour-intensive oil palm projects [19]. During this period, farmers, who were initially in the PIR-Trans schemes, managed to gain access to oil palm technology and improve their incomes. They were eventually able to access investment capital, because, once they had paid off their credit, they obtained land certificates, which could be used as collateral for borrowing money from local banks to expand production. At a time of rising oil palm prices, many of these new landowners used these accumulated assets to rapidly expand their holdings. During the later years of the oil palm boom (i.e. prior to 2008), these actors were joined by successful KKPA farmers, who were (to a limited degree) using incomes from productive oil palm holdings to invest in upgrading unproductive land into oil palm plantations. The result was spontaneous frontier development on the margins of already existing palm oil plantations. This means that the spread effect of oil palm plantation was more prevalent in this period.

Summarizing, the state agribusiness-driven policy has transformed rural areas, making Indonesia the world's largest oil palm producer. The area devoted to palm oil production had doubled to 5.5 million ha by the year 2000 [38], and by 2009 Indonesia was the world's leading oil (CPO and PKO) producer with 24.5 million tons as compared to Malaysia with 22.1 million tons [39]. The introduction of the oil palm was initially associated with direct state investments via state-owned companies. After implementing various types of schemes, the state introduced the estate-transmigration programme; both models of palm oil development led to frontier development. Settlements were officially established in newly opened areas, in the expectation that economic activities would slowly develop. During the second and the third period, both spontaneous migration and the number of independent oil palm smallholders increased in the area surrounding the existing oil palm plantations.

4. Oil palm plantation development in Riau province

Riau Province is located in the centre and on the eastern coast of Sumatra along the Strait of Malacca, the busy international shipping route connecting the Indian Ocean with the South Chinese Sea and the Pacific Ocean. Riau is also quite near Singapore, one of the biggest trading centres in Southeast Asia (see Figure 2).

Riau is rich in natural resources, particularly petroleum, natural gas, coal, forest, and rubber, oil palm and fibre plantations. There are huge deposits of oil and natural gas beneath the ground, making the province the country's largest producer of oil: 80% in the early 1970s [41] and 50% in 2006 [42, 43]. In 2010, there were 2.1 million ha of oil palm plantations [44] and also two giant pulp and paper companies. The economic potential attracts people from the surrounding areas and even from other islands (Java, Kalimantan) who hope to find a better life. Net migration to Riau has rapidly increased over time (Table 2).

Riau is divided into 10 regencies and 2 autonomous cities. Until 2004 the province included the Riau Islands, a large group of small islands located east of Sumatra and south of Singapore, but in July 2004 these islands were split off as a separate province.

According to the 2010 census, the population of Riau province was 5.54 million. Population growth was far above national average: the annual population growth rate was 4.35% in 1970-90, while the national average was only 1.49%. In 2000–10, the rate it was 3.9%, while the national figure was only 1.9%. This population growth reflects the economic potential of the province. Table 3 presents the detailed population growth by regency and city.

http://dx.doi.org/10.5772/53586

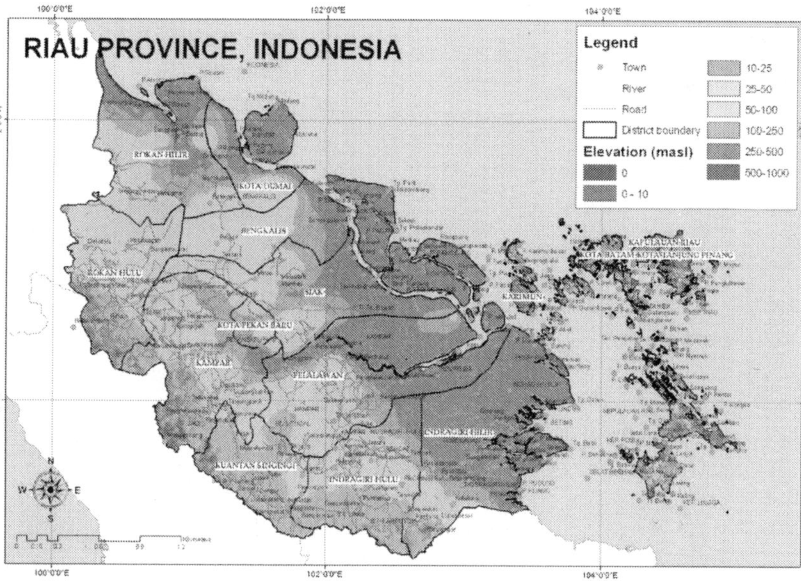

Figure 2. Topographical map of Riau. Source: ICRAF database

	1971	1980	1985	1990	1995	2000	2005
Riau	175,498 **(11%)**	269,732 **(11%)**	221,654 **(11%)**	561,364 **(22%)**	714,828 **(29%)**	1,370,491 **(72%)**	1,127,824 **(64%)**
Sumatra	1,547,401	2,384,232	2,095,077	2,552,764	2,457,043	1,903,777	1,762,728

Source: BPS 2011

Table 2. Net migration to Riau and Sumatra, 1971–2005

As shown in Table 3, the regency of Pelalawan – where timber plantations for the pulp and paper industry, and also oil palm plantations are located – had the highest population growth rate (6.8%). Siak, Kampar, Rokan Hulu and Rokan Hilir, which have significant areas of oil palm plantations, also experienced a high population growth (above 4%). Although there is no big difference in population between the districts, several districts (such as the city of Pekanbaru, Kampar Regency and Inderagiri Hilir Regency) have quite high concentrations compared to other districts. However, migrants are found almost everywhere. Many come from Java, and use the social networks that have been established through trans-migrants who settled before them [45].

Regency / city	Area (km²)	Population		
		Census 2000	Census 2010	Annual growth
Pekanbaru (city)	633	585,440	903,902	4.3%
Dumai (city)	2,039	173,188	254,337	3.8%
Kuantan Singingi Regency	5,235	216,732	291,044	2.9%
Indragiri Hulu Regency	7,611	247,306	362,961	3.8%
Indragiri Hilir Regency	13,633	555,701	662,305	1.8%
Pelalawan Regency	12,482	152,949	303,021	6.8%
Siak Regency	8,216	238,786	377,232	4.6%
Kampar Regency	10,814	447,157	686,030	4.3%
Bengkalis Regency	11,932	520,241	674,755	2.6%
Rokan Hulu Regency	7,225	265,686	475,011	5.8%
Rokan Hilir Regency	8,852	352,299	552,433	4.5%
Meranti Islands Regency*)	no data	no data	no data	–
RIAU PROVINCE	88,673	3,755,485	5,543,031	3.9%

*) Regency established in 2009; previously part of Bengkalis Regency. No data are available.

Source: BPS, 2011

Table 3. Area and population of Riau by regency and city, 2000 and 2010

Since the early 1980s, Riau Province has been the primary target for oil palm plantation development as part of Indonesia's agricultural development policy. The first large-scale oil palm plantation was established in Rokan Hulu Regency by PTPN (a state-owned plantation enterprise). The current Rokan Hulu Regency was part of Kampar Regency in the 1980s; like many other Indonesian provinces, Riau experienced the formation of new regencies in the era of post-Soeharto decentralization. This process, known as *pemekaran,* led to an increase in the number of regencies. Until 1999, Riau consisted of 5 regencies and 2 cities; it now has 10 regencies and 2 cities.

The establishment of oil palm plantations was initially realised through direct state investments. Then the private sector slowly took a more important role, especially after 1998. As mentioned, the establishment of oil palm plantations took place in the frontier area and in tandem with a resettlement programme to provide labour. Because the plantations provide higher wages and better profit, they attract people who seek to improve their livelihood and try their luck by engaging in oil palm growing. It is estimated that migrants comprise

http://dx.doi.org/10.5772/53586

around 24% of the total population (about 67% of this group arrived in the context of the transmigration programmes [46]). This migration process stimulated oil palm to expand (occupying the surrounding area), and the decision of spontaneous independent farmers to cultivate oil palm has made the oil palm sector in the province boom (Table 4). Looking at the age distribution of the plantations, the table shows that the largest proportion of newest (still immature) oil palm plantations are under the smallholder system, followed by the private estates. It means that the recent development of oil palm plantation in Riau was done by private sector and smallholder farmers.

A study on livelihoods in Riau revealed that many of the migrants are well-off and are able to buy the land of local people and plant oil palms. However, it also involves the encroachment of state forest land. For example, small-scale farmers have entered the Tesso Nilo National Park to establish oil palm plantations [46, 47].

	Area (ha)			
	Immature	**Productive stage**	**Not-productive**	**Total area**
State-owned estates	1,000	78,546	-	79,546
Private estates	147,162	758,402	385	905,949
Smallholdings	318,969	780,959	17,625	1,117,553
	467,131	1,617,907	18,010	2,103,048

Sources: Dinas Perkebunan Provinsi Riau (2011) Statistik Perkebunan Provinsi Riau 2010. Pekanbaru

Table 4. Oil palm plantations by actors in Riau, 2010

In only two decades, Riau overtook North Sumatra as the leading province in palm oil production (Figure 3). The rapid expansion of oil palm plantations has been in line with the economic attractiveness and profitability of this perennial crop. Independent smallholders are becoming increasingly dominant. They benefit from a direct link with the plantation companies, which provide technical assistance, good planting material, fertilizers and delivery contracts to the mill.

The rapid expansion of oil palm plantations is clearly illustrated in Riau's economic performance. Table 5 presents the distribution of oil palm plantations and the economic performance of the various regencies.

As shown in Table 5, each regency in Riau has oil palm plantations, with the planted area ranging from 12% in the capital city of Pekanbaru to 58.5% in Rokan Hulu Regency. In 2000–10, Riau underwent relatively high population growth. Indragiri Hilir Regency (which is dominated by peat land) had the lowest population growth rate. Rokan Hulu Regency (which has mostly mineral soil and the largest oil palm plantation in the province) had the highest population growth (see also Table 3).

Regency / city	Area (km²)	Population			Oil palm plantation (2010)		Economic performance		
		Population in 2010	Population density (per km²)	Population growth rate 2000–10	Planted area (ha)	% to total area	GRDP per capita 2010 [2]	GRDP Growth [3]	Economic growth rate [4]
Kuantan Singingi	5,235	291,044	56	2.9%	121,709	23.4%	10,649.44	8.28	7.40
Indragiri Hulu	7,611	362,961	48	3.8%	118,538	15.4%	11,088.16	7.30	6.75
Indragiri Hilir	13,633	662,305	49	1.8%	213,541	15.5%	10,157.36	7.55	7.47
Pelalawan	12,482	303,021	24	6.8%	184,110	14.8%	10,321.78	7.21	7.10
Siak	8,216	377,232	46	4.6%	232,857	28.3%	10,123.38	7.45	7.37
Kampar	10,814	686,030	63	4.3%	353,792	32.4%	6,772,80	7,40	7,29
Rokan Hulu	7,225	475,011	66	5.8%	422,613	58.5%	5,395,28	7,06	6,02
Bengkalis	11,932	674,755	57	2.6%	177,130	21.0%	6,862,21	7,68	7,40
Rokan Hilir	8,852	552,433	62	4.5%	237,743	26.5%	7,439,10	7,69	7,57
Meranti Islands [1]	no data	no data	no data	no data	-	-	8,049,62	7,29	6,29
Pekanbaru (city)	633	903,902	1.428	4.3%	8,080	12.8%	10,078,26	9,72	8,90
Dumai (city)	2,039	254,337	125	3.8%	32,935	16.2%	8,221,24	8,60	8,56
RIAU	88,673	5,543,031	63	3.9%	2,103,048	23.6%	8,782,70	7,99	8,90

Sources: adapted from the respective [Regencies' statistics from 2011]

Notes:

1. Regency established in 2009; previously part of Bengkalis Regency. No data are available.

2. Gross regional domestic product in constant prices (hence 2000 price).

3. Annual average 2004–10.

4. Annual average 2008–10.

Table 5. Oil palm plantation and economic performance in Riau, by regency

Two economic indicators are used to assess the economic performance of all regencies in Riau in relation to oil palm plantation in the region, namely gross regional domestic product (GRDP) and economic growth. GRDP is used to measure the size of a region's economy. It is basically the aggregate of gross value added of all resident producer units in the region (using GRDP per capita as an approximation of the value of goods produced per person in a region); economic growth is measured on the basis of increase in the capacity of an economy to produce goods and services from one period of time to another.

http://dx.doi.org/10.5772/53586

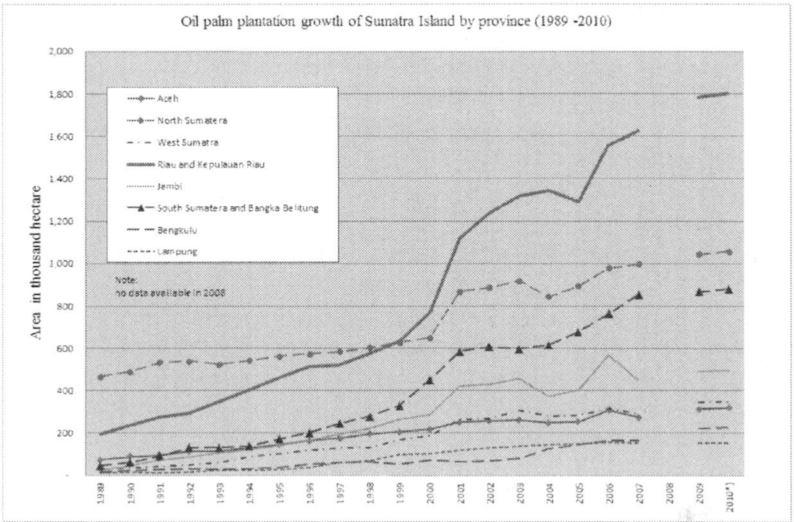

Figure 3. Oil palm plantation development in Sumatra Island by province (1989-2010)

In general, Riau performed well economically. GRDP per capita ranged from IDR 5.4 million in Rokan Hulu to IDR 11.0 million in Indragiri Hulu. The growth rate of GRDP in 2004–10 was almost 8%. In addition to oil palm, other sectors (e.g., mining in Rokan Hilir; and industry in Pelalawan and Bengkalis) also contribute to GRDP. However, looking at oil palm plantations, a recent economic assessment of small-scale oil palm cultivation revealed that returns to land ranged from IDR 92 million to IDR 143 million per ha in a 25-year cycle, and returns to labour from IDR 122,000 to IDR 178,000 per person per day. Returns to land of smallholding oil palm plantations ranged from IDR 125 million to IDR 266 million; returns to labour ranged from IDR 67,000 to IDR 297,000 [48]. It is therefore not too surprising that net migration was high, especially in sparsely populated area (where the agricultural wage rate were relatively high). Especially here, land use conversion was accelerated by the attraction of more people.

For the development of settlements, we looked at the Village Potential Statistics (PODES / *PotensiDesa*), a village-level census that is carried out three times per decade. PODES is administered by BPS (Indonesia Statistical Body) and serves to collect socio-economic information from all 69,000 Indonesian rural villages and urban neighbourhoods. The survey is based on responses of the village heads and includes a wide range of indicators, from population characteristics to infrastructure, economic activities and social life. According to PODES, from 1996 (roughly 10 years after the introduction of large-scale oil palm plantations through direct government intervention), some 100 settlement units were being prepared to become definite village administrative units in Riau. These settlements were previously part of the resettlement scheme linked to the agricultural development programme, and some of them were part of the oil palm development programme. In Kampar Regency, where large-scale oil palm plantation was first introduced, it was noted that there were 53 new settlements being prepared to become definite village administrative units. There were also 32 settlements (UPT; *Unit Permukiman Transmigrasi*) that still fall under the authority of the Transmigration Office. These settlements will also become definite village administrative unit in the near future.

Thus, the number of settlements in Riau also increased significantly. 'Settlements' in this context refers to *desa* (settlements in rural areas) and *kelurahan* (settlements in urban neighbourhoods), the *desa/kelurahan* being the lowest level of government administration in Indonesia. Using the PODES data from 1996, 2005 and 2010, the number of *desa* in ex-Kampar Regency (Rokan Hulu, Kampar and Pelalawan regencies) increased from 309 in 1996 to 492 in 2011. In the same period, the number of *kelurahan* increased from 8 to 24. The figures indicate that significant changes have taken place over a 15-year period. Even though other factors may also play a role, the multiplication and rapid growth of the *desa* and *kelurahan* cannot be explained without taking into account the rapid expansion of oil palm plantations and related activities.

Making an assessment of the socio-economic impact of oil palm adoption, we see that people living in the immediate surroundings of oil palm estates often have considerable benefits [49]. Village-level assessment showed that villages that adopt oil palm as a major source of income tend to perform well on indicators of physical, financial and human capital. At the household level, an assessment showed that more than 18% of the households had increased their income (in real terms) by 200–300 per cent after 5 years of practising oil palm cultivation. About 35% had increased their income between 400 and 1300 per cent after 5–10 years of engagement, and about 45% of the households that had practised oil palm cultivation for more than 10 years had increased their income by 2200 to over 25,000 per cent. Such positive income effects make oil palm expansion extremely difficult to stop.

In policy debates on and cost–benefit analyses of the effects of oil palm expansion, much attention is often paid to the direct effects, that is, the effects on potential incomes and/or environmental effects. However, little or no attention is paid to the indirect population effects, how to deal with the massive population inflow, and whether oil palm production will in the long run be able to sustain urban populations.

5. Concluding remarks

It is not difficult for policy makers to show that oil palms are an economically rentable crop with a huge potential for further economic growth. In addition to national demands, the growing worldwide interest in biofuels as an alternative to fossil fuels will increase demand for its feedstock and lead to the expansion of oil palm plantations in climatically suitable regions.

On the basis of a cost–benefit analysis of various crops, oil palm will probably continue to be seen as a highly profitable crop with interesting possibilities for being promoted as a source

http://dx.doi.org/10.5772/53586

of 'green' development. The Indonesian population has increasingly been attracted by this crop, as it provides them with opportunities to benefit and multiply their incomes, which will in itself provide capital for improving consumption and having a good life.

At the same time, however, Riau shows us that there are environmental costs (deforestation, invasion into peat land areas etc.) and that oil palm expansion is accompanied by rapid immigration and urbanization. Even though policy attempts are made to control land conversion or to stop deforestation, much of what is happening today cannot easily be regulated. It is not the direct effects (i.e. the expansion of plantations and production) but the indirect effects and multipliers that bring into question the long-term sustainability of the development model. The establishment of new settlements, rapid urbanization and continuing immigration will require additional employment opportunities. And along with the growing population and the conversion of rice lands into oil palm fields, food security issues will increasingly become a problem. Rather than making quick money from oil palm production, the Indonesian government should make efforts to better control the indirect effects and especially to investigate the problem how to make oil palm-based economies more sustainable and equitable in the longer term. Without interventions – and with today's laissez-faire approach – further oil palm expansion will soon lead to the depletion of natural resources and an increase in social tensions as a result of unemployment and food insecurity.

Acknowledgement

This article is based on the Agriculture beyond Food (AbF) programme, funded by Netherlands Organisation for Scientific Research (NWO) in collaboration with the Royal Dutch Academy of Sciences (KNAW).

Author details

Suseno Budidarsono[1,2], Ari Susanti[1,3] and Annelies Zoomers[1]

*Address all correspondence to: s.budidarsono@uu.nl

1 Utrecht University, Faculty of Geoscience, International Development Studies, The Netherlands

2 World Agroforestry Center (ICRAF), Southeast Asia Regional Programme, JL, CIFOR, Situ Gede, Sindang Barang, Bogor, Indonesia

3 Gadjah Mada University, Faculty of Forestry, Jl. Agro No. 1 Bulaksumur, Yogyakarta, Indonesia

References

[1] USDA (United States Department of Agriculture). 2012. Oil Seeds: World Market and Trade. Washington DC. Foreign Agricultural Services. Circular Series FOP 08–12, August 2012. Table 03.

[2] Corley, R. H. V., and Tinker, P.B.H. 2003, The Oil Palm, 4th ed. Blackwell Science. Ltd, p. 1.

[3] World Bank and IFC. 2011. The World Bank Group Framework and IFC Strategy for Engagement in the Palm Oil Sector. DRAFT FOR CONSULTATIONS

[4] Tan, K.T., Lee K.T., Mohamed, A.R., Bhatia, S., 2009. Palm oil: Addressing issues and towards sustainable development. Renewable and Sustainable Energy Reviews 13 (2009) 420–427. Elsevier.

[5] McCarthy, John F. 2010. Processes of inclusion and adverse incorporation: oil palm and agrarian change in Sumatra, Indonesia. Journal of Peasant Studies, 37: 4, 821–850. DOI: 10.1080/03066150.2010.512460 (accessed 5 January 2011).

[6] German, L., Schoneveld, G., Skutsch, M., Andriani, R.; Obidzinski, K., Pacheco, P., Komarudin, H., Andrianto, A., Lima, M.; Dayang Norwana, A.A.B. 2010. The local social and environmental impacts of biofuel feedstock expansion, A synthesis of case studies from Asia, Africa and Latin America. Infobrief No. 34. Center for International Forestry Research (CIFOR), Bogor, Indonesia.

[7] Barlow, C., Zen, Z. and Gondowarsito, R., 2003. The Indonesian Oil Palm Industry. Oil Palm Industry Economic Journal. Vol. 3. No 1. Malaysia Palm Oil Board.

[8] Tomich, T. P., van Noordwijk, M., Budidarsono, S., Gilison, A., Kusumanto, T., Mudiyarso, D., Stolle, F., and Fagi, A.M., 2001. Agricultural Intensification, Deforestation, and the Environment: Assessing Tradeoffs in Sumatra, Indonesia. In D.R. Lee and C. B. Barret (eds). Trade-offs and Synergies? Agricultural Intensification, Economic development, and Environment. Wallingford, UK: CAB International.

[9] Edwards, R., Mulligan, D. and Marelli, L. 2010. Indirect land use change from increased biofuels demand: Comparison of models and results for marginal biofuels production from different feedstock. Final Report. Luxemburg. European Commission Joint Research Centre – Institute for Energy

[10] Koh, L.P. and Ghazoul J. 2008. Biofuels, biodiversity, and people: Understanding the conflict and finding opportunities. Biological Conservation. 141. pp. 2450–2460.

[11] Germer, J. and Sauerborn, J., 2008. Estimation of the impact of oil palm plantation establishment on greenhouse gas balance. Environment, Development and Sustainability. Vol. 10, No. 6, pp. 697–716, DOI:10.1007/s10668-006-9080-1.

[12] Tilman, David, Robert Socolow, Jonathan A. Foley, Jason Hill, Eric Larson, Lee Lynd, Stephen Pacala, John Reilly, Tim Searchinger, Chris Somerville, Robert Williams.

2009. Beneficial Biofuels – the Food, Energy, and Environment Trilemma. SCIENCE. vol. 325. 2009. P. 270. AAAS.

[13] Pye, Oliver. 2009. Palm Oil as a Transnational Crisis in South-East Asia, Austrian Journal of South-East Asian Studies, 2 (2), Vienna, Austria. Society for South-East Asian Studies.

[14] Koh, L.P., 2007. Potential Habitat and Biodiversity Losses from Intensified Biodiesel Feedstock production. Conservation Biology. Vol. 21, No. 5. pp. 1373–1375.

[15] Fitzherbert, E.B., Matthew J. Struebig, A. Morel, F. Danielsen, Carsten A. Brvhl, Paul F. Donald and Ben Phalan. 2008. How will oil palm expansion affect biodiversity? Trends in Ecology and Evolution. Vol. 23, No. 10. 538–545.

[16] Buckland, H. 2005. The Oil for Ape Scandal: How Palm Oil Is Threatening Orang-Utan Survival, Friends of the Earth, the Ape Alliance, the Borneo Orangutan Survival Foundation, The Orangutan Foundation (UK) and the Sumatran Orangutan Society.

[17] Colchester, M., Jiwan, N., Andiko, Sirait, M., Firdaus, A. Y,, Surambo, A. and Pane, H. 2006. Promised Land: Palm Oil and Land Acquisition in Indonesia – Implications for Local Communities and Indigenous Peoples. Monograph. Forest People Programme (UK) and Sawit Watch (Indonesia).

[18] Marti, Serge. 2008. Losing Ground. The human rights impacts of oil palm plantation expansion in Indonesia. A Report. Friends of the Earth, LifeMosaic and Sawit Watch.

[19] Zen, Z., Barlow, C. and Gondowarsito, R., 2006. Oil Palm in Indonesian Socio-Economic Improvement – A Review of Options. Oil Palm Industry Economic Journal. Vol. 6. No 1. Malaysia Palm Oil Board.

[20] Government of Indonesia, 1981. Transmigration Programme: an Overview. Paper presented by the government of Indonesia at the IGGI meeting, Amsterdam, May 1981.

[21] Budidarsono, S., Khasanah, N., Ekadinata, A., Rahayu, S., Dewi, S., Suharto, R. and van Noordwijk, M. (eds) 2011. Reducing carbon emissions associated with oil palm plantations in Indonesia: accounting for greenhouse gas emissions over the full life cycle on peat and mineral soils and building capacity for and industry response to emerging environmental regulation in European markets. Research Report. Jakarta. Indonesian Palm Oil Commission and World Agroforestry Centre.

[22] Jelsma I. et al. (2009). Smallholder oil palm production system in Indonesia: lessons from the NESP Ophir project. Wageningen: Wageningen University.

[23] Michael Bunce. 1982. Rural Settlement in an Urban World. London. Croom Helm Ltd.

[24] Syahza, Almasdi (2004). Development Impact of Palm Oil Plantation on Rural Economic Multiplier Effect in Riau Province). Jurnal Ekonomi, Th.X/03/November/2005, Jakarta. Fakultas Ekonomi Universitas Tarumanagara, Jakarta.

[25] Poku, Kwasi. 2002. Small-scale Palm Oil Processing in Africa. FAO Agricultural Services Bulletin No 148. Rome. Food and Agricultural Organization of United Nations.

[26] Sanders, J. 1982. Palm Oil production on the Gold Coast in Aftermath of Slave Trade: Case Study of the Dante. The International Journal of African Historical Studies, Vol. 15 No. 1. pp. 49–63.

[27] Het Oostkust van Sumatra Instituut, 1938. Data Deli 1883–1938. Mededeling No. 26.

[28] Deasy, George F. 1948. Localization of Sumatra's Oil Palm Industry. Economic Geography, Vol. 18, No. 2 (Apr., 1942), pp. 153–158. Massachusetts. Clark University.

[29] Stoler, Ann L., 1985. Capitalism and Confrontation in Sumatra's Plantation Belt, 1870–1979. London. Yale University Press.

[30] Breman, Jan, 1989. Taming the Coolie Beast – Plantation Society and the Colonial Order in Southeast Asia. New Delhi. Oxford University Press.

[31] Gilbert, A. and Gugler, J. 1992. Poverty and Development: Urbanization and the Third World. 2nd edition. London, Oxford University Press, pp. 50–52.

[32] Higgins, B. 1956. Indonesia's Development Plan and Problems. Pacific Affairs, Vol. 29. No. 2 (June 1956) pp. 107–125. University of British Columbia.

[33] Graham, E. and I. Floering, 1984. The modern plantation in the Third world. Great Britain. Croom Helm Ltd.

[34] Pamin, K., 1998. A hundred and fifty years of Oil Palm development in Indonesia: from Bogor Botanical Garden to the Industry. In Proceedings of 1998 International Oil Palm Conference: Commodity of the past, today, and the future (eds A. Jatmika et al.), pp. 3–23, Indonesian Oil Palm Research Institute, Medan, Indonesia.

[35] Bangun, Derom, 2006, Indonesia Palm Oil Industry, Conference Paper. The National Institute of Oilseed Products Annual Convention, 21–25 March 2006, Sheraton Wild Horse Pass, Phoenix, Arizona, USA. Available at http://www.oilseed.org/pdf/am_2006_materials/Bangun_Text.pdf (accessed 12 July 2012)

[36] Hoshour, C. (1997). Resettlement and the politization of ethnicity in Indonesia. Bijdragen tot de Taal, Land, en Volkenkunde, Riau in Transition 153(4).

[37] Humphrey, J. and Schmitz, H., 2000. Governance and upgrading: linking industrial cluster and global value chain research. Institute of Development Studies, Sussex, IDS Working Paper 120.

[38] Casson, A., 2000. The hesitant boom: Indonesia's oil palm sub-sector in an era of economic crisis and political change. Bogor. CIFOR. P. 8.

[39] FAO. FAOSTAT – Agriculture. Interactive online statistics http://www.fao.org/crop/en/

[40] BAPPENAS (National Planning Body), 1975. Rencana Pembangunan Lima Tahun II (2nd Five Year Development Plan). Jakarta. BAPPENAS. Book D, p. 69.

[41] Statistics Indonesia. Petroleum and Natural Gas Production, 1996–2010. Jakarta. Badan Pusat Statistik Republik Indonesia.

[42] BPS dan BAPPEDA Provinsi Riau. 2008. Riau Dalam Angka 2007. Pekanbaru. BPS Provinsi Riau. Table 7.2.1.

[43] Dinas Perkebunan Provinsi Riau. 2011. Statistik Perkebunan Provinsi Riau 2010. Pekanbaru. Dinas Perkebunan.

[44] Susanti, A. and Burgers, P. 2012. Oil palm expansion in Riau Province, Indonesia: serving people, planet, profit? Background paper for European Report on Development. Confronting scarcity: managing water, energy and land for inclusive and sustainable growth. Belgium: European Union

[45] BPS dan BAPPEDA Provinsi Riau. 2011. Riau Dalam Angka 2010. Pekanbaru. Pekanbaru. BPS Provinsi Riau.

[46] Derkzen, M.L. 2011. Conserve or convert? Forest as the fuel of oil palm – an assessment of rural livelihood and their strategies to cope in an oil palm environment, Riau, Indonesia. Field Work Report. International Development Studies, Utrecht University.

[47] Heijman, S. 2010. Is the forest reduced to just an economic resources in an era of rapid oil palm expansion? Field Work Report. International Development Study, Utrecht University.

[48] Budidarsono, S., Rahmanulloh, A., Sofiyuddin, M. 2012. Economics Assessment of Palm Oil Production. Technical Brief No. 26: Oil Palm Series. Bogor, Indonesia: World Agroforestry Centre (ICRAF) Southeast Asia Regional Program. 6 pp.

[49] Budidarsono, S., Dewi, S., Sofiyuddin, M., Rahmanulloh, A. 2012. Socioeconomic Impact Assessment of Palm Oil Production. Technical Brief No. 27: Oil Palm Series. Bogor, Indonesia. World Agroforestry Centre (ICRAF) Southeast Asia Regional Program.

Section 2

Biofuels

Chapter 7

The Need for Integrated Life Cycle Sustainability Analysis of Biofuel Supply Chains

Anthony Halog and Nana Awuah Bortsie-Aryee

Additional information is available at the end of the chapter

http://dx.doi.org/10.5772/52700

1. Introduction

Climate change has been widely investigated by the scientific community, and its potential impacts are expected to affect the world's economy, ecosystem services, and societal structures within a few decades. To reduce the undesirable consequences of climate change, adaptation and mitigation technologies and policies have to be implemented. Analyzing technological advances to address sustainable development, through integrated systems level methods and approaches, is needed to predict future vulnerability to climate change and continued ecosystem deterioration [1-4]. The incorporation of sustainability notion into all sub-systems of our global society has come into full swing and must be continued to be pursued by any entity, both in private as well as public sectors. However, these challenges require integrative and transdisciplinary computational tools and methods to aid in embedding sustainability goals into corporate and government policy decision making processes [5-8]. United States Environmental Protection Agency (USEPA) has been reshaping its strategies and programs in conjunction with incorporating the triple dimensions of sustainability [9].

In the U.S. bioenergy development, recent innovations in biotechnology, genomics and complexity science have contributed to the renewed interest in converting (ligno) cellulosic biomass to valuable fuels and other bioproducts [10]. The U.S. Department of Agriculture (USDA) and Department of Energy (USDOE) are actively supporting projects to make biofuels and bioproducts economically, socially and environmentally sustainable and viable. U.S. Bioenergy Research Centers work on accelerating genomics-based systems biology research to achieve the transformational breakthroughs in basic science needed for the development of cost-effective technologies to make production of next-generation biofuels from lignocellulose, or plant fiber, commercially viable on a national scale [11]. Figure 1 shows the many potential production pathways to biomass hydrocarbons. Features of these alternative pathways include diversity in feedstocks, fuel composition, and byproducts. Integrated decision-making tools are urgently needed to support choices among these alternatives. Developing these tools effectively requires a life-cycle and dynamic perspective. Life Cycle Assessment (LCA) follows internationally accepted methods (ISO 14040 and ISO 14044) and practices to evaluate requirements and impacts of technologies, processes, and products so as to determine their propensity to consume resources and generate pollution.

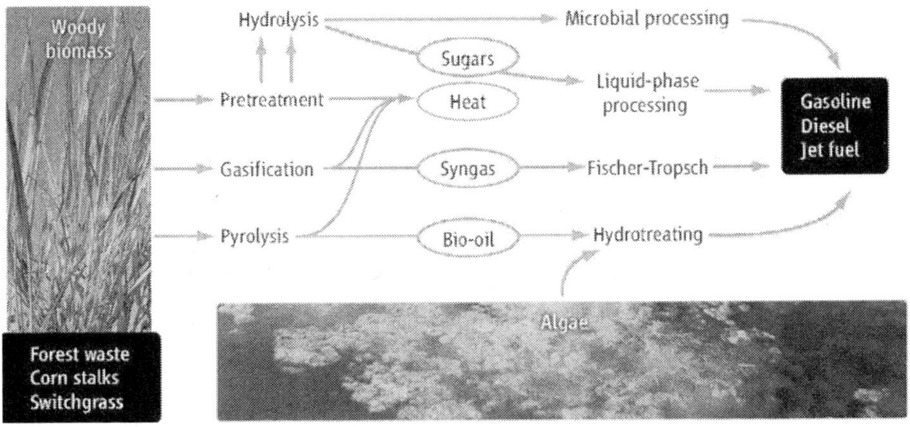

Figure 1. Pathways to biomass hydrocarbons. Reproduced with permission from [10]; published by Science, 2009

Mckone et al. [1] have identified 7 grand challenges that we must address to enable life cycle assessment (LCA) to effectively evaluate the environmental "footprint" of biofuel alternatives and to support the evolving bioeconomy. According to their work, the grand challenges for applying life cycle assessment to biofuels are:

- Understanding feedstock growers, options, and land use;
- Predicting biofuel production technologies and practices;
- Characterizing tailpipe emissions and their health consequences;
- Incorporating spatial heterogeneity in inventories and assessments;
- Temporal accounting in impact assessments;
- Assessing transitions and end states; and
- Dealing with uncertainty and variability.

The above challenges have already been addressed and analyzed in disaggregated and piecemeal fashion in several papers in various disciplines. *What we really need is to integrate the disparate methodologies systematically and computationally to obtain comprehensive and robust*

http://dx.doi.org/10.5772/52700

results (i.e. indicators and metrics) which can support decisions and policies in public and private sectors. Thus, our one grand challenge is to synthesize our current results and infer relevant and critical information either to support or not bioeconomy development. Integrated-oriented decision-making frameworks and tools with the support of information and communication technologies (ICT) and cyberinfrastructure are needed to support choices among the competing production pathways and provide information to various stakeholders. For instance, systems modeling (such as the use of DOE's GREET model) is relatively inexpensive to perform prior to physically implementing any experiments in laboratory or pilot scale. Towards developing a decision support system (DSS) tool that accounts for multi-stakeholders' interests in analyzing system's sustainability, this paper aims to highlight and describe the available tools, models and frameworks which can be used to address the 7 challenges identified and explore the possibility of their complementary attributes to result to an integrated system framework for assessing the sustainability implications of further investments in biofuels, bioenergy and bioproducts.

Thus, an in-depth analysis is needed to critically understand the re-emergence of biofuels in conjunction with sustainable development. Most of the research projects on cellulosic biofuels have not given due attention to social acceptability and economic feasibility, besides not considering the competing interests of various stakeholders [2- 3]. The interactions between environmental, social and economic impacts of biorefinery development must be analyzed in a comprehensive and integrated manner to ensure the sustainable development of a bio-industry (such as the increasing interest in "drop-in" biofuels). Concerns related to biomass harvest and its impact on soil erosion, nutrient losses, biodiversity losses, land use changes, water consumption, eutrophication and environmental impacts of auxiliaries inputs must be weighted with the benefits of cellulosic biofuels [2, 12]. Several studies (e.g. [11, 13-15]) have been conducted to identify biomass potential, assess technological efficiency and understand the environmental implications of biofuels. However, a complete and comprehensive evaluation of biofuels supply chain from a holistic and systems perspective has been lacking. Several authors [2-3; 6 8] have argued the need for further research for large scale deployment of second-generation and third generation biomass crops including their effects on land use, biodiversity and hydrology. Technological researches have focused on using cheap and easily available feedstock (e.g. woody biomass, harvest residues, agricultural residues) to advance lignocelluloses feedstock bio-refinery [16 - 17], whereas environmental concerns of biofuels have focused on carbon dioxide emissions only [18]. Though integrated forest products bio-refinery systems may result to additional revenues by producing co-products like biofuels and other biomass based chemicals in addition to the main products [16, 19], uncertainties prevail regarding the capital investments required as well as the social impacts for large-scale production [2-3].

With the above challenges in implementing bioenergy policies while carefully considering stakeholders' interests, dynamic integrated system methodologies are urgently needed to analyze the sustainability of biofuels supply chain and to better understand the overall impacts of introducing (ligno) cellulosic ethanol [5-7; 11] or even to support the development of "drop-in" biofuels.

In the subsequent sections of this paper, we elaborate on how we might support the conduct of integrated sustainability analysis and modeling of biofuel supply chains.

2. Status

2.1. Understanding stakeholders and resource requirements

The future feedstocks for biofuels may come from farms, rangelands, or forests. Because of transportation costs, harvested feedstocks are likely to be stored and processed at small- to intermediate-scale (distributed) facilities. Unlike oil companies and government agencies that have a hierarchical structure for decision-making, the first stages of biomass production might involve hundreds to thousands of decision-makers (stakeholders). Using life cycle perspective approach to influence policies that would alter the behaviors of these distributed decision-makers poses different challenges than when the decision-making authority is more highly concentrated. One expects that feedstock growers utilize land to maximize profits.

Having a large number of potential feedstocks with different characteristics in a system of distributed decision-making presents substantial challenges for current LCA approaches because of the vast scope of information needed to address so many alternatives. Multi-criteria decision analysis (MCDA), such as the Analytic Hierarchy Process (AHP), can aid in determining the most critical criteria, variables and indicators to stakeholders, which can represent their conflicting interests with respect to economic, environmental, technological and social dimensions of systems sustainability [4-5, 7-8]. The critical criteria and indicators can be ranked and identified by AHP's eigenvalues, which are calculated from stakeholders' inputs [20].

2.1.1. Multi-criteria Decision Analysis (MCDA) for Stakeholders' analysis

MCDA is a decision support system that is suitable for addressing complex problems featuring high uncertainty, conflicting objectives, different forms of data and information, multiple interests and perspectives, and including complex evolving bio-physical and socio-economic problem [21-26]. MCDA approaches have long been widely applied to economic, social, and industrial systems. An MCDA in general involves m alternatives (e.g. bioenergy systems) evaluated on n criteria (i.e. sustainability criteria), in which each of j-*th* criteria C of i-*th* alternative A has performance of xij. Each criterion is weighted, and wj is the weight of criterion j. The grouped (i.e., stakeholders) decision matrix X can be expressed as shown on Figure 2.

Wang et al. [27] present a review of MCDA methods to aid in sustainable energy decision making. In their review, the corresponding methods in different stages of multi-criteria decision-making (i.e., criteria selection, weighting, evaluation and final aggregation) are discussed. The criteria are classified into four major aspects: (i) technical (e.g., efficiency, primary energy ratio, etc.), (ii) economic (e.g., investment cost, net present value, etc.), (iii)

http://dx.doi.org/10.5772/52700

environmental (e.g., CO2 emission, NOx emission), and (iv) social (e.g., social acceptability, job creation, etc.). The weighting methods of the criteria are classified into three categories: (i) subjective weighting (e.g., pair-wise comparison, analytical hierarchy process (AHP), etc.), (ii) objective weighting (entropy method, technique for order preference by similarity to ideal solution, etc.), and (iii) combination method.

Criteria			c1	c2	⋯	cn
Weights			w1	w2	⋯	wn
Alternatives	A1	X =	x11	x12	⋯	x1n
	A2		x21	x22	⋯	x2n
	⋮		⋮	⋮	⋱	⋮
	An		xm1	xm2	⋯	xmn

Figure 2. Grouped decision matrix of MCDA

Different MCDA approaches have been applied to support different decisions including environmental and sustainable energy decision making [28-30]. We can use any of MCDA methods to aid stakeholders' analysis to find out the "most critical" criteria, indicators and metrics that represent stakeholders' interests. A combination of AHP and LCA has been used for evaluating environmental performance of pulp and paper manufacturing [31]. Halog [20] and Halog et al [29] proposed the use of analytic hierarchy process (AHP), one of MCDA methods [32-34] in stakeholders' analysis for identifying the critical criteria, indicators and metrics which represent multi-stakeholder's interests. This will provide ranking of different criteria, indicators and metrics which are important holistically. MCDA opens great applicability to support sustainability assessment of existing and emerging multi-attribute systems. For instance, AHP allows stakeholders to weigh different criteria, indicators, and metrics by calculating Eigen values. In biofuels system, where energy efficiency, investment cost, GHG emissions, land use change and social impacts are the most common criteria, MCDA is certainly applicable. Through MCDA, we can focus first on the critical ones that account stakeholder' inputs.

However, the existing life cycle thinking and MCDA methods are considered steady-state methods whereby they provide snapshots of hotspots based on historical data. They do not provide projections or trends in the future. They do not take into account the interactions of different metrics, outputs and parameters over time. To make the results more useful for decision and policy makers, we need to model the dynamic interrelationships of these variables over time. Additionally, we can explore the use of Geographic Information Systems (GIS) to assist spatial analysis when needed.

2.2. Biofuel production technologies and practices

Much of the variability among LCA results for biofuels arises from lack of knowledge about how biomass production operations and fuel production from biomass will evolve. Many alternatives exist both for production processes and for final products. An important challenge is to understand the energy, biomass, pollutant, and product mass balances of production facilities: To what extent will they be self-sufficient or even net producers of electricity? Will the facilities deliver a single product (fuel) or have multiple product streams (fuel, food, electricity, chemical commodities)? What are their waste products, air emissions, and water demands? A related challenge is accurately predicting scales of future biofuel production. For biofuels, the feedstocks are more dispersed and less dense than petroleum, which will induce biorefineries to be smaller than petroleum refineries. Ultimately, the scale of biorefining will depend on feedstock and production process choices, technological efficiency in converting feedstock to fuel, productivity of local land for feedstock production, and costs associated with feedstock production and transport, and biorefinery construction and operation. Much is unknown about this system at large scale and will remain uncertain until the system is created. Larger biorefineries may economize on refining-related impacts, but will increase transport-related impacts. Biorefinery scale has important ramifications for life-cycle impacts including the nature and the location of impacts. We can use the scenario analysis capability of the methods of agent based modeling (ABM) and system dynamics (SD) to predict potential performance of various biofuel production technologies and practices.

2.2.1. Agent Based Modeling (ABM)

Agent based model is a computer representation of the considered system (e.g. biofuels supply chain) that is comprised of multiple, interacting actors (i.e., stakeholders) [35]. ABM systems possess two distinct properties: (1) the system is composed of interacting agents; and (2) the system exhibits emergent properties, that is, properties arising from the interaction of the agents/stakeholders that cannot be deduced simply by aggregating the properties of the agents [36- 37]. ABM can be used to model the interactions of agents or sub-systems in biofuels supply chain using the metrics, variables and indicators as performance measures. Figure 3 provides schematic illustration of an agent-based system: each of the four circles represents a sub-system of agents (e.g., companies/entities) denoted by small dots and the whole arrows show how agents and sub-system of agents are interacting with each other. Interacting agents and sub-systems, though driven by only a small set of rules which govern their behavior, account for complex system behavior whose emergent dynamic properties cannot be explained by analyzing its component parts [35].

ABM aims to look at global consequences of individual or local interactions in a given geographical (e.g. regional) area. Parker et al [38] have categorized existing literatures on agent-based land use models into five categories—(i) policy analysis and planning; (ii) participatory modeling; (iii) explaining spatial patterns of land use or settlement; (iv) testing social science concepts; and (v) explaining land use functions. Fox et al [39] argue that optimization of supply chain performance is only possible when the impacts of decisions made by one agent onto another agents are understood. A systems model that captures all important interactions among different units of a supply chain would contribute to effective decision making. Julka et al [40] use petroleum refinery integrated supply chain modeler and simulator to mimic a crude refinery's supply chain to develop procurement strategies. Thus,

http://dx.doi.org/10.5772/52700

ABM can be used to look at the global consequences in developing biofuels supply chain/ network considering the individual interactions of stakeholders across all life cycle stages.

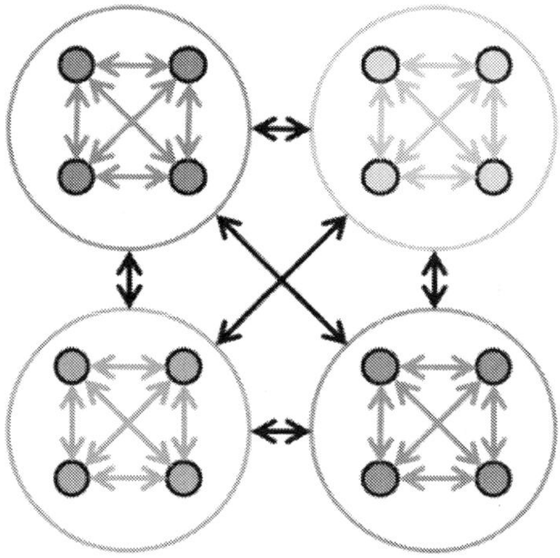

Figure 3. Illustration of an agent-based system

2.2.2. System Dynamics (SD)

SD is a well-established systems perspective/complexity science method which is originally developed by Jay Forrester at MIT [41- 42]. This has been applied in different corporate, industrial and government decisions worldwide which have the intention of modeling and understanding the interrelationships (i.e., feedbacks) of variables, indicators and metrics over time. This has been useful in modeling the interrelationships between or among subsystems which are linked by variables and aids to see how their interlinkages will produce specific overall system behavior. Before using appropriate modeling software package, it is important to draw causal loop diagrams. A causal loop diagram is a visual representation of the feedback loops in a system whereby the stocks and flows (i.e., involving different variables, parameters, indicators) are connected by either positive and negative loops. A stock (e.g., biomass, GHG, revenue, unemployment) is the term for any entity in the system that accumulates or depletes over time. A flow is the rate of change in a stock. A flow changes the rate of accumulation of the stock. The real power of system dynamics is utilized through simulation and in showing the inter-linkages between micro-, meso-, and macro-systems. SD involves computational modeling for framing, understanding, and discussing complex issues and problems [43-45]. It is recognized that the structure of any emerging system—the many circular, interlocking, sometimes time-delayed relationships among its components—is often just as important in determining its behavior as the individual components themselves. There are often properties of the whole which cannot be found among the properties-of-the-elements. The feedback loops as well as the use of stocks and flows can represent and model the critical sustainability variables, indicators, and metrics to describe how seemingly simple sub-systems display baffling nonlinearity for the whole system [46]. The modeling can be developed by sub-dividing the whole system into sub-models but we need to remember that they are interconnected by variable, parameter or a metric [4, 47]. Through SD we can create a prototype dynamic system model for the system being considered.

In SD, a system is modeled mathematically in a nonlinear, first-order differential (or integral) equation such as:

$$\frac{d}{dt}x(t) = f(x, p) \tag{1}$$

where x is a vector of levels (stocks or state variables), p is a set of parameters, and f is a nonlinear vector-valued function. Simulation of such systems is accomplished by partitioning simulated time into discrete intervals of length dt and stepping the system through time one dt at a time.

SD typically goes further and utilizes simulation to study the behavior of systems and the impacts of alternative policies [44-46]. Running *"What If"* simulations or scenarios to test certain energy and environmental policies on a prototype system model can greatly aid in understanding how an emerging system potentially evolves over time. Similar to MCDA and ABM methods, SD has been applied in a wide range of areas, for example population, ecological and economic systems, which usually interact with each other. SD has been used in the sustainability assessment of technologies in the Canadian Oil Sands Industry [4, 47] as well as in bioethanol production in Canada [48]. SD models have recently been developed in some biofuels studies. Riley et al [49] use SD model to describe the U.S. DOE biomass program. Bush et al [50] and Sheehan [51] explore the potential market penetration scenarios for first generation bio-fuels in the United States. Scheffran and BenDor [52] investigate interaction between economic conditions and land competition between different crops. Franco et al [53] use SD to understand the difficulties in fulfilling government requirements for biofuels blending and to evaluate the effect of different government policies in the production of ethanol and biodiesel.

Once a system model is validated, the procedural steps can be done iteratively depending on the scenario defined. This will also provide information about the trends of important variables over time which will give us insights and guidance on what decisions and policies to take. Eventually, this modeling procedure can support the selection and implementation of sustainable biofuel systems.

2.3. Characterizing emissions and their health consequences

LCAs of transportation fuel systems report that the fuel combustion stage makes the largest contributions to pollutant emissions and associated disease burdens. Credible and reliable impact estimates for biofuel combustion are needed. Another aspect of LCA's tailpipe challenge is the need for accurate emission factors for future fleets that cover a range of fuel alternatives and vehicle technologies. Enormous technological progress has been made in

http://dx.doi.org/10.5772/52700

controlling motor vehicle emissions, and there is strong momentum for continuing progress. In what ways and to what extent will shifts from petroleum-based fuels to biofuels affect the combustion-phase emissions of air pollutants? One historical approach to answer similar questions has been to conduct laboratory-based emissions testing. However, this approach is relatively expensive and lacks reliability for characterizing fleet-wide emissions from real drivers on real roads, and so is unlikely to provide accurate information in a timely manner. An alternative approach, used in LCA tools such as GREET [27] which assumes that vehicle emissions meet federal and state emissions standards regardless of the fuel used, and that emissions targets are the best estimate of what will happen in future years.

Since the concept of sustainability insinuates temporal and spatial connotations, it is important that variables, indicators, metrics and parameters representing stakeholders' interests are modeled over time within a geographical location, which can be done using dynamic system modeling.

2.4. Incorporating spatial heterogeneity in integrated assessment

A key challenge for integrated sustainability assessment of biofuels is to rationally select appropriate spatial scales for different impact categories without adding unnecessary complexity and data management challenges. Though methods such as LCA can address net changes across large geographical areas, it must also address how the impacts will be experienced at local or regional scales. Accurate assessments must not only capture spatial variation at appropriate scales (from global to farm-level), but also provide a process to aggregate spatial variability into impact metrics that can be applied at all geographical scales. We can use geographic information system (GIS) to support regionalized LCA. GIS has a powerful analytical ability to assess data spatially. It stores different layers of information. GIS is a tool that links location and attribute information to enable a person to visualize patterns, relationships, and trends for different parameters of a system. Previous literature and organizational reports have suggested using GIS as a tool to link the aspatial data with spatial data [54-56]. This offers at least two important contributions to the modeling of land use changes impacts on biodiversity and ecosystem goods and services caused by biofuels production. It permits analysis of spatially explicit datasets of land use and land cover, which can be used in the assessment of the areas affected by increased land occupation. It also offers an understanding of land use dynamics. The insight can aid in the predictions of land use changes caused by increased demand for biofuels [57].

2.5. Temporal accounting in integrated sustainability assessment

Similar to spatial resolution when conducting integrated sustainability analysis, selecting appropriate time scales poses challenges for biofuels system modeling efforts. Time allocations are important for comparing impacts, yet the time distribution of impacts is rarely made clear in studies such as LCAs. Many factors in systems analysis vary significantly in time. Therefore, time-based assumptions must be clearly noted and evaluated. Among the "moving targets" are population distributions, technology options, regulatory requirements, and the degree of biofuel penetration in the overall energy mix. Moreover, the inputs one uses in integrated sustainability assessments to characterize biomass and fuel production technologies as well as transportation infrastructure must capture how these systems are evolving. Different natural time scales associated with different impacts pose challenges for effective comparisons among climate-change, human health, and water use consequences. The impacts of GHG emissions are distributed over decades using integrated assessment models and are commonly discounted. Assumptions about the rate of discounting influence judgments about the relative importance of current year versus future year emissions. Similarly, impacts on water resources and soil can play out over decades. We can use dynamic system modeling and simulation.

2.6. Assessing transitions and end states in sustainability analysis

Both advocates and critics of biofuels often focus on a restricted set of scenarios that appear to reinforce their *a priori* beliefs about how biofuel production and use might function [1]. Even accomplished practitioners of LCA tend to focus attention on system end-states (related to backcasting), i.e., what biofuel production and use will be likely 20 or 30 years from now, when a proposed combination of fuels and vehicles has matured and is thoroughly deployed. This perspective ignores potentially important effects that accrue during the transition phase; the impacts from building new infrastructure, new vehicles, and integrating a new fuel into a mature and, in many respects, inelastic transportation system. This can be addressed using dynamic system modeling and simulation or agent based modeling as discussed above.

To account for transitions, LCA requires much collaboration between economists and systems engineers to address what happens during the transition phase when large-scale changes occur in many components of a complex, market driven, technological system. For example, one of many key issues is whether fuel changes will affect the performance and lifetime of vehicles or the infrastructure transporting that fuel in ways that significantly increase climate forcing, water, health, and other externalities during transition. Technology investments are needed, and these activities could cause GHG emissions to rise in the near-term as part of a longer-term effort to attain a more carbon-efficient end state. In addressing transitions, there should be recognition that emerging technologies could profoundly change the assumptions that underlie biofuel LCAs. This issue makes clear the need to support life cycle thinking methods for building scenarios from which one should learn, rather than as a tool designed to make firm predictions.

2.6.1. Scenario development and analysis for policy planning and making

Forecasting, foresighting and backcasting are approaches used for policy planning and making. These scenario development approaches have their benefits and shortfalls. Backcasting involves working backwards from a particular desired future end-point or set of goals (i.e., sustainable society) to the present state, in order to determine the physical feasibility of that future and the policy measures that would be required to reach the state [58-60]. This helps in analyzing alternative futures response to present situation and deals with problems in a different way rather than extrapolating present scenario into the future (forecasting) [61-64].

http://dx.doi.org/10.5772/52700

Backcasting makes it clear that addressing sustainability concerns requires a paradigm shift from business-as-usual attitudes. On the other hand, industries and organizations use forecasting technique as a data analysis methodology to develop future scenarios from existing information. Forecasting enables decision makers to identify reasonable estimates of various current activities. Using forecasting approach, managers and decision makers can tweak and calibrate their operations at the appropriate time in order to maximize benefits. Forecasting assists in preventing losses by taking in all relevant information and making proper judgment decisions [63]. Moreover, foresighting can be distinguished from forecasting. Forecasting is the passive attempt to diagnose or predict future events. Foresighting aims to actively change or create the future by linking it to the present. Thus, the major difference between foresighting and forecasting is that in forecasting the conclusions for today are missing. There are four major applications of foresighting: (1) assessing possible consequences of actions, (2) anticipating problems before they occur, (3) considering the present implications of possible future events; and (4) envisioning desired aspects of future societies. Foresighting which is a tool for 'decision-shaping' rather than 'decision-making' offers many benefits including: engaging policy-makers and experts in actively planning for the future; identifying potential problems early; verifying expectations and examining trends; bringing people together to create a suitable future; strengthening existing networks,; and educating the public on urgent future-related issues. It could have a positive impact on sustainable technology policy by providing a means for analyzing its broader social and economic implications.

By considering different scenarios (starting from business-as-usual to different plausible scenarios in the future (either through foresighting or backcasting), we can generate different results for our system performance measures and identify a few critical alternative systems or scenarios to be strongly considered for developing sustainable decision, policy, technology, systems, intervention, etc. Again, this is with the assumption that we can gather good quality data. We can also perform sensitivity and uncertainty analyses to improve the robustness of integrated sustainability analysis results.

2.6.2. Uncertainty and variability analysis for robust sustainability assessment

Addressing uncertainty is a major hurdle, not only for biofuels LCA, but for other integrated system modeling efforts as well. Many sources of uncertainty and variability, both inherent and epistemic, are encountered in climate-change, human-health, environmental, and economic impact assessments. Some of the uncertainties and variabilities cannot be reduced with current knowledge (i.e., through improvements in data collection or model formulation) because of their spatial and temporal scale and complexity. Effective policies are possible, but such policies must explicitly take into account uncertainty. Among those commenting on how to formally address uncertainty in impact assessments, it has been established that there are "levels" of sophistication in addressing uncertainty. In its recommendations for addressing uncertainty in risk assessment, the International Program on Chemical Safety proposed four levels, ranging from the use of default assumptions to sophisticated probabilistic assessment [1]:

- Level 0: Default assumptions; single value of result;
- Level 1: Qualitative but systematic identification and characterization of uncertainties;
- Level 2: Quantitative evaluation of uncertainty making use of bounding values, interval analysis, and sensitivity analysis; and
- Level 3: Probabilistic assessments with single or multiple outcome distributions reflecting uncertainty and variability.

Furthermore, Baumgartner [65] argued that assessing environmental and social impacts is associated with uncertainties caused by applied assessment tools, definition of assessment objectives, system boundaries of assessment and data quality. There are various ways to deal with data and model uncertainties when conducting system modeling and simulation [6, 66]. Uncertainties and variabilities come from a large number of variables and parameters considered, assumptions made, and the spatial and temporal variability in parameters or sources [67-69]. The latest LCA, MCDA, and system dynamics software packages include statistical tools to support uncertainty and sensitivity analyses. Uncertainty analysis aids to show if the model's general pattern of behavior is strongly influenced by changes in critical parameters. For system dynamics, the usual method is to perform a sensitivity analysis of the model whereby a collection of simulated experiments is performed [70]. This is done by choosing parameters, metrics and indicators that are judged to be sensitive, changing their values and then re-running the simulation model. If there is a drastic response in the results, this can show a lack of robustness in the system model. For ABM, the goal is to check whether the model addresses the right problem and provides accurate information about the system being modeled [71]. Additionally, Miller [72] proposes to use computer-based Active Nonlinear Tests (ANTs) that are capable of performing multivariate sensitivity analysis, model breaking and validation, extreme cases, and policy discovery. ANTs search across sets of parameter values and are capable of detecting important non-linear relationships among the parameters—relationships that typically go unnoticed using standard techniques [73]. Besides the probabilistic approach in handling uncertainty, fuzzy mathematics, Monte Carlo simulation, etc. can be used to address imprecision [30]. As we confront uncertainty and variability, we need to separate the "doable" and "knowable" from assumptions that are conditional components of the integrated and life cycle sustainability assessments. An informative system perspective assessment should sort out the data gaps that can be addressed with modest effort from those that would require a major undertaking. All integrated assessment efforts require tools, such as sensitivity analysis, variance propagation methods, and decision/event trees for tracking the impact of data quality and model uncertainty through all components of an assessment. A strong challenge for integrated assessments in addressing uncertainty is to provide and track metrics of data quality with respect to how data were acquired (measurements, assumptions, expert judgment, etc.), to what extent the data have been validated or corroborated, and how well the data capture technological, spatial, and temporal variations. This can be best facilitated by capitalizing cyberinfrastructure such as web-based data mining techniques.

http://dx.doi.org/10.5772/52700

3. Conclusions and discussion

Confronting the seven grand challenges, as argued by McKone et al [1], means recognizing some issues that have not been well articulated among practitioners of integrated sustainability assessment. In particular, a good balance must be attained between the needs of technology momentum and adaptive decision making. Most importantly, we must recognize that integrated assessment is an adaptive and ongoing process and not just a product. Technology momentum refers to the difficulties encountered in backing away from fixed costs (financial, institutional, and environmental) that have been sunk into one alternative pathway. Adaptive decision-making refers to learning by doing, recognizing that commodity costs and impacts can diminish as a system scales up. For biofuels we need technology momentum, but we must simultaneously maintain options for adaptive decision-making. These can be handled by the system-based approaches proposed in this paper which are being currently applied to wood-derived biofuels.

Although life cycle thinking methods can provide insights on options with the lowest impacts, results are often burdened by uncertainty such that they become more informative as technologies are deployed, making it difficult to apply life cycle thinking methods during the early phases of a major technology shift. One example is the important decision that must be confronted in a transition to (ligno) cellulosic biofuels; what end-product should be targeted among choices such as alcohols, alkanes, or a specific chemical compound such as dimethyl furan? This type of decision hinges on issues of timing, technical feasibility, and competitive advantage.

More collaboration and dialogue between basic scientists and sustainability practitioners is important for incorporating system thinking concepts into early phases of technology evaluation. Overall, approaches are needed to create more cross-talking among all members of the biofuel enterprise, which could be implemented at web-based level. Ideally, efforts toward developing the science and technology of biofuels will be continuously informed by those who are expert in integrated sustainability assessment. In this way, the biofuels community has the best opportunity to attain the overarching sustainability goals they seek. In integrated assessment of biofuels, one must recognize that no large-scale industrial product can be developed in isolation. Natural resource systems such as food, energy, water, and land are all intimately interconnected and thus integrated dynamic system modeling and analysis is absolutely needed.

To confront the uncertainty associated with sustainability analysis of biofuels combined with the irreducibility of many uncertainties, planners and policy makers must consider their role in managing uncertainty as well as managing impacts. Managing uncertainty requires addressing different aspects of the overall decision making process in the context of uncertainty. For example, decisions must be made to allocate resources among (i) investments to collect, store, and manage information; (ii) investments to improve the knowledge base (i.e., to generate new knowledge); (iii) formalization of the processes used to collect, use, and process information; (iv) formalization of processes to evaluate and communicate uncertainty; and (v) adjustment of the risk assessment process to mitigate the practical impact of the uncertainty on the analysis process. These tasks can be facilitated with the support of ICT and cyberinfrastructure.

Barriers to integrated sustainability assessments will be similar to environmental LCA (such as many stakeholders want a final answer), to be "cleared for takeoff, with no call-backs. These stakeholders view these assessments as a final exam; pass it and you are done being concerned with impacts and can proceed to technology deployment. This conceptualization serves to highlight the flaws in our thinking in sustainable technology development and commercialization. As McKone et al [1] pointed out this shortsighted perspective will fail to take advantage the true power of integrated and life cycle sustainability assessments. At its best, integrated sustainability assessment contributes to an ongoing process that organizes both information and the process of prioritizing information needs. These are opportunities for sustainability practitioners to focus attention and effort on making integrated, system perspective assessments more useful to decision makers. Integrated and life cycle sustainability analysis methods can co-evolve with a technology and provide the basis for adaptive planning. Decision makers who work in real time and often cannot wait for precise results must recognize that integrated sustainability assessment can provide valuable insights but it is not necessarily a "truth generating machine". Effective integrated and life cycle sustainability assessments can guide and inform decisions, but they cannot replace the wisdom, balance, and responsibility exhibited by effective decision-makers.

Acknowledgement

This work was carried out as part of the seed grant supported by the United States Department of Agriculture (USDA) Sustainable Bioenergy Program.

Author details

Anthony Halog[1,2*] and Nana Awuah Bortsie-Aryee[1]

*Address all correspondence to: anthony.halog@gmail.com

1 School of Forest Resources, University of Maine, Orono, Maine, USA

2 School of Geography, Planning and Environmental Management, University of Queensland, Brisbane, Queensland, Australia

References

[1] McKone T, Nazaroff W, Berck P, Auffhammer M, Lipman T, Torn M, Masanet E, Lobscheid A, Santero N, Mishra U. Grand challenges for life-cycle assessment of biofuels. Environmental Science and Technology; 2011.

[2] Rowe RJ, Terry RC, Rickart EA. Environmental change and declining resource availability for small-mammal communities in the Great Basin. Ecology 2011; 92(6), 1366-1375.

[3] van Dam J, Junginger M, Faaij A, Jurgens I, Best G, Fritsche U. Overview of recent developments in sustainable biomass certification. Biomass and Bioenergy 2008; 32(8), 749-780

[4] Halog A, Chan A. Developing a Dynamic Systems Model for Sustainable Development of the Canadian Oil Sands Industry, International Journal of Environmental Technology and Management 2008; 8 (1): 3-22.

[5] Halog A, Manik Y. Advancing integrated systems modeling framework for life cycle sustainability assessment. Sustainability 2011; 3 (2):469-499

[6] Halog A. Models for evaluating energy, environmental and sustainability performance of biofuels value chain. International Journal of Global Energy Issue 2009; 32: 87-101.

[7] Buchholz T, Luzadis V, Volk T. Sustainability criteria for bioenergy systems: Results form an expert survey. Journal of Cleaner Production 2009; 17: S86-S97.

[8] Buchholz T, Rametsteiner E, Volk T, Luzadis V. Multi criteria analysis for bioenergy systems assessments. Energy Policy 2009; 37:484-495.

[9] National Academy of Sciences. Expanding Biofuel Production. Sustainability and the Transition to Advanced Biofuels: Summary of a Workshop. National Academy Press; 2010

[10] Regalbuto J. Cellulosic Biofuels – Got Gasoline? Science 2009; 325: 822-824.

[11] USDA/USDOE. Sustainability of Biofuels: Future Research Opportunities. Report from the October 2008 Workshop, April 2009; 2009.

[12] Davis SC, Anderson-Teixeira KJ, DeLucia EH. Life-cycle analysis and the ecology of biofuels. Trends in plant science 2009; 14(3): 140-146.

[13] Farrell AE, Plevin RJ, Turner BT, Jones AD, O'Hare M, Kammen DM. Ethanol can contribute to energy and environmental goals. Science 2006; 311:506–508.

[14] Plevin RJ, O'Hare M, Jones AD, Torn MS, Gibbs HK. Greenhouse gas emissions from biofuels' indirect land use change are uncertain but may be much greater than previously estimated. Environment, Science and Technology 2010; 44: 8015–8021.

[15] Perlack R, Wright L, Turhollow A, Graham R, Stokes B, Erbach D. Biomass for a feedstock for bioenergy and bioproducts industry: the technical feasibility of a billion ton annual supply, Oak Ridge National Laboratory, TN (2005 May) 2005;78 p. ORNL/TM-2005/66, DOE/GO-102995, 2135.

[16] Mao H, Genco JM, Yoon SH, van Heiningen A, Pendse H. Technical economic evaluation of a hardwood biorefinery using the "Near-Neutral" hemicellulose pre-extraction process. Journal of Biobased Materials and Bioenergy 2008; 2(2), 177-185. doi: 10.1166/jbmb.2008.309

[17] Uihlein A, Schebek L. Environmental Impacts of a Lignocelluloses Feedstock Biorefinery System: An Assessment. Biomass and Bioenergy 2009; 33: 793-802.

[18] Dickerson K., Rubin J. Maine Bioproducts Business Pathways. Margaret Chase Smith Policy Centre, http://www.forestbioproducts.umaine.edu/library/Maine-Bioproducts-Business-Pathways/203/. Forest Bioproducts Research Initiative, School of Economics, University of Maine, Orono ME; 2008.

[19] Halog A, Mao H. Assessment of Hemicellulose Extraction Technology for Bioethanol Production in the Emerging Bioeconomy. International Journal of Renewable Energy Technology 2011; 2(3): 223 - 239.

[20] Halog A. Sustainable development of bioenergy sector: An integrated methodological framework. International Journal of Multicriteria Decision Making 2011. In press.

[21] Pór A, Stahl J, Temesi J. Decision support system for production control: Multiple criteria decision making in practice. Engineering Costs and Production Economics 1990; 20: 213-218.

[22] Spengler T, Geldermann J, Hähre S, Sieverdingbeck A, Rentz O. Development of a multiple criteria based decision support system for environmental assessment of recycling measures in the iron and steel making industry. Journal of Cleaner Production 1998; 6: 37-52.

[23] Munda G. A NAIADE based approach for sustainability benchmarking. International Journal of Environmental and Technology Management 2006; 6:65-78.

[24] Munda G. Social multi-criteria evaluation: methodological foundations and operational consequences. European Journal of Operations Research 2003; 3: 662-677.

[25] Roy B. The Outranking Approach and the Foundations of ELECTRE Methods; Bana e Costa, C.A., Ed.; Springer-Verlag: Berlin, Germany; 1991: pp155-183.

[26] Brans J, Vincke P. PROMETHEE method for multiple criteria decision-making. Management Sciences 1985; 31:647-656.

[27] Wang JJ, Jing Y-Y, Zhang CF, Zhao JH. Review on multi-criteria decision analysis aid in sustainable energy decision-making. Renewable and Sustainable Energy Review 2009; 13: 2263-2278.

[28] Pohekar S, Ramachandran M. Application of multi-criteria decision making to sustainable energy planning—a review. Renewable and Sustainable Energy Reviews 2004; 8: 365-381.

[29] Halog A, Wei B, Sagisaka M, Inaba A. A Multi-attribute assessment of environmentally-sound electric vehicle battery technologies. Journal of Industrial Engineering 2004; 1: 40-59.

[30] Halog A. Selection of Sustainable Product Improvement Alternatives. Ph.D. Dissertation. http://digbib.ubka.uni-karlsruhe.de/volltexte/242002 (accessed 16 February 2011). University of Karlsruhe, Karlsruhe, Germany, 2002.

[31] Pineda-Henson R, Culaba A, Mendoza G. Evaluating environmental performance of pulp and paper manufacturing using the analytic hierarchy process and life-cycle assessment. Journal of Industrial Ecology 2002; 6: 15-28.

[32] Saaty TL. The Analytic Hierarchy Process; McGraw Hill Company, New York, NY, USA; 1980.

[33] Saaty TL. How to Structure and Make Choices in Complex Problems. Human Systems Management. 1982; 3-4: 255-260.

[34] Saaty TL. Theory and Applications of the Analytic Network Process. RWS Publications, Pittsburgh, PA, USA; 2005.

[35] Holland JH, Miller JH. Artificial adaptive agents in economic theory. American. Economic Review 1991; 81: 365-371.

[36] Akkermans, H. Emergent supply networks 2001: System dynamics simulation of adaptive supply agents. In Proceedings of the 34th Hawaii International Conference on System Sciences Hawaii, HI, USA, 3–6 January 2001; p. 11.

[37] Axelrod, R. The Complexity of Cooperation: Agent-Based Models of Competition and Collaboration; Princeton University Press: Princeton, NJ, USA; 1997.

[38] Parker D, Manson S, Janssen M, Hoffmann M, Deadman P. Multi-agent systems for the simulation of land-use and land-cover change: A review. Annals of the Association of American Geographers 2003; 93: 314-337.

[39] Fox MS, Barbuceanu M, Teigen R. Agent-oriented supply-chain management. International Journal of Flexible Manufacturing Systems 2000; 12:165-188.

[40] Julka N, Karimi I, Srinivasan R. Agent-based supply chain management: A refinery application. Computers and Chemical. Engineering 2002; 26:1771-1781.

[41] Sterman JD. System dynamics modeling: Tools for learning in a complex world. California. Management Review 2001;43: 8-25.

[42] Borshchev A, Filippov A. From system dynamics and discrete event to practical agent based modeling: Reasons, techniques, tools. In Proceedings of the 22nd International Conference of the System Dynamics Society, 2004. Oxford, UK, 25–29 July.

[43] Radzicki M, Taylor R. Origin of System Dynamics: Jay, W. Forrester and the History of System Dynamics. http://www.systemdynamics.org/DL-IntroSysDyn/start.htm (accessed 22 February 2011). U.S. Department of Energy's Introduction to System Dynamics; 2011.

[44] Forrester JW. Industrial Dynamics. John Wiley & Sons Inc, New York, NY, USA; 1961.

[45] Forrester JW. Counterintuitive behavior of social systems. Technology Review 1971;73: 52-68.

[46] Saeed K. Slicing a complex problem for systems dynamics modeling. System Dynamics Review 1992; 8: 251-262.

[47] Halog A, Chan A. Toward sustainable production in the Canadian oil sands industry. In Proceedings of the 13th CIRP International Conference on Life Cycle Engineering, European Union, 31 May–2 June 2006; pp. 131-136.

[48] Chan A, Hoffman R, McInnis B. The role of systems modeling for sustainable development policy analysis: The case of bio-ethanol. Ecology and Society 2004; 9, Article 6.

[49] Riley C, Wooley R, Sandor D. Implementing systems engineering in the US. Department of Energy Office of the Biomass Program. http://www.nrel.gov/docs/fy07osti/41406.pdf (accessed 16 February 2011). In Proceedings of International Conference on System of Systems Engineering: SOSE in Service of Energy and Security, San Antonio, TX, USA, 16–18 April 2007; 2007

[50] Bush B, Duffy M, Sandor D, Peterson S. Using system dynamics to model the transition to biofuels in the United States. In Proceedings of the Third International Conference on Systems of Systems Engineering Monterey, California, CA, USA, 2–4 June 2008; NREL/CP-150-43153. http://www.nrel.gov/docs/fy08osti/43153.pdf (accessed 16 February 2011).

[51] Sheehan JJ. Biofuels and the conundrum of sustainability. Current Opinion in Biotechnology 2009; 20: 318-324.

[52] Scheffran J, BenDor T. Bioenergy and land use: A spatial-agent dynamic model of energy crop production in Illinois. International Journal of Environmental Pollution 2009; 39: 4-27.

[53] Franco C, Ochoa MC, Florez AM. A system dynamics approach to biofuels in Colombia. In Proceedings of the 27th International Conference of the System Dynamics Society 2009, Albuquerque, NM, USA, 26–30 July 2009. http://www.systemdynamics.org/conferences/ 2009/proceed/papers/P1189.pdf (accessed 16 February 2011).

[54] Keam S. McCormick N. Implementing sustainable bioenergy production: A compilation tools and approaches. IUCN, Gland, Switzerland; 2008.

[55] Geyer R, Lindner JP, Stoms DM, Davis FW, Wittstock B. Coupling GIS and LCA for biodiversity assessments of land use. International Journal of Life Cycle Assessment 2010; 15(7): 692-703. doi: 10.1007/s11367-010-0199-9

[56] Neupane B. Incorporating biodiversity impact into environmental life cycle assessment of woodchips for bioethanol production, FTY2011-001; http://www.library.umaine.edu/theses/theses.asp?Cmd=abstract&ID=FTY2011-001 (accessed 16 February 2011) University of Maine; 2011

[57] Kloverpris J, Wenzel H, Banse M, Milai Canals L, Reenberg A. Global land use implications of biofuels: state-of-the-art. Conference and workshop on modeling global land use implications in the environmental assessment of biofuels. International Journal of Life Cycle Assessment 2008; 13 (3): 178-183

[58] Holmberg J, Robert JH. Backcasting from non-overlapping sustainability principles—A framework for strategic planning. International Journal of Sustainable Development and World Ecology 2000; 7:1-18.

[59] Holmberg J, Lundqvist U, Robert K, Wackernagel M. The ecological footprint from a systems perspective of sustainability. International Journal of Sustainable Development and World Ecology 1999; 6:17-33.

[60] Kuisma J. Backcasting for Sustainable Strategies in the Energy Sector. IIIEE Report. Lund, Sweden, September 2000; Volume 18.

[61] Robinson J. Future subjunctive: Backcasting as social learning. Futures 2003; 35: 839-856.

[62] Geurs K, Wee VB. Backcasting as a tool to develop a sustainable transport scenario assuming emission reductions of 80–90%. Innovations 2000; 13:47-62.

[63] Dreborg KH. Essence of backcasting. Futures 1996; 28:813-828.

[64] Borbely M, Muirhead JR, Graniero PA. Backcasting and forecasting biological invasions of inland lakes. Ecological Applications 2004; 14:773-783.

[65] Baumgartner, R. Dealing with uncertainty—integrated sustainability assessment based on Fuzzy Logic Sustainable Development and Planning II; WIT Press: Billerica, MA, USA; 2006 Volumes 1, 2.

[66] Law A, Kelton W. Simulation Modeling and Analysis, 3rd ed.; McGraw-Hill: New York, NY, USA; 2002.

[67] Kempner R, Beck J, Petrie J. Design and analysis of bioenergy networks: a complex adaptive system approach. Journal of Industrial Ecology 2009; 13: 284-305.

[68] Halog A, Sagisaka M, Inaba A. Modeling uncertainties in assessing waste gasification technology. MACRO Review Special Issue 2003; 16:251-255.

[69] Huijbregts M. Dealing with parameter uncertainty and uncertainty due to choices in life cycle assessment. International Journal of Life Cycle Assessment 1998; 3:343-351.

[70] Ford A. Modeling the Environment: An Introduction to System Dynamics Modeling of Environmental Systems; Island Press: Washington, DC, USA 1999; p3

[71] North MJ, Collier NT, Vos JR. Experiences creating three implementations of the repast agent modeling toolkit. ACM Transaction on Modeling and Computer Simulation 2006; 16:1-25.

[72] Miller J. Active nonlinear tests (ANTs) of complex simulation models. Management Science 1998; 44: 620-630.

[73] Scholl H. Agent-based and system dynamics modeling: A call for cross study and joint research. http://projects.ischool. washington.edu/jscholl/Papers/DTABS01.PDF (accessed 16 February 2011). In Proceedings of 34th Annual Hawaii International Conference on System Sciences, Maui, HA, USA, 3–6 January 2011; 2011.

Chapter 8

The Logistics of Bioenergy Routes for Heat and Power

Antonio M. Pantaleo and Nilay Shah

Additional information is available at the end of the chapter

http://dx.doi.org/10.5772/52651

1. Introduction

This chapter aims to overview the logistics of bioenergy systems, focusing on the economic and sustainability implications of the different transport, processing and energy conversion systems for heat and power generation. The main research trends of biomass processing, decoupling of treatment and energy conversion, integration into existing infrastructures and energy systems, and optimal location and sizing of bioenergy facilities are reviewed. For this purpose, a description of supply chains modelling and research trends, technical options and related cost figures for the various steps of the biomass supply chains are overviewed. Moreover, the opportunities to integrate bioenergy into existing energy systems are explored, investigating the use of biofuels in combination with fossil fuels into existing plants and networks. Finally, the main research trends in the optimization of scale and location of the different steps of bioenergy routes are overviewed.

2. Biomass supply chains modelling and key issues

The term "biomass" includes several typologies of organic based materials that can be processed in a variety of methods to produce biofuels and bio-products suitable for several markets, such as energy, industry and food. An overview of bioenergy pathways is reported in Figure 1.

When evaluating bioenergy routes, a system perspective has to be taken, encompassing components such as biomass resource, supply management, processing and conversion systems, energy services. In fact, developing sustainable bioenergy from a economic, environmental and social point of view requires an optimization of the structure and functioning of the supply chain/networks, adjusted to the specific conditions of the production sytems (climate and topology, feedstock, technologies, infrastructures, energy end uses, etc). Steps such as biomass harvesting, storage, refining and transport are particularly relevant, and should be facilitated by suitable logistics of supply chains and operations management techniques.

Bioenergy models, as energy models in general, are useful in problems such as projecting future energy demand and supply, assessing the impacts of different energy technologies and energy efficiency measures, optimizing the operations of energy generators. In recent years, the total number of available energy models has grown tremendously, and various classification schemes that provide insight in the differences and similarities between energy models are available in literature, as reported in Table 1 [1-3]. One of the problems with classifying energy models is that there are many possible categories, while there are only few models that fit into one distinct category. In general, model design requires a trade-off between representational fidelity, model performance, and flexibility to multiple contexts. It is also evident that there is no energy tool that addresses all issues, but instead the 'ideal' energy tool is highly dependent on the specific objectives that must be fulfilled.

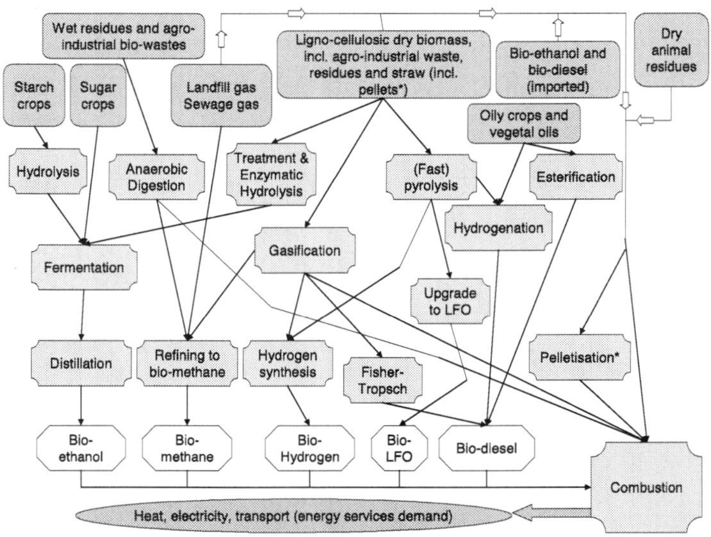

Figure 1. Simplified bioenergy conversion systems patrhways

Classification criterion	Description
1. Purposes of Energy Models	General: potentials assessment, forecasting Specific: energy demand, biomass supply, impacts, appraisal, integrated approach, modular build-up
2. The Model Structure: Internal & External Assumptions	Degree of endogenization, description of non-energy sectors, description end-uses, description supply
3. The Analytical Approach	Top-Down or Bottom-Up
4. The Underlying Methodology	Econometric, Macro-Economic, Economic Equilibrium, Optimization, Simulation, Spreadsheet/Toolbox, Backcasting, Multi-Criteria
5. The Mathematical Approach	Linear programming, mixed-integer programming, dynamic programming
6. Geographical Coverage	Global, Regional, National, Local, or Project
7. Sectoral Coverage	Energy sectors or overall economy
8. The Time Horizon	Short, Medium, Long Term
9. Data Requirements	Qualitative, quantitative, aggregated/disaggregated

Table 1. Classification of energy models for bioenergy

In Table 2 the key factors in bioenergy modelling and biomass supply chains optimization are proposed. In particular, these factors include: (i) the biomass/biofuel chemical-physical properties (moisture, bulk density, LHV, ashes, metal contents, total solids and volatile solids percentages, etc), processing/handling properties (hydrofobicy, storability, grinding, odours, etc) and their influence on transport, storage, drying, conditioning and processing steps, (ii) the biomass seasonality and economic factors such as the relationships between quantity and timing of withdrawal and unitary supply costs. The integration of GIS based tools allows to assess the location over the territory of biomass potentials, transport, storage and processing infrastructures, and final energy demand sites. When estimating biomass potentials in bioenergy models, the factors that are commonly taken in account are the land uses, existing and competing uses of biomass, yield estimates and influence of environmental conditions (such as weather conditions). Moreover, sustainability issues such as direct and indirect land use change, energy inputs in biomass production, harvesting and processing steps and food vs no-food dynamics should also be accounted for. Logistics and infrastructure aspects are also crucial factors. In particular, both the various biomass/biofuel transport modes (ship, road, rail) and biofuel/energy distribution options (pipelines, networks, road) should be taken in account. Moreover, biomass storage and processing infrastructures should be considered, both in the case of existing and new facilities. In the processing and energy conversion steps, both the biomass to biofuel and the biofuel to energy technologies should be modelled. In order to take in account the trade-offs between large/small biomass supply radius (and related transport costs) and large/small biomass processing and conversion facilities, including the potentials of decentralized small scale plants, factors such as scale economies and influence of size on process efficiencies at various conversion technologies should be considered. Moreover, the presence of existing energy infrastructures and the options for biomass co-refining or biomass co-firing in existing fossil fuel plants should be considered, in order to evaluate the opportunities of integration of bioenergy into existing energy systems. Bioenergy modelling should also take in account the options of coupling vs decoupling of processing and energy confersion plants, as discussed in next section. When investigating these integration opportunities, an accurate modelling of biofuel properties and their suitability for dual-fuelling in conventional plants is particularly important. Finally, in order to favourite bioenergy plants locations near to the energy demand, thus maximizing the energy, environmental and economic benefits of these routes, a proper modelling of the energy demand and its suitability for biomass/biofuel uptake is very important. The assessment of potential energy demand regards both stationary applications (heat/cool/power) and fuels for transports. In the first case, the optimization of biomass fired cogeneration or trigeneration (heat/cool/power) plants (in terms of size, locations and technologies) requires, other than the previously mentioned factors, a proper modelling of: (i) energy demand patterns (daily and seasonal variation of energy demand), (ii) quality of heat demand (temperature of heat/cool required), (iii) existing energy supply systems and related costs (baseline scenarios), (iv) subsidy regimes for bioenergy.

In order to address the specific issues of bioenergy, several methods have been used to model and analyse different aspects of the agricultural and forestry biomass logistics system. A number of basic models have been developed in literature to calculate the costs and compare different handling chains and strategies [4-6]. The recent development of advanced computational tools strongly contributed to the improvement of mathematical models for analysis and optimization of such complex supply and logistic systems [7-12], even if the contribution of these methods in biomass logistics could be limited by the high complexity and dynamic environment of bioenergy.

Biomass	**Territory and potentials**	**Infrastructures and logistics**	**Processing-energy conversion**	**End uses**
Temporal biomass availability (seasonality)	Biomass supply location over the territory	Transport systems	Biomass to biofuel technologies	End use typology
Biomass quality	Accessibility issues and available transport modes	Storage and processing infrastructures	Biofuel to energy technologies	Baseline energy scenario
Handling properties	Land uses and biomass yields estimates	Energy infrastructures and integration options (district heating, gas networks, pipelines)	Economies of scale, efficiencies	Energy demand patterns
cost vs quantity biomass	influence of environmental conditions		Processing-conversion coupling vs decoupling	Quality of energy demand
	Alternative and competing uses of biomass		Biofuel suitability for conversion processes	Subsidy regimes for bioenergy

Table 2. Classification of key factors in bioenergy modelling

Moreover, although many researches have an energy system approach, few actually use models that account for the many trade offs and the alternative handling options in the design of whole biomass supply chains. A detailed dynamic simulation program for harvesting, storage, pre-processing and transport of biomass, the IBSAL model, is proposed in [13]. It assumes time and space dependent availability of biomass under the influence of weather conditions and predicts the number, size and location of equipment needed to meet a certain demand. It also calculates the biomass supply costs, energy inputs and emissions, taking in account factors such as the operational parameters of the machines and storage constraints. One of the major innovations consists on the use of non-linear equations to describe these dependencies, e.g. a third-degree polynomial to represent the moisture content as a function of number of days since the start of harvest, or a gamma distribution to simulate the time dependent biomass availability during the harvesting period. However, the methodology is applied to corn stover supply and the implementation to different typologies of feedstocks and agricultural machinery systems would require specific experimental data to inform the model. Moreover, the model is only focused on the supply side and does not include any biomass to energy conversion process or final end uses. To partially overcome these limits, an evolution of the IBSAL model is proposed in [14]. The improved model assesses the logistics of multi-biomass supply and related storage issues to fed a cellulosic ethanol production plant, by a stochastic model with variable input data, such as weather, yields and machine breakdowns. The specific research problem is, in this case, to evaluate how the feedstocks daily demand of the plant can be met throughout the year, what is the cost of the agricultural logistic system, and what are the possible bottlenecks of the supply chains. However, the research does not propose an explicit storage and transport optimization strategy, that could be useful in order to minimize the supply area to meet a given demand, define the optimal location and sizing of storage facilities or scheduling for transport operations. Moreover, the research is focused on a single end-user facility and tailored for a very large straw supply chain and ethanol plant (capacity of 70 million litres/year). Specific issues arising from dispersed and small scale farming techniques, tortuosity of transport networks, land accessibility and ground slope, different storage techniques or other techno-economic factors should be captured when implementing this approach in different agricultural scenarios.

In [12] the storage and transport issues of biomass are assessed and the application to relevant case studies is proposed. In particular, the storage problem and the advantages of a multi-biomass supply chain on the logistic costs are evaluated. The use of intermediate storage locations between the fields and the power plant is often required for several logistic, economic, agronomic and environmental reasons. On the other side, the option of settling the storage facility next to the biomass power plant requires a storage layout with biomass drying capability using dumped heat from the power plant. This concept aims at reducing faster the biomass moisture content and prevents material decomposition as well as fungus and spores formation. In [12] three biomass storage solutions are compared, in terms of total system cost. The concept of multi-biomass is adopted in its simplest form, since two locally available biomass types are considered. The biomass supply chain modelling considers the seasonal availability of the resource, which requires very large storage of biomass for a significant time period, if year-round operation of the power plant is desired. The limited time frame for collecting a large amount of biomass leads also to significant seasonal need of resources, both equipment and workforce. This seasonal demand may increase the cost of obtaining these resources, while leading to suboptimal utilization of resources, particularly of the storage space. The multi-biomass approach may reduce these problems significantly, if the biomass availability is properly shifted over the time. Another characteristic of the biomass supply chain is that it has to deal with low-density materials. As a result, there is increased need for transportation and handling equipment, as well as storage space. This problem is enhanced by the low heating value, which is partly due to the moisture of most agricultural biomass types. The low density of biomass increases further the cost of collection, handling, transport and storage stages of the supply chain. Finally, several biomass types require customized collection and handling equipment, leading to a complicated structure of the supply chain.

In [15], a linear mixed-integer model is proposed, that includes resources, handling/processing, storage and end uses. It is based on the wider *eTransport* model [16], developed for expansion planning in generic energy systems where several alternative energy carriers and technologies are considered simultaneously. The model is based on a network-node system approach, where both the topology and geographic distance of multiple energy infrastructures and the technical and economic properties of different investment alternatives are considered. The model minimises total energy system cost (investments, operation and emissions) of meeting predefined energy demands of energy (electricity, gas, heating) within a geographical area and over a given planning horizon, including alternative supply infrastructures for multiple energy carriers. The model is baed on a nested optimisation, calculating both the optimal diurnal operation of the energy system (operational model) and the optimal expansion plan over a 20–30 years horizon (investment model). In the specific case of bioenergy flows, the amount of energy (and specific operating cost) at any point in the supply chain depends both on the volume and the moisture content in the biomass, and can be defined as a function of two main properties of the biomass [17]: the appearance (biomass in chips, pellets, logs) and the quality (moisture content). Since the moisture content has large influence on the efficiency of various biomass conversion processes, one of the main focus of the research is to represent the relationships between moisture and energy content of various biomasses and to handle long-term processes in the optimization, such as passive drying effects. As an example, the model allows choosing between cheap/free long-term passive drying during storage or spending fuel for forced and fast drying. Biomass density and heating value are also influenced by the processing and storage technologies.

In [17] another methodology for optimization of agricultural supply chains by dynamic programming is described, to find the lowest cost from harvest to end use. The model explicitly deals with the product properties (quality and appearance), which are influenced by handling, processing, transport and storage actions. In particular, agricultural commodities are described according to the appearance states (describing if a product is (un)packed, (un)wrapped, (un)labelled or cut into pieces) and quality states (describing the quality which can be expressed as microorganism infestation, ripeness, moisture content, colour, taste). The types of actions in agrichains are thus: i) handling (actions which modify the appearance states of a product, such as wrapping, cutting and labelling); ii) processing (actions which modify the quality states of a product, such as cooling and drying); iii) transport and storage (actions which alter the quality states of a product. Chain optimisation refers to the construction of routes defining which actors should perform which actions (handling, processing, transportation and storage) at which process conditions, in order to achieve minimum total chain costs while achieving targets.

Another MILP model for the optimal design and operation of biofuel supply chains is proposed in [18] and applied to biodiesel supply chains in Greece. The model incorporates both the optimization of raw materials-feedstocks and biofuel production plants location. It includes the possibility to choose between the domestic biomass production and the import of biomass and-or biofuels to meet given bioenergy targets. However, the model is tailored for a single biofuel production process, it does not take in account storage, transport and environmental issues and costs and it represents the demand side as a fixed quantity of biodiesel to be produced in the whole investigation area.

The work presented in [19] describes an environmental decision support system based on three modules: a GIS-based interface for the characterization of the problem and for the determination of the parameters involved in the formulation of the problem; a database where data characterizing the problem is stored; the optimization module, subdivided into strategic planning, tactical planning and the operational level. The necessity of taking into account different levels derives from the different time scales to be considered and from the different decisions to be performed. Long-term decisions refer to plant sizing, location, and selection among the various technology options. Tactical level decisions refer to planning over a medium- short-term horizon, and are generally considered within a discrete-time setting, with the assumption that the plant capacity and the facilities are known. Finally, the operational level is based on the explicit modelling of the supply-chain process as an ordered sequence of the operations that should be performed from biomass collection to energy conversion. In this case, a non-linear mixed-integer programming optimization is proposed. The main focus is the optimal planning of forest biomass use for energy production.

Another non-linear decision support model is proposed in [20]. The problem considered is optimal exploitation of biomass resources with several harvesting sites and a few centralized combustion plants on a regional level. The aim is to find the optimal capacity of heat and power generation as well as the optimal utilization of biomass resources and transport options. The time horizon considered is one year so that the model is capable of giving long-term decision support.

Another decision support system (DSS) for bioenergy applications, with special reference to harvesting wood for energy from conventional forestry and short rotation forestry, is proposed in [21]. In particular, the work addresses the calculation of delivery costs for wood fuel from conventional forest in the UK. Moreover, an exhaustive review of topics related to the problems of modelling bioenergy supply systems is provided. The same research group proposed other DSSs: the Coppice decision support system (CDSS), a spread- sheet model that can be used to model the costs of growing short rotation coppices under UK conditions, and the Coppice harvesting decision support system (CHDSS), which models the supply chain from the standing Coppice crop through harvesting, storage and transport. These DSSs, as well as other models, have been linked together to produce a bioenergy assessment model (BEAM), which is a comprehensive biomass to electricity model.

3. Bioenergy transport systems

Biomass transport modelling is essential to optimize bioenergy supply chains, plant size and locations. Various typologies of biomass transport models are available in literature. A first type is a simple continuous model [22,23], which is suitable for idealized situations; a second type is a discrete model with defined grid road systems [24,25]; a third type is a complete discrete model incorporating GIS [26,27]. Road tortuosity in the first and second type of models are generally based on assumptions without carrying out road system evaluations. In the last type, the road network is rasterised and then continuous grids of distance and transportation costs to the plant sites are computed using functions of Euclidean distance and allocation. Moreover, in case of on-farm biomass transport, previous studies [28] show that the haulage cost is also dictated by farm landscape attributes and infrastructure.This section overviews the biomass and biofuel transport systems and related costs with different supply route scenarios. The available handling, loading and transport technologies for the various categories of biomasses are assessed. The selection of transport modes is influenced by the typology of biomass feedstocks and supply chain dimension, and a possible biomass/biofuel classification for this purposes can be as follows: (i) forestry products and urban green; (ii) agricultural energy crops and by-products, (iii) urban and agro-industrial bio-wastes with high moisture content; (iv) waste vegetable oils and liquid biomass; (v) long distance transport of solid and liquid biomass; (vi) gaseous biofuels, including biogas, syngas, biomethane. The main trade-offs of road, rail, ship, pipeline transport systems are investigated in the following, and the key factors influencing the optimal choice of the transport mode are discussed.

3.1. Transport systems for solid biomass

Transportation is a cost element in any energy project, but this is especially true for biomass because of the lower energy and bulk density compared with fossil fuels. Several studies have shown that truck transport cost of agricultural residues biomass ranges from 20% to greater than 40% of total delivered cost, depending on distance traveled and mode of transportation [22]. Long-distance transport of biomass including the use of trucks and ships has been addressed in literature [23,24], proving that, despite the long shipping distance, the costs of Latin America wood chips in the receiving European harbour can be as low as 40 Eur/ t or 2.1 Eur/GJ, and the crop's costs account for 25–40% of the delivered costs. The relatively expensive truck transport from production site to gathering point restricts the size of the production area, so that a high biomass yield per hectare is vital to enable large-scale systems.

Many studies have shown that the optimum size of biomass processing and conversion plants is large when abundant biomass is available, and low-cost transport systems are

http://dx.doi.org/10.5772/52651

used; on the contrary, when the specific biomass transport cost increases, because of low energy density of the feedstock and long transport distances, and scale economies and conversion efficiencies are less influenced by the size, the optimal plant size tends to be lower [25-29].In addition, many field sources of biomass are, by their nature, remote from the population centers that will use the produced energy. Thus, developers of such biomass projects will have the alternative of moving the biomass to a plant near the energy consumer, or moving the produced energy from a remote biomass processing plant, and the selection of optimal plant location is based on the relative costs and energy losses of biomass, biofuels and energy transport and intermediate storage. Moreover, both at a large scale and in urban areas, biomass transport by truck may not be physically possible owing to traffic congestion and resulting community opposition. Rail transport of biomass reduces the frequency of loads and offers better environmental performances in comparison to road transport. A specific comparison of rail vs truck transport of biomass is proposed in [30], and the minimum shipping distance for rail transport above which lower costs/km offset the incremental fixed cost in comparison to truck is estimated in the range of 145-170 km for wood chips and straw in a North American setting. Pipeline transport would deliver biomass with minimum ongoing community impact, but is feasible only for liquid and gaseous biomass [31], and will be discussed in the next section.

In [32] the relative cost of transportation by truck, rail, ship, and pipeline for three biomass feedstocks, by truck and pipeline for ethanol, and by transmission line for electrical power is assessed, for various plant sizes. Distance fixed costs and distance variable costs (including power losses during transmission), are calculated for each biomass type and mode of transportation. The results show that pipelining is competitive only at large scale, while transhipment is feasible for distances higher than 1,000-3,000 km, on the basis of the typology of biomass.

In [33] the delivery cost of different combinations of multiple forms of lignocellulosicfeedstocks including agricultural and woody biomass is analysed. In particolar, three types of biomass i.e., wheat, straw, corn stover and forest biomass were considered in different forms such as loose biomass, bales/bundles, chopped/chipped and pellets. It was found that the delivery cost of a combination of woody and agricultural biomass feedstocks is lower than that for a single type of biomass, and traffic congestions resulting from biomass supply to a large facility could be significantly reduced by increasing the density of biomass.

However, selection of a transportation mode cannot be based on only one issue. Economical, environmental, social, and technical parameters should be integrated to select the best system [34].

Transportation costs for biomass and its products have a distance fixed component (DFC) that is incurred regardless of the distance travelled, and includes loading-unloading costs depreciation, insurance, interests and the administrative cost of biomass transport, and a distance variable component (DVC) that includes costs of fuels, repair, tire, lubrication and labor. DFC depends on the type of biomass being transported and the equipment and contractual arrangements involved, which are both case specific, and vary based on the specific form of biomass to a far greater extent than DVC. For example, large round bales of stover or straw would require different treatment for transhipment from truck to rail than woodchips or pellets. The impact of DFC on overall transportation cost diminishes with increasing distance. Moreover, biomass transportation costs are often referred to the total number of actual metric tons as road limits, and in this case the calculated transport cost per dry metric ton will vary for every biomass source. For truck, rail, and ship transport, mass is the primary factor setting the cost of shipment, although for low density loads volume can become the limiting factor. For pipelines transporting a single phase liquid, for example ethanol, liquid volume is the primary factor, whereas for two- phase slurry pipelines carrying biomass the amount of dry matter is the primary factor, because moisture level reaches equilibrium during transport. For both ship, road and rail transport modes, the DFC for low density biomass (straw) is significantly higher than for chips, pellets or TOP. Infact, chips and pellets lend themselves to bulk handling by methods such as conveying or pneumatic transfer, whereas straw/stover is moved as a large bale.

	Transport mode	Fuel	Capacity range	Capacity range	Fixed cost	Variable cost	Main drawbacks	Sources
1	Truck-small	Solid-liquid biomass	15 m3	5 t	2-4 Eur/t	0.2 Eur/km m3	emission levels, traffic congestions, road suitability (for large trucks)	[23,24,30,32, 35
	Truck-medium		35 m3	25 t		0.15 Eur/km m3		
	Truck-large		100 m3	40 t		0.1 Eur/km m3		
	Liquid-tank truck	Bio-oil	30 m3	35 t	5.7 $/m3	0.18-0.07 $/ km*m3		[23,32,41]
	Liquid tank trailer		60 m3	70 t	5.6 $/m3	0.15-0.05 $/ km*m3		
2	Rail	Solid-liquid biomass	2,500 m3	1000 t	5-14 $/t	0.02-0.03 $/km t	Rail network availability	[30,32]
3	Ship		6,700-105,000 m3	4,000-63,000 t	11-34 $/t	0.01 $/km t	Large scale storage capacity, long distance emission levels, ships avalability	[23,24,32]
4	Pipeline-1	Bio-oil, biodiesel	156 m3/day		0.1	0.29	Investment costs, refurbishment costs in case of existing infrastructures, energy losses (DH)	[41,42]
	Pipeline-2		469 m3/day		0.04	0.12		
	Pipeline-3		1000 m3/day		0.02	0.07		
	Pipeline-4	ethanol	1000 m3/day		0	4.13 C-0.5885 $/km t		[32]
5	Gas network	gas	Highly variable on the basis of pipeline diameter		50-150 kEur/km			[61-63]
6	District heating	90* / 120° heat			350-450 kEur/km			[43-46,50,64 ,65,99,112]

Notes: Variable transport cost figures are composed by fuel cost, transport manteinance and spare parts costs, personnel costs; fixed costs are given by loading-unloading costs and all the other costs that are not dependent on the transport distance;

Pipeline-1: capacity bio-oil plant 250 t/day, density 1,2 t/m³, transport capacity 156 m³/day, pipeline diameter 5.1 cm, distance between booster 9.1 km; 65 MW capacity delivered energy; Pipeline-2: capacity bio-oil plant 750 t/day, density 1,2 t/m³, transport capacity 469 m³/day, pipeline diameter 7.6 cm, distance between booster 9.4 km; 195 MW capacity delivered energy; Pipeline-3: capacity bio-oil plant 1600 t/day, density 1,2 t/m³, transport capacity 1000 m³/day, pipeline diameter 9.9 cm, distance between booster 8.1 km; 416 MW capacity delivered energy; C = capacity of bio-ethanol pipeline t/day

Pipeline costs include installation costs

Table 3. Biomass, biofuels and bioenergy transport modes: technical parameters and cost figures

The techno-economic parameters reported in Table 3 are obtained from an overview of literature data on capacities and costs of various biomass, biofuels and energy transport routes. However, cost figures are affected by a relevant range of uncertainties. As regards truck transport of wood chips and straw, as an example, fixed and variable transport costs range between 3.8-4.9 $/dry t and 0.11-0.15 $/t km in the Northern America scenario, as discussed in [35], while data for wood chips in Brazil [36] and Sweden [37] and mixed agricultural and forest residues in Thailand [38] present cost variations in the range of 50%.

The truck operating cost can vary because most of the cost components are region specific, and influenced by fuel taxation. A small change in the equipment use would have large impact on the costs [39]. Driver and fuel costs have wider range of tolerance within them [40]. The firm size from where truck or trailer are rented also affect the cost. Some costs are lower for small farms (such as wages, administrative costs) but these are offset by economics of scales of costs for equipment, tire and consumables which lead to large variations of total costs. There are also many different sizes and types of trucks available. In the specific case of small transport distances, which is typical of the integration of bioenergy in urban areas, the data are obtained from official prices of transports from operators in Italy. The data for medium and large truck are also referred to the Italian scenario (fuel taxation level and fixed costs).

3.2. Transport systems for liquid biomass

Liquid biomass, both in the form of pyrolysis bio-oil, row vegetable oil, bio-ethanol, biodiesel or other BTL fuel, present an higher energy density in comparison to solid biomass and can be transported by trucks, rail, ship and pipelines. Specific transport issues arise in case of high viscosity and corrosive bio-oils, such as pyrolysis oils, that require stainless steel tanks with an average 14% increase in transport costs [23]. Transport of conventional liquid fuels (per tonne) is also assumed to be 25% higher than for solid fuels [23]. Costs for liquid biomass by trucks are reported in Table 3, according to [41] and considering pyrolysis bio-oil. In case of biodiesel and bioethanol these costs could be reduced, because of the lower viscosity (that means quickier loading/unloading rate) and absence of corrosive materials for tanks.

Pipeline transport can be an economically interesting option for large scale transport of bio-oil and over long distances. Today, most of the crude oil is transported by pipeline, and the transport costs benefit from economy of scale in capital cost. Traffic congestion problems are also mitigated. Pipeline transportation of liquid fuels has been used over several decades. Recently, several studies have been carried out on the pipeline transport of raw biomass in the form of a slurry [31,32,35]. Bio-oil and liquid biofuels in general can be transported by pipeline in larger capacities and over longer distances. Current practice is to transport bio-oil by trucks from the production plant. An important characteristic of bio-oils is their high viscosity, that decreases when increasing temperature. In the case of pyrolysis bio-oil, at about 45 °C, its viscosity for pipeline transportation is 15 cSt which is similar to crude oil. To maintain the bio-oil in the pipeline over 45 °C, the pipeline has to be insulated. In the case of low pH bio-oil, the corrosion to carbon steel requires the use of high density polyethylene (HDPE). Similar to truck transportation cost, pipeline transportation cost has both fixed cost (FC) and variable cost (VC). Fixed cost of pipeline transport includes capital cost of inlet and outlet stations. Inlet station refers to the terminal where bio-fuel moves from the storage tank to the pipeline through pumps. Outlet station refers to the terminal where it moves from the pipeline to the storage tank. The inlet station costs include: capital cost of storage tank, building and foundation cost, fittings and valves cost, inlet pump cost and access road cost. Similarly, the outlet station costs include storage tank cost, fittings, valve and small distribution pump cost and building cost. In [42], investment cost figures for inlet and outlet station for a bio-oil pipeline at a transport capacity in the range of 156-2,000 m^3 per day (corresponding to a bio-oil plant using 250-3200 dry tonnes of biomass per day and a pipeline energy transport capacity of about 65-830 MW) are reported. Variable cost of pipeline transport includes capital cost of pipeline, installation and construction cost, operating cost of pipeline, booster station cost, maintenance cost of pipeline and pumps, communication line cost, insulation costs and road access cost. Operating cost of the pipeline includes labor required for running the system and electricity required for pumps. For transport of bio-oil over longer distances, booster stations are required to overcome the frictional losses during the transport. The variable cost for the same bio-oil pipeline capacity range, including the booster station and a length of 100 km are proposed in [42]. These cost figures have been used to inform a detailed techno-economic model based on discounted cash flow analysis, in order to calculate the cost of pipeline transport ($/m3) of bio-oil for different capacities of pipeline (m3/day) at various lengths of pipeline. These cost figures are reported in Table 3. The results report that the pipeline transport cost decreases with the increase in capacity of pipeline and is directly proportional to the distance of transport. Although the pump power increases with the increase in the capacity, the total cost of pipeline transport of bio-oil ($/m3) decreases with the capacity, predominantly due to the benefits from the economy of scale in the capital cost of pipeline. Because of the lower fixed transport costs of pipeline in comparison with truck systems, for short distances and large quantity of delivered fuels, the pipeline option could be more promising. For long distances, the bio-oil heating requirements to mantain the viscosity and the power consumption of the pumps due to the fiction losses should be carefully assessed. However, it should be noted that pipeline costs are highly influenced by the specific installation area, since in densely popolate urban areas, where most of the energy demand is concentrated, the costs can be even 5 times higher than in rural areas.

In [43] the life cycle assessment of transportation of bio-oil by pipeline and by truck are compared. The scope of the work includes the transportation of bio-oil by truck or pipeline from a centralized plant to an end-user. Two cases are studied for pipeline transport of bio-oil: the first case considers a coal based electricity supply for pumping the bio-oil through a pipeline; the second case considers an electricity supply from a renewable resource. The two cases of pipeline transport are compared to two cases of truck transport (truck trailer with capacity 30 m^3 and super B-train truck with capacity 60 m^3). The results report values of 345 and 17 g of CO_2/m^3 km, respectively in the case of coal based and renewable electricity, and similar values for transport by trailer and super B-train truck are 89 and 60 g of CO_2/m^3 km,

http://dx.doi.org/10.5772/52651

respectively. Energy input for bio-oil transport is 3.95 MJ/ m^3 km by pipeline, 2.59 MJ/m^3 km by truck and 1.66 MJ/ m^3 km by super B-train truck.

In the case of liquid biofuels, other than the previous transport systems, pipelines can be used. In the case of high viscosity bio-oils, the pipelines should be probably heated in order to achieve acceptable transport yields. The advantages of pipeline systems are in terms of avoided congestion during delivery, avoided air emissions from trucks, and reduced operational costs. However, sometimes it is not possible to install pipelines, in particular in urban areas with planning constraints or high refurbishment costs. The solution of centralized biomass processing facilities and decentralized energy conversion plants is based on the concept that the high density biofuel can be easily stored and transported to the CHP plants near to the loads by means of efficient distribution systems as pipelines, eventually integrated into existing ones. The costs and the energy losses of biofuels distribution networks would be in most cases lower than that one of district heating networks.

4. Biomass storage, drying and pre-treatment systems

The biomass handling, storage and pretreatment are crucial steps for an optimal development of bioenergy supply chains. Different biomasses require specific treatments and the seasonality of supply increases the complexity of dimensioning and optimal operation of these facilities.

The storage requirements of various biomass and biofuel typologies and the technical options currently adopted are reviewed in the following, together with cost figures of different storage systems. These costs could be particularly relevant when low energy density biomasses, with high seasonality and particularly complex storage requirements have to be stored.

The biomass supply chain presents several distinctive characteristics that diversify it from a typical supply chain. One of them is the need to store the biomass in a proper way, because of its seasonal availability and the necessity of continuous operation of biomass conversion plants. Moreover, in case of imported biomass (wood chips, bio-oils) the transport logistics constraints and the possibility to purchase and hence store large quantities of biomass are crucial issues in order to favourite trading and achieve good market prices. The biomass storage is a particularly important task, both for the relevant investment costs of some storage technologies and for the biomass and energy losses and safety issues related to the selection of poor storage systems. Since most of the biomass-to-energy applications to date concern single biomass use, there is a need of storing very large amounts of biomass for a significant time period, if year-round operation of the power plant is desired. The limited time frame for collecting a large amount of biomass leads also to significant seasonal need of resources, both equipment and workforce. This seasonal demand may increase the cost of obtaining these resources, while leading to their suboptimal utilization, particularly as regards storage space. The problems introduced by the seasonality of biomass availability may be avoided, if a biomass that is available year- round is used, which is very rare in practice. The multi-biomass approach may smooth significantly these problems and is quite often applied in real cases. Another characteristic of the biomass supply chain is that it has to deal with low-density materials. As a result, there is increased need for transportation and handling equipment, as well as storage space. This problem is enhanced by the low heating value, which is partly due to the increased moisture of most agricultural biomass types. The low density of biomass increases further the cost of collection, handling, transport and storage stages of the supply chain. Finally, several biomass types require customized collection and handling equipment, leading to a complicated structure of the supply chain. For example, there are different requirements on handling and transportation equipment and storage space configuration if biomass is procured in the forms of sticks, chips, round bales, plastic bags, etc. Moreover, in case of wet biomass for biogas plants, storage issues are particularly relevant since the mass and energy losses during a not accurate storage can be very relevant. Other typologies of biomass can not be easily stored without a preliminary pre-treatment (drying), because of odour problems and health and safety regulations (i.e. wet olive cake). Liquid biomass (bio-oils) should be also stored in a proper way in order to avoid acidification and deterioration of the biofuel. Therefore, the typology of biomass and the form in which the biomass will be procured often determines the investment and operational costs of the respective bioenergy exploitation system, as it affects the requirements and design of the biomass supply chain.

In case of solid biomass for thermochemical applications, on-field storage is a low-cost option, with the drawback of high biomass losses, difficult control of moisture content, risk of auto-ignition, health and safety issues, and finally land occupation that can hinder next cropping. The use of intermediate storage between field and energy conversion plant is also an option, that implies double biomass transport and often higher total delivery costs [57]. In case of long distances, the use of road-rail transport systems could be integrated with intermediate storage [22]. Storage location at the premises of biomass upgrading and biofuel conversion plants could facilitate the drying process, by means of dumped heat from the process plants, thus preventing material decomposition and health and safety risks.

As regards solid biomass for termochemical conversion systems, three typologies of storage are assessed in [11]: i) closed warehouse with biomass drying capability, by hot air injection generated by dumped heat of the CHP plant which helps to avoid quality degradation of the biomass while simultaneously increasing the energy content of the biofuel; ii) covered storage facility of a pole-frame structure having a metal roof without any infrastructure for biomass drying where a 0.5% material loss/month rate has been assumed; iii) ambient storage of biomass, covered only with a plastic film presenting the highest material loss rate, which is assumed to be 1% material loss/month.

In Table 4 the main characteristics and costs of the available storage systems are described.

Biomass drying provides significant benefits in case of thermochemical conversion systems, such as increased boiler efficiency,lower air emissions, improved boiler operations. The three main options for lignocellulosic biomass drying are rotary dryers, flash dryers and superheated steam dryers. The first types of dryers are less sensitive to biomass size and are the most common option, even presenting the greates fire hazard. Flash dryers are more compact and easier to control, but require small particle size, while superheated steam dryers present the best energy efficiency performances with very low air emission levels. The dryer selection is dependent on the biomass typology, opportunity of integration into biomass processing systems, required air emission levels, availability of waste heat. The biomass drying technologies required in case of thermochemical energy conversion processes are reviewed in [66-68]. In particular, in [66] a detailed description of dryer technologies and heat recovery systems for biomass drying are provided. Guidelines about optimal selection of drying technology and size on the basis of the specific process and feedstocks are also provided, including cost figures, environmental performances and safety issues for each option under investigation.

Storage typology	Material loss (%/month)	Investment cost	O&M costs (% investment/ yr)	Maximum height (m)	Suitable biomass	Note
Open storage	1-3%	20-50 Eur/m²	4	3-4	Solid biomass	Risks of ignition
Covered storage	0,5-1%	100-150 Eur/m²	4	6-8	Solid biomass	
Closed wharehouse	negligible	200-300 Eur/m²	5	6-8	Solid biomass	Possible integration with drying systems and biomass treatments
Plastic covered storage	0,5-2%	50-100 Eur/m²	4	6-8	Wet biomass for biogas	
Depressurized wharehouse	negligible	300-500 Eur/m²		6-8	Solid biomass	Required to minimize odours emissions of biomass
Silos	Negligible	25-35 Eur/m³		6-8	Liquid-solid biomass	
Storage tank	negligible	40-50 Eur/m³		6-8	Wet solid-liquid biomass	Required to minimize pre-fermentation of wet biomass in biogas plants

Table 4. Main characteristics of biomass storage [11, 58-60]

In case of wet biomass, overall efficiency can often be improved by dewatering prior to thermal drying. On the downside, mechanical dewatering equipment itself can consume a large amount of energy and have high maintenance requirements, which must be weighed against the reduction in drying energy. Dewatering equipment includes drying beds, filters and screens, presses, and centrifuges. Depending on the material and the specific type of equipment, mechanical dewatering equipment may reduce moisture content to as little as approximately 50% [67]. Passive dewatering methods, such as using filter bags that are impervious to rain but allow moisture to seep out, can achieve moisture contents as low as 30% at low cost, but long periods of time – on the order of two to three months – may be required. An overview of dewatering and drying technologies on the basis of biomass properties is proposed in [67,68], incuding cost analyses, energy performances, health and environmental ssues.

Technologies such as natural drying, solar drying, gas or biomass fired rotating kilns, drying systems coupled to CHP plants with heat recovery systems are compared.

The biomass treatment and upgrading processes are required to obtain high energy density biofuels, which can be easily transported, stored, and that are suitable for high efficiency energy conversion processes, possibly at the premises of the energy demand. In Table 5, the commercially available and the most promising biomass treatment processes are described, to produce solid, liquid and gaseous biofuels. In most cases, these processes are implemented near to the biomass production sites, in order to minimize the transport costs, facilitate the trade on the market and the storage issues. However, when integrating biomass routes into existing energy systems, the specific logistics, economic and environmental constraints of energy demand in tertiary and residential sectors imply the necessity to locate these processing facilities in industrial areas, eventually decoupling them to the final energy conversion of biofuels near to the loads. Moreover, locating these processes in industrial areas could facilitate the implementation of biorefineries approaches and the integration of multiple processes.

The most promising biofuels are pellets (and in particular torrefied pellet with higher LHV), bio-oils (both from FAME and 2nd gen thermochemical processes on lignocellulosic biomass) and bio-methane (from AD biogas upgrading or 2nd gen FT processes on lignocellulosic biomass).

n	Biofuel	Treatment	Input biomass	References
Solid biofuel				
1	Pellet	Chipping-drying-pelletization	Lignocellulosic biomass	[69-71]
2	TOP (torrefied pellet)	Torrefaction-pelletization	Lignocellulosic biomass	[24,72-74]
3	Chip	Chipping-drying	Lignocellulosic biomass	
4	TOP (torrefied pellet)	Hydrotreatment-drying/dewatering	Wet lignocellulosic biomass	[75-77]
Liquid biofuel				
5	Bio-oil	Mechanical or chemical refining / oil hydrotreatments	Vegetable oils and fat oils	[78-80]
6	Pyrolysis oil (BTL)	Pyrolysis and thermochemical processes on lignocell biomass	Lignocellulosic biomass	[47,81-83]
7	Biodiesel	Esterification of FAME (fatty acid methyl esters)	Vegetable oils and fats	
8	Biodiesel-FT	Gasification coupled to FT biodiesel process	Lignocellulosic biomass	[84-86]
9	Bioethanol	2nd gen process from lignocellulosic biomass	Lignocellulosic biomass	[87-89]
Gas biofuel				
10	Syngas	Gasification of lignocellulosic biomass	Lignocellulosic biomass	[90-92]
11	Biogas	Anaerobic Digestion	Wet fermentable biomass	[93,94]
12	Biomethane-AD	AD and biogas upgrading	Wet fermentable biomass	[95,96]
13	Biomethane-FT	Gasification+syngas upgrading	Lignocellulosic biomass	[97,98]
14	Bio-hydrogen	Dark fermentation-AD processes	Wet fermentable biomass	[89,100-102]
15	Bio-hydrogen-FT	Catalytic synthesis from FT processes	Lignocellulosic biomass	[103-106]

Table 5. Biomass processing technologies for heat and power generation

5. Energy conversion and integration with existing infrastructures

The biofuels can then be converted into energy for stationary applications by means of several technologies. The heat generation is the cheapest and most profitable conversion system for solid biomass and in absence of specific incentives for bio-electricity. The district heating (DH) option is interesting in case of high heat demand density (i.e. new buildings or refurbishment of existing ones), and possibility to increase the networks load factor by district cooling with adsorption chillers. The CHP option with solid biomass can be attractive in case of high electricity costs, incentives for biomass electricity, favourable rules for on-site generation and net metering, presence of suitable heat/electricity demand and possibility to manage the logistic constraints of the biomass transports and storage. The technological options are ORC plants up to 1-2 MWe [107, 108] and ST, possibly in cofiring, for higher size [109]. In the case of liquid and gaseous biofuels, the options of internal combustion engines (ICE) and gas turbines (GT) [46], also in cofiring with natural gas, are available and allow minimizing the biomass transport, storage and air emission constraints which are typical of large solid biomass boilers and make their diffusion difficult in urban areas. In perspective, the use of small scale ICE, but also microturbines (MT) [110] and fuel cells (SOFC) [111], fired by high quality biofuels (bioethanol, biomethane, biohydrogen [89,106]) for heat and power, could be a very promising option, in particular if connected to a centralized biofuel distribution network, and integrated with the gas network.

One of the key issues when implementing competitive and sustainable bioenergy routes is the integration with existing energy systems and infrastructures.

In this context, there are several promising opportunities of repowering existing fossil fuel plants (brownfield plants) for biomass cofiring, both in the case of CHP and district heating systems [113-115]. Moreover, new power plants can be installed in dual-fuel configurations, in order to increase plant operation flexibility, reduce the problems of biomass storage, handling, seasonality, transport of relevant quantities of biofuels, that are typical of single fuel plants. On the contrary, when a power plant is designed to fire both biofuels and fossil fuels, the typical technical and economic problems of only biomass-fired power plants can be drastically reduced, and large scale (and hence higher conversion efficiencies) can be achieved avoiding the use of huge quantities of biomass. ICEs are typical technologies that can be fed by multi-fuels; in particular Diesel engines are suitable for diesel/gas operation with a maximum gas (or biogas) quantity of 75% [116] and a slight efficiency reduction. Also gas turbines can be fed by natural gas in combination to bio-oils, biodiesel, or bio-ethanol. As an example, GE's LM6000-PC aeroderivative gas turbine can be fired by natural gas, ethanol, biodiesel fuels size 35-60 MWe.

Another interesting energy systems integration opportunity regards the use of existing infrastructures for biofuels and fossil fuels processing (co-refining) and the transport. In the latter case, the potentials to use existing natural gas to transport biomethane from thermochemical synthesis or anaerobic digestion processes are particularly promising.

6. Decoupling of biomass processing and optimal sizing

The biomass processing and pre-treatment facilities are influenced by scale economies and in most cases large processing plants can minimize the biofuel production costs, in particular when efficient biomass transport systems are implemented and the variable component of transport cost (dependent on the biomass collection distance) is not dominant. Moreover, biomass processing plants require large sites for biomass storage and handling, and the amenity issues related to the presence of these industrial facilities are often not compatible with residential areas. On the contrary, the final biofuel energy conversion should be located at the premises of the energy demand, and in particular where it presents the highest costs, such as in residential areas, in order to minimize the energy distribution costs. This is particularly relevant in the case of heat (and eventually combined cool generation by adsorption chillers) or CCHP plants. For this reason, several researches on bioenergy are focused on decoupling of biomass processing and biofuel energy conversion, to favourite the integration of bioenergy into urban and peri-urban energy systems.

As an example, in [47] it is described how systems de-coupling applied to fast pyrolysis and diesel engines can distinguish itself from the other conversion technologies, since several remote generators are much better served by a large fast pyrolysis plant that supplies fuel to de-coupled diesel engines than by constructing an entire close- coupled system at each generating site. Another advantage of de-coupling is that the fast pyrolysis conversion step and the diesel engine generation step can operate independently, with intermediate storage of the fast pyrolysis liquid fuel, increasing overall reliability. Peak load or seasonal power requirements would also benefit from de-coupling since a small fast pyrolysis plant could operate continuously to produce fuel that is stored for use in the engine on demand. A similar approach, but related to Fisher-tropsh liquids production at a centralized catalytic synthesis facility with the two options of direct biomass transportation and gasification to centralised plant or preliminary distributed processing of biomass by fast pyrolysis to bio-oil is proposed in [27]. The results show that, for large biomass collection radius, the intermediate and distributed processing of biomass to bio-oil presents lower total production costs, because of the lower biomass delivery costs that offsets the higher operation and biomass costs. A similar approach, related to torrefactionvs fast pyrolysis bio-oil vs wood pellets pretreatment and long distance transport to FT liquid or power plants is proposed in [24], including a detailed assessment of overall chain efficiency in long distance biofuel transport, increased energy conversion efficiency of high quality biofuels, and sensitivity to the main techno-economic parameters. The results report that torrefaction coupled to pelletization to feed BIGCC or cofiring power plants allows minimizing the energy production costs. Another research proposed in [48] compares the production of wood thinning chips, pellet, fast pyrolysis bio-oil and bio-methanol with the further options of cofiring or cogeneration, in order to define the best biomass convesrion strategies, and the benefits of densification in case of long distance transport are enhanced. Finally, in [49] the options of HTC treatment of lignocellulosic biomass vspelletization and coupling these facilities to CHP plants are investigated; the results show that HTC can be a very interesting option for wet biomass, competitive to drying-pelletization.

http://dx.doi.org/10.5772/52651

Bioenergy plants can be also conducted at a wide range of capacities. The problem of optimal size calculation of biomass-to-energy conversion plants has been widely addressed in literature, on the basis of the trade off between the high conversion efficiencies and economies of scale of large size plants and the low biomass collection radius, transport costs and feedstocks collection and management requirements of small size plants [27, 29,51-54]. Factors such as feedstock availability and spatial distribution, terrain and road conditions, biomass transport specific costs, storage costs, existing energy infrastructures, biomass seasonality issues, conversion plant scale factors and efficiencies influence this optimization problem. Logistic aspects are particularly relevant when low energy density and highly dispersed feedstocks are used. Moreover, small scale plants can facilitate the use of excess heat generated, that can match local loads, if a cogeneration configuration is selected. In [55,56], two generic analytical frameworks are proposed, to calculate the optimal conversion plant size for biogas plants.

7. Conclusions

This chapter overviewed the logistic issues of bioenergy routes for stationary applications, discussing the supply chain modelling approaches proposed in literature, the various options for storage, transport, processing and energy conversion of the biomass, and the research trends in order to improve the sustainability and economics of biomass for heat and power.

One of the most interesting research areas regards the optimal location and sizing of biomass processing and conversion facilities, on the basis of the biomass resource, the logistics of supply and conditioning, the final energy end-user typology and existing energy infrastructures for bioenergy integration.

The following main considerations can be drawn: i) high quality biofuels (pellet, bio-oils, biomethane) should be used in order to minimize transport, storage and environmental issues and facilitate the energy coinversion of biomas at the premises of energy loads by CHP plants; ii) decoupling of biomass upgrading and biofuel energy conversion near to the loads is a very promising option; iiii) small boilers are suitable for rural areas and low heat density zones, while DH is feasible with high energy density loads or when cooling distribution can be introduced to increase the network load factor; iv) integration into existing infrastructures is a key factor (i.e. possibility to use existing gas networks for bio-methane); v) solid biomass CHP implies large storage, transport and air emission issues, should be integrated into DH schemes and localized where space and logistics of transport are not a constraint; vi) large CHP plants should be integrated as possible into brownfield plants and using cofiring options to maximize energy conversion efficiencies while limiting the amounts of biomass required; vii) the most reliable technological option currently available for small scale biomass CHP in urban and periurban areas are ORC plants fed by solid biofiuels and ICE fed by liquid or gaseous biofuels, while promising technologies for small scale on site biofuel CHP are microturbines and fuel cells. In conclusion, the economic competitiveness of bioenergy routes in CHP schemes is strongly influenced by the subsidies available for bio-electricity, while biomass heating and cooling can be, at some extent, competitive with fossil fuels even without incentives.

Author details

Antonio M. Pantaleo[1*] and Nilay Shah[2]

*Address all correspondence to: a.pantaleo@imperial.ac.uk

1 Department of Agricultural and environmental sciences, University of Bari, Italy

2 Centre for Process Systems Engineering, Imperial College London, UK

References

[1] S. Jebaraj and S. Iniyan, "A review of energy models," Renewable and Sustainable Energy Reviews, vol. 10, no. 4, pp. 281-311, Aug. 2006.

[2] N. V. Beeck, Classification of Energy Models, Tilburg University and Eindhoven University of Technology May. 1999, p. 25.

[3] D. Connolly, H. Lund, B. V. Mathiesen, and M. Leahy, "A review of computer tools for analysing the integration of renewable energy into various energy systems," Applied Energy, vol. 87, no. 4, pp. 1059-1082, Apr. 2010.

[4] Allen, J., Browne, M., Hunter, A., Boyd, J., & Palmer, H. (1998). Logistics management and costs of biomass fuel supply. International Journal of Physical Distribution and Logistics Management, 28

[5] Clegg, J. M., & Noble, D. H. (1987). A cost model of straw conservation systems. Agricultural Engineering, 42

[6] Mantovani B, Gibson HA 81995). Simulation model for analysis of harvesting and transport costs for biomass based on geography, density and plant location. Energy in World Agriculture 5:253–80

[7] Tatsiopoulos IP, Tolis AJ. (2003) Economic aspects of the cotton-stalk biomass logistics and comparison of supply chain methods. Biomass and Bioenergy 24(3):199–214.

[8] Cundiff, J. S., Dias, N., &Sherali, H. D. (1997). A linear programming approach for designing a herbaceous biomass delivery system. Bioresource Technology, 59,47-55

[9] Eksioglu, S. D., Acharya, A., Leightley, L. E., &Arora, S. (2009). Analyzing the design and management of biomass-to- biorefinery supply chain. Computers & Industrial Engineering, 57, 1342-1352

[10] Nagel, J. (2000). Determination of an economic energy supply structure based on biomass using a mixed-integer linear optimization model. Ecological Engineering, 16, 91-102

[11] Rentizelas, A. A., Tatsiopoulos, I. P., &Tolis, A. J. (2009). An optimization model for multi-biomass tri-generation energy supply. Biomass and bioenergy, 33, 223-233.

[12] A. Rentizelas, A. J. Tolis, and I. P. Tatsiopoulos, "Logistics issues of biomass: The storage problem and the multi-biomass supply chain," Renewable and Sustainable Energy Reviews, vol. 13, no. 4, pp. 887-894, May. 2009.

[13] Sokhansanj S, Kumar A, Turhollow AF. Development and implementation of integrated biomass supply analysis and logistics model (IBSAL). Biomass and Bioenergy 2006;30:838–47.

[14] Ebadian, M., Sowlati, T., Sokhansanj, S., Stumborg, M., &Townley-Smith, L. (2011). A new simulation model for multi-agricultural biomass logistics system in bioenergy production. Biosystems Engineering, 1-11. IAgrE.

[15] S. van Dyken, B. H. Bakken, and H. I. Skjelbred, "Linear mixed-integer models for biomass supply chains with transport, storage and processing," Energy, vol. 35, no. 3, pp. 1338-1350, Mar. 2010.

[16] B. H. Bakken, H. I. Skjelbred, and O. Wolfgang, 82007) "e-Transport: Investment planning in energy supply systems with multiple energy carriers," Energy, vol. 32, no. 9, pp. 1676-1689,

[17] Gigler JK, Hendrix EMT, Heesen RA, van den Hazelkamp VGW, Meerdink G. On optimization of agri-chains by dynamic programming. European Journal of Operational Research 2002;139:613–25.

[18] Papastopulou C., Kondili E., Kaldellis J, Development and implementation of an optimization model for biofuels supply chain, Energy, 2011, doi:10.1016/j.energy.2011.08.013

[19] Frombo F, Minciardi R, Robba M, Sacile R. A decision support system for planning biomass-based energy production. Energy 2009;34:362–9.

[20] Freppaz D, Minciardi R, Robba M, Rovatti M, Sacile R, Taramasso A. Optimizing forest biomass exploitation for energy supply on a regional level. Biomass and Bioenergy 2004;26:15–25.

[21] C. P. Mitchell, "Development of decision support systems for bioenergy applications," Biomass and Bioenergy, vol. 18, no. 4, pp. 265-278, Apr. 2000.

[22] Kumar, A., Sokhansanj, S., & Flynn, P. C. (2006). Development of a Multicriteria Assessment Model for Ranking Biomass Feedstock collection and transportation systems. Applied Biochemistry And Biotechnology, 129, 71-88.

[23] Hamelinck, C. N., Suurs, R. A. A., &Faaij, A. P. C. (2005). International bioenergy transport costs and energy balance. Biomass and Bioenergy, 29, 114-134. doi:10.1016/j.biombioe.2005.04.002

[24] Uslu, a, Faaij, a, & Bergman, P. (2008). Pre-treatment technologies, and their effect on international bioenergy supply chain logistics. Techno-economic evaluation of torrefaction, fast pyrolysis and pelletisation. Energy, 33(8), 1206-1223. doi:10.1016/j.energy.2008.03.007

[25] Searcy, E., & Flynn, P. (2009). The impact of biomass availability and processing cost on optimum size and processing technology selection. Applied biochemistry and biotechnology, 154(1-3), 92-107. doi:10.1007/s12010-008-8407-9

[26] Jenkins, B. M. (1997). A comment on the optimal sizigi of a biomass utilization facility under constant and variable costs scalings. Biomass and Bioenergy, 13, 1-9.

[27] Wright, M., Brown, R., &Boateng, A. (2008). Establishing the optimal sizes of different kinds of biorefineries. Biofuels, Bioprod. Bioref., 2, 229-238. doi:10.1002/bbb

[28] R. P. Overend, The average haul distance and transportation work factors for biomass delivered to a central plant. Biomass 2, 75-79 (1982).

[29] Nguyen, M. H., & Prince, R. G. H. (2006). A simple rule for bioenergy conversion plant size optimization: bioethanol from sugar cane and sweet sorghum Biomass and Bioenergy, 10(3), 361-365.

[30] Mahmudi, H., & Flynn, P. C. (2006). Rail vs truck transport of biomass. Applied biochemistry and biotechnology, 129-132, 88-103.

[31] Kumar, A., Cameron, J. B., & Flynn, P. C. (2004). Pipeline transport of biomass. Applied biochemistry and biotechnology, 113-116(1), 27-39. Retrieved from http://www.ncbi.nlm.nih.gov/pubmed/15054194

[32] Searcy E, Flynn P, Ghafoori E, Kumar A. The relative cost of biomass energy transport. ApplBiochemBiotechnol 2007;137–140:639–52

[33] Sultana, A., & Kumar, A. (2011). Optimal configuration and combination of multiple lignocellulosic biomass feedstocks delivery to a biorefinery. Bioresource technology, 102(21), 9947-56. Elsevier Ltd. doi:10.1016/j.biortech.2011.07.119

[34] Gold, S., &Seuring, S. (2011). Supply chain and logistics issues of bio-energy production. Journal of Cleaner Production, 19(1), 32-42. Elsevier Ltd. doi:10.1016/j.jclepro.2010.08.009

[35] Kumar, A., Cameron, J. B., & Flynn, P. C. (2005). Pipeline transport and simultaneous saccharification of corn stover. Bioresource technology, 96(7), 819-29. doi:10.1016/j.biortech.2004.07.007

[36] Marrison, C.I., Larson, E.D., 1995. Cost versus scale for advanced plantation-based biomass energy systems in the US and Brazil. In: Proceeding of Second Biomass Con-

ference of the Americas: Energy, Environment, Agriculture, and Industry, 1995, pp. 1272– 1290

[37] Hankin, A.C., Stokes, B., Twaddle, A., 1995. The transportation of fuel wood from forest to facility. Biomass Bioenergy 10 (2–3), 141– 151

[38] Junginger, M., Faaij, A., Broek, V.D., Koopmans, A., Hulscher, W., 2001. Fuel supply strategies for larger-scale bio-energy projects in developing countries. Electricity generation from agricultural and forest residues in Northeastern Thailand. Biomass and Bioenergy 21 (4), 259–275

[39] Berwick, M.D., 1997. Truck Costs for Owner/Operators. Upper Great Plains Transportation Institute, North Dakota State University, Fargo. North Dakota, USA

[40] By Ray Barton Associates Ltd. 2008. Operating costs of trucking and surface intermodal transportation in Canada

[41] Pootakham, T., & Kumar, A. (2010). Bio-oil transport by pipeline: a techno-economic assessment. Bioresource technology, 101(18), 7148-54. Elsevier Ltd. doi:10.1016/j.biortech.2010.03.136

[42] Pootakham, T., & Kumar, A. (2010). A comparison of pipeline versus truck transport of bio-oil. Bioresource technology, 101(1), 414-21. Elsevier Ltd. doi:10.1016/j.biortech.2009.07.077

[43] Vallios, I., Tsoutsos, T., &Papadakis, G. (2009). Design of biomass DH systems. Biomass and Bioenergy, 33(4), 659-678

[44] Tassi E, Fiore A, Soluzioni per la produzione e distribuzionedell'energiatermica in un centrourbano, report, CESI 2004

[45] C. Weber, N. Shah, 2011, Optimisation based design of a district energy system for an eco-town in the UK, Energy 36(2)

[46] Lazzarin, R., Noro, M. (2006). Local or district heating by natural gas: Which is better from energetic, environmental and economic point of views? Applied Thermal Engineering, 26(2-3), 244-250

[47] Bridgwater, a. V., Toft, a. J., &Brammer, J. G. (2002). A techno-economic comparison of power production by biomass fast pyrolysis with gasification and combustion. Renewable and Sustainable Energy Reviews (Vol. 6, pp. 181-246).

[48] Polagye, B., Hodgson, K., &Malte, P. (2007). An economic analysis of bio-energy options using thinnings from overstocked forests. Biomass and Bioenergy, 31(2-3), 105-125.

[49] Erlach, B., Wirth, B., &Tsatsaronis, G. (2011). Co-Production of Electricity, Heat and Biocoal Pellets from Biomass: A Techno-Economic Comparison with Wood Pelletizing. Proceedings World Renewable Energy Congress 2011, 508-515.

[50] H Zinko, B Bohm, H Kristjansson, U Ottosson, M Rama, K. S. (2008). District heating distribution in areas with low heat demand intensity (p. 134), IEA RD Programme on "Dh and Cooling including the integration of CHP

[51] B. M. Jenkins. AA comment on the optimal sizing of a biomass utilization facility under constant and variable cost scaling. Biomass and bioenergy, 13(1/2) pp 1-9, 1997.

[52] Yu, Y., Bartle, J., Li, C.-Z., & Wu, H. (2009). Mallee Biomass as a Key Bioenergy Source in Western Australia: Importance of Biomass Supply Chain. Energy & Fuels, 23(6)

[53] Gan, J. (2007). Supply of biomass, bioenergy, and carbon mitigation: Method and application. Energy Policy, 35(12), 6003-6009.

[54] Overend, R. (1982). The average haul distance and transportation work factors for biomass delivered to a central plant, Biomass, 2, pp75-79

[55] Sliz-Szkliniarz, B., & Vogt, J. (2011). A GIS-based approach for evaluating the potential of biogas production from livestock manure and crops at a regional scale: A case study for the Kujawsko-PomorskieVoivodeship. Renewable and Sustainable Energy Reviews, 16(1)

[56] JianbangGan, C.T. Smith. Optimal plant size and feedstock supply radius: A modeling approach to minimize bioenergy production costs Biomass and Bioenergy 35 (8) p. 3350-3359 2011

[57] Allen J, Browne M, Hunter A, Boyd J, Palmer H. Logistics management and costs of biomass fuel supply. Int J PhysDistrib Logistics Manage 1998;28:463–77

[58] Il condizionamentodellebiomasselignocelluyosiche, Veneto Agricoltura, accessed july 2012, http://www.venetoagricoltura.org/upload/pubblicazioni/Progetto%20Biocolt/Cap%205.pdf

[59] web site www.biomassenergycentre.org.uk accessed july 2012

[60] Research into the Storage of Woody Biomass Fuel for Heating Equipment, Maurice Millar NIFES Consulting Group (2006), available at http://www.scotland.gov.uk/Resource/Doc/217736/0090841.pdf, accessed july 2012

[61] Parker N, Using natural gast transmission pipeline costs to estimate hydrogen pipeline costs, 2004, available at http://escholarship.org/uc/item/9m40m75r

[62] Gas network official costs for public works in Italy, http://www.regione.abruzzo.it/osservatorioAppalti/prezzario/index.asp?modello=sottocategoriaU&servizio=dt&stileDiv=sottocategoria&b=sezioneU9&tom=9 accessed july 2012

[63] Capitolato e prezziario per interventi di manutenzione, variant, nuoviinterventiedemergenzesumetanodotti, n°7910, (2011) available at www.edison.it

[64] IEA Report Contract: 1704-05-02-01-005, Project: 8DHC-08-03, District heating distribution in areas with low heat demand density, 2008, www.iea-dhc.org

[65] comparison of three wood gasification systems for energy balance and overall efficiency. Biomass and Bioenergy, 34(3)

[66] Amos, W. A. (1998). Report on Biomass Drying Technology (p. 35).

[67] Roos, C. J. (2008). Biomass Drying and Dewatering for Clean Heat & Power. Water Resources, 3165(September).

[68] Wimmerstedt, Roland, "Recent advances in biofuel drying," Chemical Engineering and Processing, Issue 38, pp. 441-447, 1999.

[69] Mani, S., Sokhansanj, S., Bi, X., &Turhollow, A. (2006). Economics of producing fuel pellets from biomass. Applied Engineering in Agriculture, 22(3), 421–426.

[70] Larsson, S. H., Thyrel, M., Geladi, P., &Lestander, T. a. (2008). High quality biofuel pellet production from pre-compacted low density raw materials. Bioresource technology, 99(15), 7176–82. doi:10.1016/j.biortech.2007.12.065

[71] Bergman, P. C. A. (2005). Combined torrefaction and pelletisation The TOP process. Analysis, (July).

[72] 72. van der Stelt, M. J. C., Gerhauser, H., Kiel, J. H. a., &Ptasinski, K. J. (2011). Biomass upgrading by torrefaction for the production of biofuels: A review. Biomass and Bioenergy, 35(9), 3748-3762.

[73] Prins, M., Ptasinski, K., & Janssen, F. (2006). More efficient biomass gasification via torrefaction. Energy, 31(15), 3458–3470.

[74] Deng, J. (2009). Pretreatment of agricultural residues for co-gasification via torrefaction. Journal of Analytical and Applied Pyrolysis, 86(2), 331–337.

[75] Hoekman, S. K., Broch, A., & Robbins, C. (2011). Hydrothermal Carbonization (HTC) of Lignocellulosic Biomass. Energy & Fuels, 25(4), 1802 1810. doi:10.1021/ef101745n

[76] Stemann, J., & Ziegler, F. (2011). Hydrothermal Carbonisation (HTC): Recycling of Process Water. Proceedings 19th European Biomass Conference and Exhibition, 6-10 June 2011, Berlin, Germany.

[77] B. Wirth, B.Eberhardt, H. Lotze-Campen, B. Erlach, S. Rolinski, P. R. (2011). Hydrothermal Carbonization: Influence of Plant Capacity, Feedstock Choice and Location on Product Cost, B. Wirth. Proceedings 19th European Biomass Conference and Exhibition, 6-10 June 2011, Berlin, Germany (pp. 2001–2010).

[78] Tyson, K. S., Bozell, J., Wallace, R., Petersen, E., &Moens, L. (2004). Biomass Oil Analysis: Research Needs and Recommendations. Contract (p. 116).

[79] Bezergianni, S., Dimitriadis, A., Kalogianni, A., &Pilavachi, P. a. (2010). Hydrotreating of waste cooking oil for biodiesel production. Part I: Effect of temperature on product yields and heteroatom removal. Bioresource technology, 101(17), 6651–6.

[80] Swain, P. K., Das, L. M., &Naik, S. N. (2011). Biomass to liquid: A prospective challenge to research and development in 21st century. Renewable and Sustainable Energy Reviews, 15(9), 4917–4933.

[81] Bridgwater, a. V. (2011). Review of fast pyrolysis of biomass and product upgrading. Biomass and Bioenergy, 1–27.

[82] Mohan, D., Pittman,, C. U., & Steele, P. H. (2006). Pyrolysis of Wood/Biomass for Bio-oil: A Critical Review. Energy & Fuels, 20(3), 848–889.

[83] Tock, L., Gassner, M., &Maréchal, F. (2010). Thermochemical production of liquid fuels from biomass: Thermo-economic modeling, process design and process integration analysis. Biomass and Bioenergy, 34(12), 1838–1854.

[84] Demirbas, A. (2009). Progress and recent trends in biodiesel fuels. Energy Conversion and Management, 50(1), 14–34.

[85] FatihDemirbas, M. (2009). Biorefineries for biofuel upgrading: A critical review. Applied Energy, 86, S151–S161.

[86] Boerrigter H, van der Drift A. Liquid fuels from solid biomass: the ECN concept(s) for integrated FT-diesel production systems. Energy research Centre of the Netherlands Report, RX–03-060, Petten, The Netherlands; October 2003. 18 p.

[87] O. J. Sánchez and C. a Cardona, "Trends in biotechnological production of fuel ethanol from different feedstocks.," Bioresource technology, vol. 99, no. 13, pp. 5270-95, Sep. 2008.

[88] C. a Cardona and O. J. Sánchez, "Fuel ethanol production: process design trends and integration opportunities.," Bioresource technology, vol. 98, no. 12, pp. 2415-57, Sep. 2007.

[89] R. C. Saxena, D. K. Adhikari, and H. B. Goyal, "Biomass-based energy fuel through biochemical routes: A review," Renewable and Sustainable Energy Reviews, vol. 13, no. 1, pp. 167-178, Jan. 2009

[90] Kirubakaran, V., Sivaramakrishnan, V., Nalini, R., Sekar, T., Premalatha, M., & Subramanian, P. (2009). A review on gasification of biomass. Renewable and Sustainable Energy Reviews, 13(1), 179–186.

[91] Wang, L., Weller, C., Jones, D., & Hanna, M. (2008). Contemporary issues in thermal gasification of biomass and its application to electricity and fuel production. Biomass and Bioenergy, 32(7), 573–581.

[92] Bacovsky, D., Dallos, M, Manfred W, "Status of 2 nd Generation Biofuels Demonstration Facilities," IEA Bioenergy Task 39 report, 2010

[93] Zhang, R., El-Mashad, H. M., Hartman, K., Wang, F., Liu, G., Choate, C., & Gamble, P. (2007). Characterization of food waste as feedstock for anaerobic digestion. Bioresource technology, 98(4), 929–35.

[94] Monson, K.D., S.R. Esteves, A.W. Guwy, R. D. (n.d.). Anaerobic Digestion of Biodegradable Municipal Wastes: A Review. (University of Glaymorgan, Ed.)Review Literature And Arts Of The Americas (May 2007.).

[95] E. Ryckebosch, M. Drouillon, and H. Vervaeren, "Techniques for transformation of biogas to biomethane," Biomass and Bioenergy, vol. 35, no. 5, pp. 1633-1645, May. 2011.

[96] Wellinger A, Lindberg A. Biogas Upgrading and Utilisation. [Internet] IEA Bioenergy Task 24: Energy From Biological Conversion of Organic Waste; 2005 [cited January 2011] Available from: http://www.biogasmax.eu/media/biogas_upgrading_and_utilisation__018031200_1011_24042007.pdf

[97] C. N. Hamelinck and A. P. C. Faaij, "Outlook for advanced biofuels," Energy Policy, vol. 34, no. 17, pp. 3268-3283, Nov. 2006

[98] van der Meijden, C. M., Veringa, H. J., &Rabou, L. P. L. M. (2010). The production of synthetic natural gas (SNG): A comparison of three wood gasification systems for energy balance and overall efficiency. Biomass and Bioenergy, 34(3), 302–311.

[99] Carl-Johan Fogelholm, AGebremedhin, S Kim, L Pedersen, T Savola, J Stang, T Tveit. and H. Zinko (2008). Improved Cogeneration and Heat Utilization in DH Networks (p 105) IEA report

[100] N. Martinezperez et al., "The potential for hydrogen-enriched biogas production from crops: Scenarios in the UK," Biomass and Bioenergy, vol. 31, no. 2-3, pp. 95-104, Feb. 2007.

[101] Hawkes FR, Dinsdale R, Hawkes DL, Hussy I. Sustainable fermentative hydrogen production: challenges for process optimisation. International Journal of Hydrogen Energy 2002;27(11):1339–47.

[102] Woodward J, Orr M, Cordray K, Greenbaum E. Enzymatic production of biohydrogen. Nature 2000;405:1014–5.

[103] Demirbas MF. Hydrogen from various biomass species via pyrolysis and steam gasification processes. Energy Source Part A 2006;28:245–52.

[104] Wu W, Kawamoto K, Kuramochi H. Hydrogen-rich synthesis gas production from waste wood via gasification and reforming technology for fuel cell application. J Mater Cycles Waste Manage 2006;8:70–7

[105] Rapagna S, Jand N, Foscolo PU. Catalytic gasification of biomass to produce hydrogen rich gas. Int J Hydrogen Energy 1998;23:551–7.

[106] Hamelinck, C. N. (2002). Future prospects for production of methanol and hydrogen from biomass, 111, 1–22.

[107] Chinese, D., Meneghetti, A., &Nardin, G. (2004). Diffused introduction of Organic Rankine Cycle for biomass-based power generation in an industrial district: a systems analysis. International Journal of Energy Research, 28(11)

[108] Rentizelas, a, Karellas, S., Kakaras, E., &Tatsiopoulos, I. (2009). Comparative techno-economic analysis of ORC and gasification for bioenergy applications. Energy Conversion and Management, 50(3), 674-681

[109] Berndes, G., Hansson, J., Egeskog, A., &Johnsson, F. (2010). Strategies for 2nd generation biofuels in EU – Co-firing to stimulate feedstock supply development and process integration to improve energy efficiency and economic competitiveness. Biomass and Bioenergy, 34(2), 227-236.

[110] Kaikko, J., &Backman, J. (2007). Technical and economic performance analysis for a microturbine in combined heat and power generation. Energy, 32(4), 378-387

[111] Lin, C., Wu, S., Lee, K., Lin, P., & Chang, J. (2007). Integration of fermentative hydrogen process and fuel cell for on-line electricity generation. International Journal of Hydrogen Energy, 32(7), 802-808. doi:10.1016/j.ijhydene.2006.09.047

[112] Rentizelas, a., Tolis, a., &Tatsiopoulos, I. (2008). Biomass District Energy Trigeneration Systems: Emissions Reduction and Financial Impact. Water, Air, & Soil Pollution: Focus, 9(1-2), 139-150.

[113] Al-Mansour F, ZuwalaJ.Anevaluation of biomass co-firing in Europe. Biomass Bioenergy 2010;34:620–9.

[114] Hughes EE. Biomass co-firing: economics, policy and opportunities. Biomass Bioenergy 2000;19:457–65.

[115] R. V. Ree and R. Korbee, "Biomass Cofiring Potential and Experiences in The Netherlands," Energy Policy, no. November, 2000

[116] P. Quaak, H. Knoef, and H. Stassen, Energy from Biomass A Review of Combustion and Gasification Technologies. World bank technical paper – energy series1999, pp. 78.

Chapter 9

Sustainable Multipurpose Biorefineries for Third-Generation Biofuels and Value-Added Co-Products

Stephen R. Hughes, William R. Gibbons,
Bryan R. Moser and Joseph O. Rich

Additional information is available at the end of the chapter

http://dx.doi.org/10.5772/54804

1. Introduction

The transition to third-generation biofuels is driven by the need to integrate biomass-derived fuels more seamlessly into the existing petroleum based infrastructure. Ethanol, whether derived from corn or sugarcane in first-generation processes or biomass in second-generation facilities, has limited market access due its dissimilarity to conventional petroleum-derived fuels. Limitations include restrictions on ratios in which ethanol can be blended with gasoline, lack of compatibility with diesel and jet engines, inability to transport ethanol through existing pipeline network, and propensity to absorb water. While it is clear that biomass can provide a sustainable and renewable source of carbon to replace a significant portion of petroleum resources currently used to generate fuel, power, and chemicals [1,2], it is also obvious that technologies must be developed to convert biomass into direct replacements for petroleum products. This transition from first- and second-generation biofuels to third-generation biofuels will involve numerous facets [3], the centerpiece likely being a multipurpose biorefinery that utilizes many inputs and produces an even greater number of outputs. The first steps to incorporating each of the individual platforms into one integrated sustainable operation are well underway [4], and this transition promises to be a continuing evolution.

First-generation biorefineries use feedstocks such as corn starch or sugar cane that are renewable, but that also have feed/food uses. As production levels have increased, along with human populations, concerns about competition with food needs have arisen [5,6]. Nevertheless, over the past 30 years these first-generation feedstocks have paved the way for production of biofuels via a more sustainable system without negative impacts on the environment or food supplies [3]. Second-generation biorefineries are based on biomass feedstocks that are more widely available and that are not directly used as food, although some are used as livestock feed. Technologies are under development to efficiently convert biomass into ethanol as well as valuable co-products. These are leading the way to sustainably meeting energy needs while also supplying materials for chemical and manufacturing industries [3]. Biomass has the unique advantage among renewable energy sources that it can be easily stored until needed and provides a liquid transportation fuel alternative for the near term. However, cellulosic ethanol can displace only the 40% of a barrel of crude oil that is used to produce light-duty gasoline. Research, development, and demonstration on a range of technologies are needed to replace the remaining 60%, which is primarily converted to diesel and jet fuel. About 15% of our current crude oil consumption is used to produce solvents, plastics, cleaners, and adhesives [7]. Thus, cost-effective technologies are needed to produce biofuels that are suitable for drop-in use in cars, trucks, and jet planes. These advanced biofuels can be sustainably produced from cellulosic and algal feedstocks. Biomass conversion technologies are also needed to produce chemical intermediates and high-value chemicals. Compatibility with the existing infrastructure will aid in process integration and increase profitability of biorefineries [7].

Biorefining has been defined as the sustainable processing of biomass into a spectrum of marketable products and energy [3,5]. The biorefinery of the future will conduct many types of processes, including those producing advanced biofuels, commodity chemicals, biodiesel, biomateials, power, and other value-added co-products such as sweeteners and bioinsecticides. With the tools provided by molecular biology and chemical engineering, the types of co-products, chemicals and biofuels that can be derived from biomass may be almost limitless. Biorefineries combine the necessary technologies for fractionating and hydrolyzing biological raw materials with conversion steps to produce and then recover intermediates and final products. The focus is on the precursor carbohydrates, lignins, oils, and proteins, and the combination of biotechnological and chemical conversion processes of the substances [8]. Most of these processes are being developed individually, but have the potential to be more efficient and economical when combined in multi-process crossover regimens using by-products or waste materials from one process to produce advanced animal feeds, human nutritional supplements, high-value peptides, or enzymes needed in other processes [4]. Use of existing infrastructure would significantly decrease the ramp-up time for economical large-scale production of advanced biofuels [7,9].

To fully meet the requirement for safe and sustainable energy production, third-generation biorefineries must be better integrated, more flexible, and operate with lower carbon and economic costs than second-generation facilities [5]. The main areas that must be addressed are biomass production and supply, process optimization and integration, and overall sustainability [10-12]. Technology is developing rapidly in these areas. A major task is to identify the most promising bio-based products, in particular food, feed, value-added materials, and chemicals to be co-produced with energy to optimize overall process economics and minimize overall environmental impact [13]. Challenges to achieving the promise of advanced biofuels include: overcoming biomass recalcitrance, addressing logistics of transportation of raw feedstock and finished products, providing fair prices for crops or agricultural residues, and tailoring crops and production to specific environments and cultures [14].

2. Biomass production and supply logistics

Feedstock costs represent a large part of biorefinery operating costs, therefore availability of an affordable feedstock supply is crucial for the viability of every biomass processing facility. Economics of biomass production vary with location, feedstock type, political policies, current infrastructure, and environmental concerns. Biofuels may be derived from forestry (thinning and logging), agriculture (residues or dedicated biomass crops), municipal wastes, algal-based resources, and by-products or waste products from agro-industry, food industry, and food services [15]. A key factor is to identify biomass resources that are sustainable because they require minimal water, fertilizer, land use, and other inputs. Feedstocks must be high in energy content, be easy to obtain in large quantities, and be amenable to the conversion processes. Intensive research is in progress on technologies to deliver high-quality, stable, and infrastructure-compatible feedstocks from diverse biomass resources to biorefineries [7].

A strategic analysis was performed in 2005 [16] and updated in 2011 [1] that identified sufficient biomass feedstock availability across the United States to meet near-term and potentially long-term bioenergy goals. The assessment took into consideration environmental sustainability and identified likely costs, assuming a farm-gate or roadside feedstock price of $40-$60 per dry ton. The study did not include additional costs for preprocessing, handling, and transporting the biomass, as these are specific to the feedstock, its condition and form, the type of handling system, and storage conditions. The analysis also did not account for feedstock density or proximity to potential processing facilities. The feedstocks evaluated were those that are currently produced from agriculture and forestry sources, including grain crops (mainly corn for ethanol, sorghum, and barley), sugarcane, sugar beets, oil crops (primarily soybeans for diesel), canola, sunflower, rapeseed, municipal solid wastes, fuelwood, mill residues, pulping liquors, and urban wood wastes, as well as potential forest and agricultural biomass and waste resources. Under conservative assumptions, the combined resources from forests and agricultural lands, assuming feedstock prices of $40-$60 per dry ton, will increase from 138 258 million dry tons in 2012 to 243-767 million dry tons by 2030 [1]. Total energy crops, including perennial grasses such as switchgrass and miscanthus, woody crops such as poplar, willow, southern pine, and eucalyptus, and annual energy crops such as high-yield sorghum, are projected to contribute significantly to this increase, going from less than 4 million dry tons in 2012 to 34-400 million dry tons in 2030. Energy crops have the potential advantages of being produced on marginal lands not used for growing food, requiring essentially no fertilizers or irrigation, and, especially if perennial, requiring little or no tilling [1].

Although sufficient biomass supply is potentially available, continued improvements in biomass feedstocks worldwide are required to achieve viable third-generation biorefineries [5]. Feedstock production improvements include maximizing yield, nutrient (N, P, and K) and water efficiency, and sustainability of production (an area with high potential for rapid gains). Screening of plant species and plant breeding is critically important to increase efficiency of biomass production while minimizing inputs, maintaining soil fertility, managing water balance, and controlling invasiveness. Techniques to estimate the biomass production potential and to evaluate the impacts and sustainability of production in a given location are required. Logistic-related improvements include increasing efficiency of harvest, addressing the issue of seasonality to provide continuous supply, and ensuring that biomass cultivation helps drive regional development. Costs in transporting biomass to the biorefinery can be reduced by using optimized harvesting equipment, appropriate preparation for shipment, and efficient collection, storage, and transfer networks, especially for multi-feedstock biorefineries [5]. Processing improvements include optimizing the composition and properties of biomass for handling and transport to meet downstream quality requirements, along with imparting traits such as greater digestibility for ease of conversion (an area where basic research has made inroads). New technologies are reducing the cost of preparing biomass for conversion. Each step of the preparation is designed to develop next-generation feedstocks. Mechanical treatments reduce the size of the feedstock, providing fractionation and separation. Thermal and chemical processes control moisture content, remove contaminants, and improve digestibility and stability to reduce fouling in process equipment. Treated or untreated biomass is typically blended or mixed in specific proportions, often with additives to improve conversion efficiency. Temperature and pressure are used to form a high-density, stable feedstock for efficient storage and transport [7].

3. Third-generation biofuels

Advanced biofuels were defined by the Final Rule from the United States Environmental Protection Agency (EPA) Renewable Fuel Standard (RFS) Program as being renewable fuels, other than ethanol derived from corn starch, for which lifecycle greenhouse gas emissions are at least 50% less than the gasoline or diesel fuel it displaces [17]. Advanced biofuels may include any of the following: 1) ethanol derived from cellulose, hemicellulose, or lignin; 2) ethanol derived from sugar or starch (other than corn starch); 3) ethanol derived from waste material, including crop residue, other vegetative waste material, animal waste, food waste, and yard waste; 4) biomass-based diesel; 5) biogas (including landfill gas and sewage waste treatment gas) produced through the conversion of organic matter from renewable biomass; 6) butanol or other alcohols produced through the conversion of organic matter from renewable biomass; 7) other fuel derived from cellulosic biomass. Typically, advanced biofuels are used for transportation, although some may be used in generators to produce electricity and others may eventually replace propane and heating oils [14]. Alcohol can substitute for gasoline in spark ignition engines; biodiesel, green diesel, and dimethyl ether can be used in compression ignition engines; Fischer-Tropsch process produces a variety of hydrocarbon fuels, the main one is a diesel-like fuel for compression ignition engines [15].

Third-generation biofuels, also referred to as drop-in biofuels, are considered advanced biofuels [17]. Third-generation biofuels are direct replacements for gasoline, diesel, and jet fuels currently produced from petroleum, and are, in fact, chemically identical to their petroleum-derived counterparts. This allows third-generation biofuels to be directly substituted for petro-fuels without any alterations to pipelines and infrastructure used to deliver the fuel, nor modifications to the engines in which the fuel will be burned. These infrastructure-compatible fuels are derived from biomass or algae, typically through thermochemical processes, although some biochemical processes are being developed as well. These fuels deliver more energy per gallon than ethanol, and conversion processes also yield a range of co-products that help to enhance the economic and environmental sustainability of biorefineries. The knowledge gained and technological advances made through research on cellulosic ethanol have accelerated advances for third-generation biofuels. The previous research on cellulosic feedstock supply, pretreatment, and logistics has helped to improve feedstocks for third-generation biofuels. Similarly, technologies to break down biomass for further processing can be applied to the processing of advanced biofuels. Research on subsequent processing of intermediates and by-products into high-value biological products and chemicals is critical to improving the profitability of third-generation biorefineries [7].

4. Biomass conversion technologies

Lignocellulose is the least expensive and most abundant form of biomass and is cheaper than crude oil on an energy basis. Technically it is possible to convert cellulosic materials and organic wastes into biofuels. Commercialization is limited because low-cost processing technologies that efficiently convert a large fraction of the lignocellulosic biomass energy into liquid fuels have not been developed to date. Thus, it is essential to continue focused research on processes to efficiently and economically convert lignocellulosic biomass into liquid fuels. Three basic routes for this conversion are 1) gasification of biomass to syngas ($CO + H_2$) and further conversion of syngas to liquid fuels, 2) fast pyrolysis or liquefaction of biomass to produce bio-oils followed by upgrading or blending for use as fuels, and 3) hydrolysis of biomass into sugar and lignin monomer units for conversion to targeted products [18]. The conversion technologies for producing liquid biofuels from biomass are outlined in Figure 1.

4.1. Hydrolysis of biomass and fermentation of sugars

Sugar streams for fermentation to biofuels can be obtained directly as sucrose from crops such as sugar cane, sweet sorghum, or energy beets. Alternatively, hydrolysis of starch crops yields glucose and hydrolysis of biomass yields glucose, xylose, and small amounts of other five carbon sugars. Lignocellulosic biomass is composed of cellulose, hemicelluloses, lignin, and proteins bound together in a complex structure that is recalcitrant to enzymatic hydrolysis by cellulase and hemicellulase enzymes. A pretreatment step is required to render the lignocellulosic biomass susceptible to the action of these hydrolytic enzymes [19]. Many factors, such as lignin content, crystallinity of cellulose, and particle size affect the digestibility of biomass. In theory, the ideal pretreatment process produces a disrupted, hydrated substrate that is easily hydrolyzed, but avoids formation of sugar degradation products and fermentation inhibitors. Various pretreatments have been proposed including comminution; delignification by white-rot fungi; chemical pretreatment with acids, alkali, organic solvents or ionic liquids; combined thermal/chemical pretreatment with steam, dilute acid, ammonia, or lime; organosolv-based fractionation treatments; and carbon dioxide or steam explosion [19,20]. Steam pretreatment,

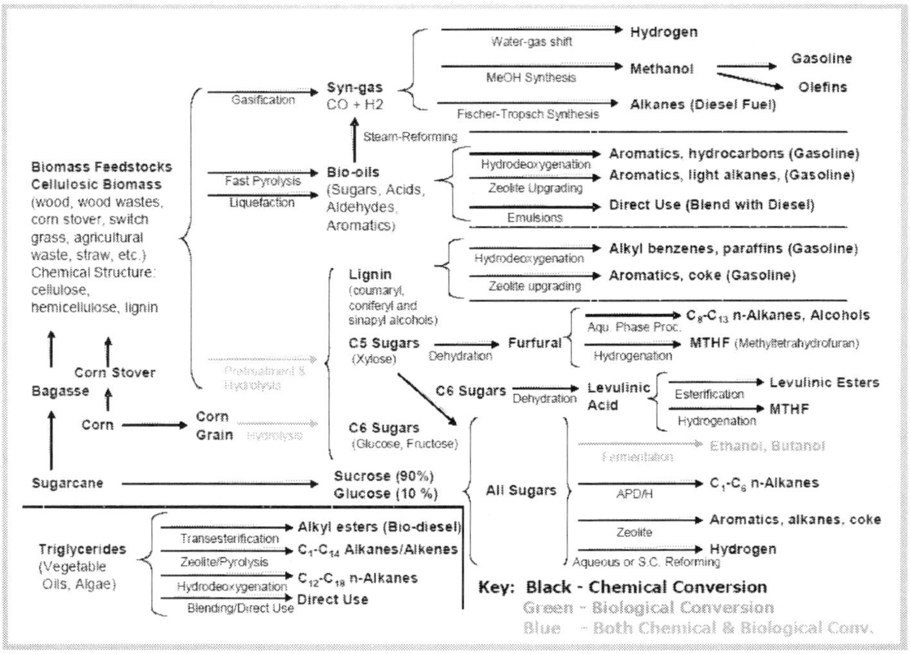

Liquid biofuel production processes use either biochemical or chemical catalysts. Biological catalysts, such as yeast to produce ethanol, are homogenous catalysts in the same liquid phase as the biomass feed. Chemical catalysts range from homogeneous acids to solid heterogeneous catalysts. (Reproduced from [9]; Adapted from [18])

Figure 1. Production of Liquid Biofuels

lime pretreatment, liquid hot water pretreatments and ammonia based pretreatments appear to have the most advantages for biorefinery applications. The main effects are dissolving hemicellulose and altering lignin structure, providing improved accessibility for hydrolytic enzymes [21]. Because the pretreatment process is typically the second most expensive unit cost in the conversion of lignocellulosic biomass, careful analysis and optimization of this process has the potential to significantly reduce biorefinery costs [19].

Genetic engineering of industrial microbes so they are capable of using lignocellulosic feedstocks of variable composition and without catabolite repression is crucial for development of third-generation biorefineries. Since most microbes possess carbon catabolite repression, mixed sugars derived from the lignocellulose are consumed sequentially, reducing the efficacy of the overall process. To overcome this barrier, microbes that exhibit the simultaneous consumption of mixed sugars have been isolated or developed and evaluated for the lignocellulosic biomass utilization. Specific strains of *Escherichia coli, Saccharomyces cerevisiae,* and *Zymomonas mobilis* have been engineered for simultaneous glucose and xylose utilization via mutagenesis or introduction of a xylose metabolic pathway. Other microbes, such as *Lactoba-*

cillus brevis, L. buchneri, and *Candida shehatae* possess a less stringent carbon catabolite repression mechanism and show simultaneous consumption of glucose and xylose. Using these phenotypes, various integrated processes have been developed that incorporate both enzyme hydrolysis of lignocellulosic material and mixed sugar fermentation, thereby enabling greater productivity and fermentation efficacy [22,23]. In addition to utilizing multiple substrates, these microbes must tolerate toxic substrate impurities such as by-products from feedstock pretreatment and hydrolysis, as well as potentially inhibitory products produced by the fermentation reaction itself.

At the present time, most recombinant strains for biorefinery applications are based on *E. coli* and *S. cerevisiae* because these organisms have been extensively studied and are relatively easy to engineer with well-developed genetic tools and established physiology [24]. However, the limited range of materials that can be fermented remains an obstacle to cost-effective bioethanol production in spite of substantial investments over the last 30 years in worldwide efforts to engineer xylose utilization in these strains [25,26]. Although several genetically engineered strains of *S. cerevisiae* have been developed that will ferment xylose to ethanol [27-29], further optimization is needed. It will require the simultaneous expression at sufficiently high level of all the enzymes and proteins needed to allow industrial yeast strains to efficiently metabolize pentose as well as hexose sugars under anaerobic conditions. In addition, for cost-effective industrial ethanol production from biomass it will be necessary to express the enzymes required to saccharify the lignocellulosic feedstocks that are the source of hexose and pentose sugars. Genes considered necessary for complete fermentation of xylose and arabinose, the two major pentose sugar constituents of lignocellulosic biomass, include those encoding xylose isomerase (XI), xylulokinase (XKS), arabinose A, arabinose B, and arabinose D [27,29,30]. These genes may be obtained from microorganisms naturally capable of metabolizing these sugars. Saccharification of lignocellulosic feedstocks also requires utilization of hydrolytic enzymes including cellulases and hemicellulases after initial chemical pre-treatment [31,32]. The cost-effectiveness of the fuel ethanol fermentation process could be further enhanced by obtaining high-value co-products and by-products from the process, such as monomers for polymer production and commercially important proteins and peptides. Genes for these proteins and peptides can be mutagenized, placed in an expression system capable of producing high levels of functional proteins or peptides, and screened in high throughput to optimize desired characteristics. [33-36].

Although extensive efforts have been made to engineer *E. coli* and *S. cerevisiae* to use both hexose and pentose sugars [22,37-40], substrate versatility remains a significant issue. Therefore, other strains are being investigated. For example, *Clostridia* strains possess exceptional substrate diversity, utilizing simple and complex carbohydrates, such as cellulose, as well as CO_2/H_2 or CO. In addition, they contain a wide variety of extracellular enzymes to degrade large biological molecules (cellulose, xylans, proteins, lipids) and produce a broad spectrum of chemicals that can be used as precursors to, or directly as, biofuels and industrial chemicals [22,41-43]. *Clostridia* are found in virtually all anaerobic habitats containing organic matter and thus have developed the ability to ferment mono- and disaccharides as well as complex polysaccharides like cellulose and hemicellulose, which makes them ideal platforms for fermenting biomass feedstocks. They produce metabolites such as butyrate, acetate, lactate, caproate, butanol, acetone, acetoin, ethanol, and many more [24]. *Clostridia* are anaerobic microbes producing a large array of metabolites by utilizing simple and complex carbohydrates, such as cellulose, as well as CO_2/H_2 or CO. Efforts are underway to develop genetic and genomic tools for these microbes, and recent efforts to metabolically engineer *Clostridia* demonstrate their potential for biofuel and biorefinery applications. Pathway engineering to combine established substrate-utilization programs with desirable metabolic programs could lead to modular design of strains suitable for many applications. Engineering complex phenotypes--aerotolerance, abolished sporulation, and tolerance to toxic chemicals--could lead to superior bioprocessing strains [24].

Another significant challenge in using wild-type microbes to convert feedstocks into advanced biofuels is to overcome their endogenous regulation of biofuel-producing pathways that limits yields and productivities. Reconstruction of advanced biofuel pathways in specific heterologous hosts has worked, but use of data-driven and synthetic-biology approaches could further optimize both the host and the pathways to maximize biofuel production from a broader range of substrates. Research will undoubtedly lead to the creation of additional metabolic engineering techniques that can be used to improve pathway flux, and to additional synthetic-biology approaches to optimize microbial hosts for successful commercialization of third-generation biofuels [44,45].

4.2. Gasification

Ethanol and third-generation biofuels can also be produced by a process called gasification. Gasification systems use high temperatures and a low-oxygen environment to convert biomass into synthesis gas, a mixture of hydrogen and carbon monoxide. The synthesis gas, or "syngas," can then be chemically converted into biofuels using the Fischer-Tropsch process or newer advanced catalytic processes. For example, Schmidt and co-workers [46] have combined the three reactions of older thermal gasification processes into a single, small reactor in which gasification takes place over a catalyst to directly produce third-generation biofuels. Synthesis gas can also be microbially converted into biofuels, although the low product tolerance of the microbes has been a limiting factor.

4.3. Pyrolysis

Biomass pyrolysis is the thermal depolymerization of biomass at moderate temperatures in the absence of added oxygen. The biomass is initially converted to a mixture of liquid (pyrolysis oil), solid (biochar), and gaseous fractions that can be used in the production of fuels and chemicals. Fractionation of the pyrolysis oil results in various qualities of oil needed for further upgrading into fine chemicals, automotive fuels, and energy [13]. An updated pyrolysis approach developed by Huber and co-workers uses catalysts to convert biomass into high-octane gasoline-range aromatics in a single step [47,48]. Pyrolysis conditions can be adjusted to optimize the production or chemical composition of a given fraction [49,50].

4.4. Transesterification of oils and fats to biodiesel

Esterification and transesterification have been used for more than a decade to produce biodiesel from plant or animal-derived lipids. Any feedstock that contains free fatty acids and/or triglycerides such as vegetable oils, waste oils, animal fats, and waste greases can be converted to biodiesel. However, the product must meet stringent quality standards [51]. Consequently, fuel standards such as ASTM D6751 in the United States and EN 14214 in Europe have been implemented to ensure that only high quality biodiesel reaches the marketplace. Similar standards have been adopted elsewhere. Acquisition of refined commodity oils such as soybean oil may account for more than 80% of the cost to produce biodiesel. As a consequence, inexpensive, non-food feedstocks are critically important to improve process economics. Such low-value feedstocks often contain contaminants such as moisture and free fatty acids that render them incompatible with simple, homogeneous, alkaline-catalyzed transesterification. In such cases, alternative methods such as heterogeneous acid catalysis are needed for efficient conversion to biodiesel. An economic comparison between different conversion methods utilizing low-value feedstocks revealed that the heterogeneous acid catalyst process had the lowest total capital investment and manufacturing cost. For biodiesel to expand and mature in the market a number of key issues must be addressed, such as improving production efficiency through development of cost-effective catalysts capable of converting low-quality feedstocks into biodiesel, enhancing availability of low cost feedstocks, and managing agricultural land and water. In addition, biodiesel will require continuous improvement in producing cleaner emissions and reducing environmental impacts, although some of these issues are addressed by exhaust after-treatment technologies such as exhaust gas recirculation (EGR) and selective catalytic reduction (SCR) [52].

4.5. Technologies to convert carbohydrates into mixed hydrocarbons

Dissolved sugars can also be converted into hydrocarbons through routes that resemble petroleum processing more than fermentation. Researchers have developed several technologies in which dissolved sugars react in the presence of solid-phase catalysts under carefully controlled conditions (to avoid unwanted by-products) to produce targeted ranges of hydrocarbons for use as fuels or chemical feedstocks [53,54]. Genetically altered microorganisms have also been developed that ferment sugars into hydrocarbons instead of alcohols [55]. Genes were isolated that, when expressed in *Escherichia coli*, produce alkanes, the primary hydrocarbon components of gasoline, diesel and jet fuel. If commercialized, this single step conversion of sugar to fuel-grade alkanes by a recombinant microorganism would lower the cost of producing drop-in hydrocarbon fuels that are low carbon, sustainable, and compatible with the existing fuel distribution infrastructure. The process does not require elevated temperatures, high pressure, toxic catalysts, or complex operations. The recombinant *E. coli* secretes the hydrocarbons from the cell, so it is not necessary to rupture the cell. In addition, because the hydrocarbons are insoluble in water, they will form a separate organic phase and the microbes are not poisoned by the accumulating fermentation product as occurs with alcohol [56].

4.6. Renewable diesel and gasoline

Traditional petrochemical refinery operations such as catalytic cracking and hydroprocessing (HP) can be applied with modifications to biological feedstocks such as bio-oils and triglycerides to produce non-ester renewable hydrocarbon gasoline and diesel fuels [57]. Fluid catalytic cracking (FCC) may be viewed as continuous pyrolysis (400+ °C) at atmospheric pressure in the presence of heterogeneous acid catalyst and is used to produce gasoline. During FCC, long-chain hydrocarbons are cracked into smaller molecules, most of which fall within the gasoline boiling range. Among the reactions that occur (both parallel and consecutive) during FCC include protolytic cracking, dehydrogenation, decarboxylation, decarbonylation, scission, cyclization, oligomerization, coking, and hydride transfer [58]. Zeolite-based catalysts have been used for industrial FCC for over 40 years. These catalysts contain a faujasite-type zeolite as the major active component, which is embedded in a silica and/or alumina matrix. This matrix acts a binder, serves as a diluting medium, provides large mesopores for diffusion to the active zeolite crystal and facilitates heat transfer during cracking reactions [58,59].

HP utilizes both high temperature and pressure along with hydrogen and heterogeneous catalysts to remove heteroatoms (such as oxygen, sulfur, nitrogen, and metals) and unsaturation and yields principally diesel and jet fuels [60]. Sulfur in diesel fuels is limited to 10 ppm in Europe (EN 590) and 15 ppm in the United States (ASTM D975). Consequently, an important process that occurs during HP is hydrodesulfurization (HDS), as crude oil may contain up to 2% (by weight) of sulfur [61]. During HDS, chemically bound sulfur is eliminated as H_2S [62]. A two-stage HDS unit is typically employed whereby a Co-Mo/Al_2O_3 catalyst is first used followed by Ni-Mo/Al_2O_3 (or Ni-W/Al_2O_3). HDS over Co-Mo primarily removes sulfur from aliphatic hydrocarbons. The more active Ni-Mo facilitates hydrogenation of aromatic sulfur as well as saturation of aromatic hydrocarbons. The two-stage deep desulfurization needed to produce ultra-low sulfur (<15 ppm S) diesel (ULSD) fuel has caused changes to the chemical composition of ULSD relative to its low sulfur (<500 ppm) diesel (LSD) fuel predecessor, which was historically prepared in only one HDS stage utilizing a Co-Mo catalyst. The resulting ULSD fuel contains fewer aromatics and heteroatom-containing hydrocarbons relative to LSD [60,63,64]. A drawback to applying existing commercial HP catalysts to biological feedstocks is that the lack of heteroatoms (especially sulfur and nitrogen) in the biological feedstocks causes the catalysts to rapidly lose activity. Therefore, to maintain catalyst activity the feedstocks must be doped with dimethyl disulfide (DMDS) and tetrabutylamine (TBA) [65]. This is of course an undesirable solution, especially in an integrated biorefinery setting where substances such as DMDS and TBA represent non-biological inputs. Recently, zeolites such as Pt/H-ZSM5 have shown promise as HP catalysts for triglycerides such as jatropha oil to yield C15-C18 hydrocarbons directly [66].

Both FCC and HP require atmospheric distillation post-production to yield fuels with the appropriate boiling ranges. Current technology for production of biofuels from these processes involves comingling of biological feedstocks with traditional petroleum feeds to produce a hydrocarbon fuel whose carbons are primarily derived from petroleum [58,67]. Direct production of renewable gasoline and diesel fuels from FCC and HP without comingling requires development of new catalysts with higher tolerance of biological feedstocks. If such processes are to be performed independently from the classic petroleum refinery, then process economics improvements are needed to reduce production costs. Stand alone units or those integrated into a multi-product integrated biorefinery may become more economically competitive as scale of production increases. Demonstration facilities utilizing a patented Universal Oil Products (UOP) process for conversion of triglycerides to renewable diesel using HP have been reported [68]. Important advantages of renewable hydrocarbon gasoline and diesel fuels relative to ethanol and biodiesel are that the former are indistinguishable from their petroleum counterparts, they have greater storage and oxidative stability, they can be transported via existing pipeline infrastructure, they are not hygroscopic, and they can be blended in any proportion with conventional petroleum-derived fuels [57,68].

4.7. Algal biofuel production

As a result of the interest in developing additional biomass feedstocks, research into the production of liquid transportation fuels from microalgae, is reemerging. These microorganisms use the sun's energy to combine carbon dioxide with water to create biomass more efficiently and rapidly than terrestrial plants. Oil-rich microalgae strains are capable of producing the feedstock for a number of transportation fuels—biodiesel, "green" diesel and gasoline, and jet fuel—while mitigating the effects of carbon dioxide released from sources such as power plants [56]. Research and demonstration programs are being conducted worldwide to develop the technology needed to commercialize algal lipid production. Algae store chemical energy in the form of biological oils, such as neutral lipids or triglycerides, when subjected to stresses such as nutrient deprivation [69]. The oil can be extracted from the organisms and converted into biodiesel by transesterification with short-chain alcohols such as methanol or ethanol [70] or by catalytic deoxygenation/hydrogenation of fatty acids into linear hydrocarbons [71]. Another approach is to engineer algae or cyanobacteria to directly produce fuel compounds, instead of oil [72]. These biofuel replacements for gasoline, diesel, and jet fuel will give higher fuel efficiency than ethanol and biodiesel, and will work in existing engines and fuel distribution networks.

4.8. Anaerobic digestion

Anaerobic digestion is the use of microorganisms in oxygen-free environments to convert organic material into methane and carbon dioxide. This biogas is currently produced from crop residues, food scraps, and manure. Anaerobic digestion is also frequently used in the treatment of wastewater and to reduce emissions from landfills. When functioning well, the bacteria can convert about 90% of the biomass feedstock into biogas (containing about 55% methane), which is a readily useable energy source (combusted for thermal energy and/or used to power electrical generators). Solid remnants of the original biomass input, which are left after the digestion process, are typically used as a fertilizer (although it should be chemically assessed for toxicity and growth-inhibiting factors). Biogas production can be part of sustainable biochemicals and biofuels-based biorefinery platform, since it can derive value from low-value by-product or waste streams. Value can be increased by optimizing methane yield and economic efficiency of biogas production [13,73].

5. Platform integration

A key factor in achieving a successful biomass-based economy will be the development of biorefinery systems allowing efficient and cost-effective processing of biological feedstocks into a range of bio-based products that integrate seamlessly into the existing infrastructure [12]. Within the operation of a biorefinery, significant opportunities exist to produce commodity and high-value chemicals in conjunction with the production of bioenergy and biofuels [13]. From a technical point of view, almost all industrial materials made from petroleum resources could be replaced by their bio-based counterparts. However, the bio-based products must be no more expensive, perform at least as well, and have lower environmental impacts. Production of these materials in integrated multi-purpose biorefineries offers the most cost-effective approach to achieving this goal. In general, biofuels, both conventional and advanced, can be produced sustainably in the future only with a significant reduction in costs, which potentially can be accomplished by integrated co-production of value-added products [13]. By producing multiple products, a biorefinery can take better advantage of the intrinsic chemical complexity of biomass components and intermediates to maximize the value derived from the biomass feedstock. A biorefinery might, for example, produce one or several low-volume, but high-value, chemical products as well as a low-value, but high-volume liquid transportation fuel, while generating electricity and process heat for its own use and perhaps enough for sale of electricity. The high-value products enhance profitability, while the high-volume fuel provides economies of scale and helps meet national energy needs, and the power production reduces costs and avoids greenhouse-gas emissions [74].

The development of promising and innovative bio-based chemicals and polymers depends on the feedstock and the resulting process stream or platform [13]. These platforms include: 1) single carbon molecules such as biogas or syngas that can give rise to methanol, dimethylether, ethanol, or Fischer-Tropsch diesel, 2) six carbon carbohydrates from starch, sucrose, or cellulose and mixed streams with five and six carbon carbohydrates from hemicelluloses that can potentially produce succinic, itaconic, adipic, glutamic, and aspartic acids, and 3-hydroxypropionic acid or aldehyde, isoprene, and farnesene, plus more from the chemical processing of glucose, 3) lignin whose structure suggests it could form supramolecular materials and aromatic chemicals, 4) oils (plant-based or algal) that produce glycerol for propylene glycol, epichlorohydrin, and 1,3-propanediol and that are being developed for manufacture of polymers (polyurethanes, polyamides, and epoxy resins), 5) organic solutions from grasses such as clover or alfalfa that contain proteins, amino acids, carbohydrates, and 6) pyrolytic liquids that are expected to produce phenols, organic acids, furfural, hydroxymethyl furfural, and levoglucosan [13]. The continued growth in biobased chemicals and materials will give impetus to the cost-effective production of biofuels in a biorefinery setting. Given the expanding range of feedstocks, platform technologies, and co-products, numerous combinations for third-generation biorefineries are possible [12].

http://dx.doi.org/10.5772/54804

5.1. Multipurpose biorefinery based on starch and cellulosic biofuel platforms

Multipurpose advanced biorefineries that hydrotreat plants oils and animal fats into renewable fuels can be combined with cellulosic ethanol production via fermentation by optimized yeast strains. Concomitant production of high-value bio-based products and advanced animal feeds would be accomplished from by-products from the facility. Cellulosic n-butanol could be produced from mutant strains of *Clostridium acetobutylicum* and *C. beijerinckii* developed to tolerate high concentrations of butanol. Furthermore, engineered algae could be used for urea and ammonia production for emissions control technologies for diesel-operated trucks, for fertilizer, and for production of sucrose and algal oils. The multipurpose biorefinery would require construction of support areas, including research and pilot facilities, a strain collection building, and distillation and post-fermentation processing facilities. Unusable waste streams would be utilized as pyrolysis or biomethane feedstocks to power the biorefinery. Ideally, the biorefinery would produce third-generation biofuels that would be distributed through existing infrastructure. A high-volume animal feed station could be established for distribution to local farms. A possible arrangement of the components of a multipurpose biorefinery combining several of these platforms is shown in Figure 2.

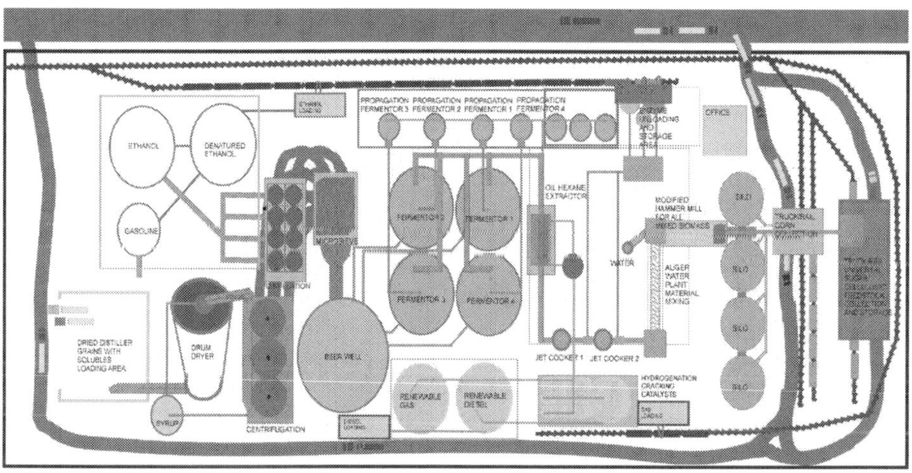

Cellulosic ethanol or n-butanol production by optimized yeast or *Clostridium* strains is combined with an existing starch ethanol production facility. Concomitant production of high-value bio-based products and advanced animal feed is also accomplished from by-products at the integrated facility.

Figure 2. Multipurpose Biorefinery Combining Starch and Cellulosic Biofuel Platforms

5.2. Multipurpose biorefinery based on sugar and syngas platforms

Another example of a multipurpose biorefinery is built on two different platforms, sugar and syngas, to promote different product slates [74]. The sugar platform is based on biochemical conversion processes and focused on fermentation of sugars extracted from biomass feedstocks. The syngas platform is based on thermochemical conversion processes and focused on gasification of biomass feedstocks and by-products from conversion processes. A diagram of this integrated biorefinery is shown in Figure 3.

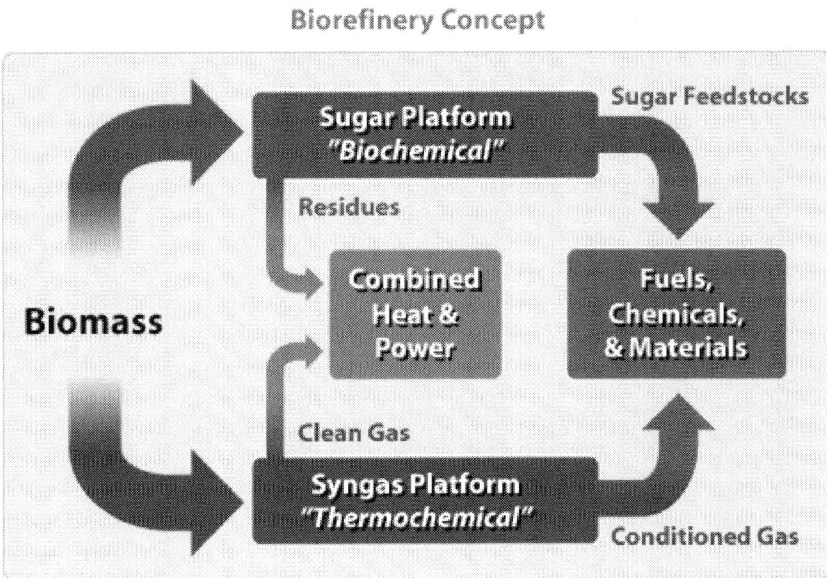

The sugar platform uses biocatalysts such as yeast to produce liquid biofuels from fermentation of sugars. The syngas platform uses high temperatures and a low-oxygen environment to convert biomass into synthesis gas that can then be chemically converted into biofuels. (Reproduced from [74])

Figure 3. Integration of Sugar and Syngas Platforms

Projects developing this concept are focused on new technologies for integrating the production of biomass-derived fuels and other products in a single facility. The emphasis is on using new or improved processes to derive products such as ethanol, 1,3 propanediol, polylactic acid, isosorbide, and various other chemicals. These projects include facilities to develop and validate process technology and sustainable agricultural systems to economically produce sugars and chemicals such as lactic acid and ethanol. These facilities will also develop 1) a novel biomass technology to utilize distiller's grain and corn stover blends to achieve significantly higher ethanol yields while maintaining the protein feed value, 2) a biobased technology to produce a wide variety of products based on 3-hydroxypropionic acid, produced by fermentation of carbohydrates, and 3) an integrated process for recovery of the hemicellulose, protein, and oil components from corn fiber for conversion into value-added products [74].

5.3. Conversion of biomass sugars to hydrocarbon chemicals and fuels

Figure 4 shows an example of an integrated biorefinery that produces third-generation biofuels via chemical catalysis to convert plant-based sugars into a full range of hydrocarbon products identical to those made from petroleum, including gasoline, diesel, jet fuel, and chemicals for plastics and fibers. The biofuels are drop-in replacements that enable full utilization of existing processing, pipeline, storage, and transportation infrastructure. The process converts aqueous carbohydrate solutions into mixtures of hydrocarbons and has been demonstrated with conventional sugars obtained from existing sugar sources (corn wet mills, sugarcane mills, etc.) as well as with a wide variety of cellulosic biomass from nonfood sources. The process can accommodate a broad range of compounds derived from biomass, including C5/C6 sugars, polysaccharides, organic acids, furfurals and other degradation products generated from the deconstruction of biomass. The soluble carbohydrate streams are processed through the aqueous phase reforming (APR) step. The APR step utilizes heterogeneous catalysts at moderate temperatures and pressures to reduce the oxygen content of the carbohydrate feedstock. The reactions in the APR step include: (1) reforming to generate hydrogen, (2) dehydrogenation of alcohols / hydrogenation of carbonyls, (3) deoxygenation, (4) hydrogenolysis, and (5) cyclization [75].

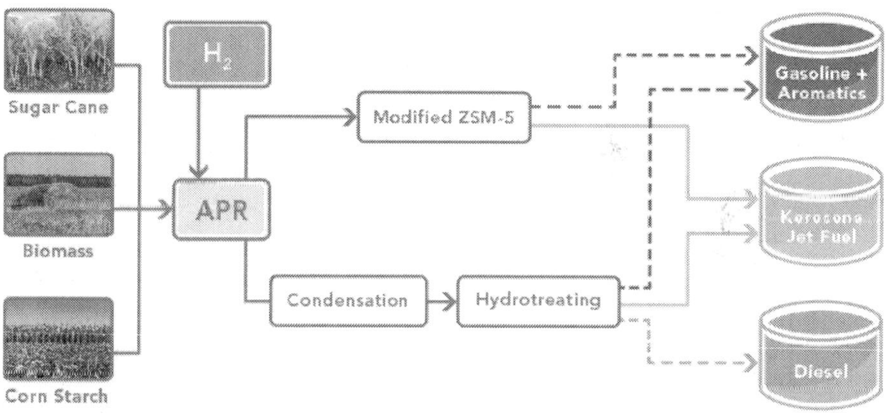

Catalytic chemistry converts plant-based sugars into a full range of hydrocarbon products identical to those made from petroleum, including gasoline, diesel, jet fuel, and chemicals for plastics and fibers. (Reproduced from [75])

Figure 4. Biomass Conversion to Hydrocarbon Chemicals and Fuels

An advantage to this process is the ability to produce hydrogen in-situ from the carbohydrate feedstock or utilize other sources of hydrogen such as natural gas for higher yields and lower costs. The product from the APR step is a mixture of chemical intermediates including alcohols, ketones, acids, furans, paraffins and other oxygenated hydrocarbons. Once these intermediate compounds are formed they can undergo further catalytic processing to generate a cost-effective mixture of nonoxygenated hydrocarbons. A modified ZSM-5 catalyst is used to convert the chemical intermediates from the APR step to a high-octane gasoline blendstock that has a high aromatic content similar to a petroleum-derived reformate stream. The chemical intermediates from the APR step can also be converted into distillate range hydrocarbon components through a condensation step followed by conventional hydrotreating [75].

5.4. Integrated forest biorefinery

An integrated forest biorefinery is diagrammed in Figure 5. In ths example, a facility processing biomass to syngas to biofuels is integrated into a pulp and paper mill [76]. The biomass feedstocks for this biorefinery are forest and agricultural residuals. The biomass is dried and sized prior to gasification and then fed into the fluidized bed stream reformer through a screw feed system. It is gasified to produce syngas with the correct hydrogen to carbon ratio for gas-to-liquids processing. The syngas passes through a conventional heat recovery and gas clean-up train. The gas-to-liquids technology is the Fischer-Tropsch (FT) process, a mature technology. In the reactor the syngas, under pressure and temperature, with the FT catalyst is converted to straight chain hydrocarbons that range from light gases to heavy waxes, including gasoline, naphtha, and diesel [76].

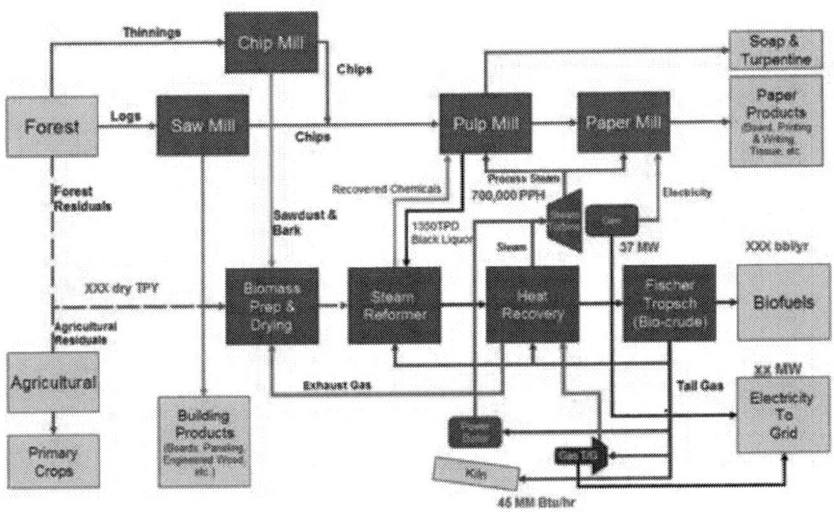

Biomass feedstocks from an existing pulp and paper mill are used to create new revenue streams by producing high-value products such as biofuels and biochemical and at the same time improving the efficiency of the core paper-making operations (Reproduced from [76])

Figure 5. Integration of Paper Mill with Biomass Gasification for Biofuels

The gasification process is ideal for use in a forest products biorefinery because it is configured for high-performance integration with pulp and paper facilities and is capable of handling a wide variety of cellulosic feedstocks, including mill by-products (spent liquor), woodchips, forest residuals, agricultural wastes, and energy crops. The syngas can be used as a substitute for natural gas and fuel oil and as a feedstock for the production of value-added products such as biodiesel, ethanol, methanol, acetic acid, and other biochemicals [76].

6. Conclusion

A crucial step in developing a worldwide bio-industry is to establish integrated third-generation biorefineries that are capable of efficiently converting a broad range of biomass feedstocks into commercially viable biofuels, biopower, commodity and high-value chemicals, and other bioproducts. Integrated biorefineries are similar to conventional refineries in that they produce a range of products to optimize both the use of the feedstock and production economics. Third-generation biorefineries will use novel technologies and diverse biomass feedstocks - requiring significant investments in research, development, and deployment to reduce costs and improve performance to achieve competitiveness with petroleum fuels. These biorefineries will employ various combinations of feedstock and conversion technologies to produce a variety of products, with the main focus on producing biofuels. Co-products can include chemicals (or other materials), animal feed, and heat and power. As pretreatment, conversion, and integration technologies continue to improve, sustainable third-generation biorefineries will become a reality.

Author details

Stephen R. Hughes[1*], William R. Gibbons[2], Bryan R. Moser[3] and Joseph O. Rich[1]

*Address all correspondence to: stephen.hughes@ars.usda.gov

1 USDA, ARS, NCAUR, Renewable Product Technology Research Unit, Peoria, IL, USA

2 Biology/Microbiology Department, South Dakota State University, Brookings, SD, USA

3 USDA, ARS, NCAUR, Bio-Oils Research Unit, Peoria, IL, USA

References

[1] Perlack, R. D, & Stokes, B. J. U. S. billion-ton update: biomass supply for a bioenergy and bioproducts industry. A study sponsored by U. S. Department of Energy, Energy Efficiency and Renewable Energy, Office of the Biomass Program (2011).

[2] Raschka, A, & Carus, M. Industrial material use of biomass: basic data for Germany, Europe and the world (2012). nova-Institute for Ecology and Innovation Study Report. nova-Institute GmbH. Germany. 28pp.

[3] International Energy AgencyIEA Bioenergy Task 42. Biorefineries: Adding Value to the Sustainable Utilisation of Biomass. September (2009).

[4] United States Department of EnergyBiomass Program. Integrated Biorefineries. http://www1.eere.energy.gov/biomass/integrated_biorefineries.htmlaccessed 27 September (2012).

[5] Strategic Research Targets for (2020). Collaboration Initiative on BiorefineriesStar-COLIBRI. European Commission's 7th Framework Programme. European Biorefinery Joint Strategic Research Roadmap for 2020.

[6] Field, C. B, Campbell, J. E, & Lobell, D. B. Biomass energy: the scale of the potential resource. Trends in Ecology and Evolution (2007). doi:j.tree.2007.12.001, 23(2), 65-72.

[7] U.S. Department of Energy, Replacing the whole barrel to reduce U.S. dependence on oil Energy Efficiency and Renewable Energy, Report DOE/EE-0762. July 2012.

[8] Kamm, B, & Kamm, M. International biorefinery systems. Pure and Applied Chemistry (2007). doi:pac200779111983, 79(11), 1983-1997.

[9] National Science FoundationChemical Bioengineering, Environmental, and Transport Division. Breaking the chemical and engineering barriers to lignocellulosic biofuels: next generation hydrocarbon biorefineries. Ed. George W. Huber. March (2008). pp.

[10] Taylor, G. Biofuels and the biorefinery concept. Energy Policy (2008). , 36-4406.

[11] Lynd, L. R, Laser, M. S, Bransby, D, Dale, B. E, Davison, B, Hamilton, R, Himmel, M, Keller, M, Mcmillan, J. D, Sheehan, J, & Wyman, C. E. How biotech can transform biofuels. Nature Biotechnology (2008). , 26(2), 169-172.

[12] Kokossis, A. C, & Yang, A. On the use of systems technologies and a systematic approach for the synthesis and the design of future biorefineries. Computers and Chemical Engineering (2010). , 34(9), 1397-1405.

[13] International Energy AgencyIEA Bioenergy Task 42. Bio-based Chemicals: Value Added Products from Biorefineries. February (2012).

[14] Advanced Biofuels USATruly sustainable renewable future. April 17, (2012). http://advancedbiofuelsusa.info/truly-sustainable-renewable-futureaccessed 29 September 2012)

[15] Nigam, S. N, & Singh, A. Production of liquid biofuels from renewable resources. Progress in Energy and Combustion Science (2011). , 37-52.

[16] Perlack, R. D, Wright, L. L, Turhollow, A. F, Graham, R. L, Stokes, B. J, & Erbach, D. C. Biomass as a feedstock for a bioenergy and bioproducts industry: the technical

feasibility of a billion-ton annual supply. A joint study sponsored by U. S. Department of Energy and U. S. Department of Agriculture (2005).

[17] Federal RegisterMarch 26, (2010). Final Rule 40 CFR 80; 75 FR 14670; Regulation of fuels and fuel additives: changes to renewable fuels program., 14670-14904.

[18] Huber, G. W, Iborra, S, & Corma, A. Synthesis of transportation fuels from biomass: chemistry, catalysts, and engineering. Chemical Reviews (2006). , 106-4044.

[19] Agbor, V. B, Cicek, N, Sparling, R, Berlin, A, & Levin, D. B. Biomass pretreatment: fundamentals toward application. Biotechnology Advances (2011). , 29-675.

[20] Kumar, D, & Murthy, G. S. Impact of pretreatment and downstream processing technologies on economics and energy in cellulosic ethanol production. Biotechnology and Biofuels (2011). DOI:

[21] Hendriks, A. T, & Zeeman, G. Pretreatments to enhance the digestibility of lignocellulosic biomass. Bioresource Technology (2009). DOI:j.biortech.2008.05.027, 100(1), 10-18.

[22] Kim, J. H, Block, D. E, & Mills, D. A. Simultaneous consumption of pentose and hexose sugars: an optimal microbial phenotype for efficient fermentation of lignocellulosic biomass. Applied Microbiology and Biotechnology (2010). , 88-1077.

[23] Den Haan RMcBride JE, La Grange DC, Lynd LR, Van Zyl WH. Functional expression of cellobiohydrolases in *Saccharomyces cerevisiae* towards one-step conversion of cellulose to ethanol. Enzyme and Microbial Technology (2007). , 40-1291.

[24] Tracy, B. P, Jones, S. W, Fast, A. G, Indurthi, D. C, & Papoutsakis, E. T. Clostridia: the importance of their exceptional substrate and metabolite diversity for biofuel and biorefinery applications. Current Opinion in Biotechnology (2012). DOI:j.copbio. 2011.10.008, 23-364.

[25] Farrell, A. E, Plevin, R. J, Turner, B. T, Jones, A. D, Hare, O, & Kammen, M. DM. Ethanol can contribute to energy and environmental goals. Science (2006). , 311(5760), 506-508.

[26] Saha, B. C. Hemicellulose bioconversion. Journal of Industrial Microbiology and Biotechnology (2003). , 30(5), 279-291.

[27] Karhumaa, K. Garcia Sanchez R, Hahn-Hägerdal B, Gorwa-Grauslund M-F. Comparison of the xylose reductase-xylitol dehydrogenase and the xylose isomerase pathways for xylose fermentation by recombinant *Saccharomyces cerevisiae.* Microbial Cell Factories (2007).

[28] Sedlak, M. Ho NWY. Production of ethanol from cellulosic biomass hydrolysates using genetically engineered *Saccharomyces* yeast capable of cofermenting glucose and xylose. Applied Biochemistry and Biotechnology (2004).

[29] Wisselink, H. W, & Toirkens, M. J. Berriel M del RF, Winkler AA, van Dijken JP, Pronk JT, van Maris AJA. Engineering of *Saccharomyces cerevisiae* for efficient anaerobic alcoholic fermentation of L-arabinose. Applied and Environmental Microbiology (2007). , 73(15), 4881-4891.

[30] Jin, Y. S, Ni, H, Laplaza, J. M, & Jeffries, T. W. Optimal growth and ethanol production from xylose by recombinant *Saccharomyces cerevisiae* require moderate D-xylulokinase activity. Applied and Environmental Microbiology (2003)., 69(1), 495-503.

[31] Rudolf, A, Baudel, H, Zacchi, G, Hahn-hägerdal, B, & Lidén, G. Simultaneous saccharification and fermentation of steam-pretreated bagasse using *Saccharomyces cerevisiae* TMB3400 and *Pichia stipitis* CBS6054. Biotechnology and Bioengineering (2007)., 99(4), 783-790.

[32] Saha, B. C, Iten, L. B, Cotta, M. A, & Wu, Y. V. Dilute acid pretreatment, enzymatic saccharification, and fermentation of rice hulls to ethanol. Biotechnology Progress (2005)., 21(3), 816-822.

[33] Hughes, S. R, Riedmuller, S. B, Mertens, J. A, Li, X-L, Bischoff, K. M, Cotta, M. A, & Farrelly, P. J. Development of a liquid handler component for a plasmid-based functional proteomic robotic workcell. Journal of the Association for Laboratory Automation (2005)., 10(5), 287-300.

[34] Hughes, S. R, Riedmuller, S. B, Mertens, J. A, Li, X-L, Bischoff, K. M, Qureshi, N, Cotta, M. A, & Farrelly, P. J. High-throughput screening of cellulase F mutants from multiplexed plasmid sets using an automated plate assay on a functional proteomic robotic workcell. Proteome Science (2006).

[35] Hughes, S. R, Dowd, P. F, Hector, R. E, Riedmuller, S. B, Bartolett, S, Mertens, J. A, Qureshi, N, Liu, S, Bischoff, K. M, Li, X-L, & Jackson, J. S. Jr., Sterner, D., Panavas, T., Rich, J.O., Farrelly, P.J., Butt, T.R., Cotta, M.A. Cost-effective high-throughput fully automated construction of a multiplex library of mutagenized open reading frames for an insecticidal peptide using a plasmid-based functional proteomic robotic workcell with an improved vacuum system. Journal of the Association for Laboratory Automation (2007)., 12(4), 202-212.

[36] Hughes, S. R, Dowd, P. F, Hector, R. E, Panavas, T, Sterner, D. E, Qureshi, N, Bischoff, K. M, Bang, S. S, Mertens, J. A, Johnson, E. T, & Li, X-L. Jackson JS Jr, Caughey RJ, Riedmuller SB, Bartolett S, Liu S, Rich JO, Farrelly PJ, Butt TR, LaBaer J, Cotta MA. Lycotoxin-1 insecticidal peptide optimized by amino scanning mutagenesis and expressed as a coproduct in an ethanologenic *Saccharomyces cerevisiae* strain. Journal of Peptide Science (2008)., 14(9), 1039-1050.

[37] Laluce, C. Schenberg ACG, Gallardo JCM, Coradello LFC, Pombeiro-Sponchiado SR Advances and developments in strategies to improve strains of *Saccharomyces cerevisiae* and processes to obtain the lignocellulosic ethanol–a review Applied Biochemistry and Biotechnology (2012). doi:s12010-012-9619-6, 166(8), 1908-1926.

[38] Hughes, S. R, Sterner, D. E, Bischoff, K. M, Hector, R. E, Dowd, P. F, Qureshi, N, Bang, S. S, Grynaviski, N, Chakrabarty, T, Johnson, E. T, Dien, B. S, Mertens, J. A, Caughey, R. J, Liu, S, & Butt, T. R. LaBaer J, Cotta MA, Rich JO. Engineered *Saccharomyces cerevisiae* strain for improved xylose utilization with a three-plasmid SUMO yeast expression system. Plasmid (2008)., 61(1), 22-38.

[39] Van Maris AJAWinkler AA, Kuyper M, de Laat WTAM, van Dijken JP Pronk JT. Development of efficient xylose fermentation in *Saccharomyces cerevisiae*: xylose isomerase as a key component. Advances in Biochemical Engineering/Biotechnology (2007). , 108-179.

[40] Hahn-hägerdal, B, Karhumaa, K, Fonseca, C, Spencer-martins, I, & Gorwa-grauslund, M. F. (2007). Towards industrial pentose-fermenting yeast strains. Applied Microbiology and Biotechnology 2007;, 74(5), 937-953.

[41] Ezeji, T. C, Qureshi, N, & Blaschek, H. P. Bioproduction of butanol from biomass: from genes to bioreactors. Current Opinion in Biotechnology (2007). DOI:j.copbio. 2007.04.002, 18-220.

[42] Qureshi, N, Saha, B. C, & Cotta, M. A. Butanol production from wheat straw hydrolysate using *Clostridium beijerinckii*. Bioprocess and Biosystems Engineering (2007). , 30-419.

[43] Qureshi, N, Li, X. L, Hughes, S. R, Saha, B. C, & Cotta, M. A. Butanol production from corn fiber xylan using *Clostridium acetobutylicum*. Biotechnology Progress (2006). , 22-673.

[44] Peralta-Yahya, PP, & Zhzng, . , Del Cardayre SB, Keasling JD. Microbial engineering for the production of advanced biofuels. Nature 2012;488 320-328. doi:10.1038/ nature11478

[45] García, V, Päkkilä, J, Ojamo, H, Muurinen, E, & Keiski, R. L. Challenges in biobutanol production: how to improve the efficiency? Renewable and Sustainable Energy Reviews (2011). , 15(2), 964-980.

[46] Dauenhauer, P. J, Dreyer, B. J, Degenstein, N. J, & Schmidt, L. D. Millisecond reforming of solid biomass for sustainable fuels. Angewandte Chemie- International Edition in English (2007). , 46(31), 5864-5867.

[47] Carlson, T. R, Vispute, T. P, & Huber, G. W. Green gasoline by catalytic fast pyrolysis of solid biomass derived compounds. Chem Sus Chem (2008). , 1(5), 397-400.

[48] Huber, G. W, & Dale, B. E. Grassoline at the pump. Scientific American (2009). , 301(1), 52-59.

[49] Mohan, D, Pittman, C. U, & Steele, P. H. Pyrolysis of wood/biomass for bio-oil: a critical review. Energy &Fuels (2006). , 20(3), 848-889.

[50] Compton, D. L, Jackson, M. A, Mihalcik, D. J, Mullen, C. A, & Boateng, A. A. Catalytic pyrolysis of oak via pyroprobe and bench scale, packed bed pyrolysis reactors. Journal of Analytical and Applied Pyrolysis , 201190-174.

[51] Knothe, G. Biodiesel and renewable diesel: a comparison. Progress in Energy and Combustion Science (2010). , 36-364.

[52] Janaun, J, & Ellis, N. Perspectives on biodiesel as a sustainable fuel. Renewable and Sustainable Energy Reviews (2010). , 14-1312.

[53] Kunkes, E. L, Simonetti, D. A, West, R. M, Serrano-ruiz, J. C, Gärtner, C. A, & Dumesic, J. A. Catalytic conversion of biomass to monofunctional hydrocarbons and targeted liquid-fuel classes. Science (2008). DOI:10.1126/science.1159210, 322(5900), 417-421.

[54] Chheda, J. N, Huber, G. W, & Dumesic, J. A. Liquid-phase catalytic processing of biomass-derived oxygenated hydrocarbons to fuels and chemicals. Angewandte Chemie- International Edition in English (2007). , 46(38), 7164-7183.

[55] Lee, S. K, Chou, H, Ham, T. S, Lee, T. S, & Keasling, J. D. Metabolic engineering of microorganisms for biofuels production: from bugs to synthetic biology to fuels. Current Opinion Biotechnology (2008). , 19(6), 556-563.

[56] Schirmer, A, Rude, M. A, Li, X, & Popova, E. Del Cardayre SB. Microbial biosynthesis of alkanes. Science (2010). , 329(5991), 559-562.

[57] Huber, G. W, & Corma, A. Synergies between bio- and oil refineries for the production of fuels from biomass. Angewandte Chemie- International Edition in English (2007). , 46(38), 7184-7201.

[58] Al-sabawi, M, Chen, J, & Ng, S. Fluid catalytic cracking of biomass-derived oils and their blends with petroleum feedstocks: a review. Energy & Fuels (2012). , 26-5355.

[59] Lavrenov, A. V, Bogdanets, E. N, Chumachenko, Y. A, & Likholobov, V. A. Catalytic processes for the production of hydrocarbon biofuels from oil and fatty raw materials: contemporary approaches. Catalysts in Industry (2011). , 3(3), 250-259.

[60] Elliot, D. C. Historical developments in hydroprocessing bio-oils. Energy & Fuels (2007). , 21-1792.

[61] Pawalec, B, Navarro, R. N, & Campos-martin, J. M. Fierro JLG. Towards near zero-sulfur liquid fuels: a perspective review. Catalysis Science & Technology (2011). , 1-12.

[62] Stanislaus, A, Marafi, A, & Rana, M. S. Recent advances in the science and technology of ultra low sulfur diesel (ULSD) production. Catalysis Today (2010). , 153-1.

[63] Kulkarni, P. S. Afonso CAM. Deep desulfurization of diesel fuel using ionic liquids: current status and future challenges. Green Chemistry (2010). , 12-1139.

[64] Hanafi, S. A, & Mohamed, M. S. Recent trends in the cleaning of diesel fuels via desulfurization processes. Energy Sources Part A: Recovery, Utilization, and Environmental Effects (2011). , 33-495.

[65] Bezergianni, S, Dimitriadis, A, Kalogianni, A, & Knudsen, K. G. Toward hydrotreating of waste cooking oil for biodiesel production. Effect of pressure, H_2/oil ratio, and liquid hourly space velocity. Industrial & Engineering Chemistry Research (2011). , 50-3874.

[66] Murata, K, Liu, Y, Inaba, M, & Takahara, I. Production of synthetic diesel by hydrotreatment of jatropha oils using Pt-Re/H-ZSM-5 catalyst. Energy & Fuels (2010). , 24-2404.

[67] Melero, J. A, Clavero, M. M, Calleja, G, Garcia, A, Miravelles, R, & Galindo, T. Production of biofuels via the catalytic cracking of mixtures of crude vegetable oils and nonedible animal fats with vacuum gas oil. Energy & Fuels (2010). , 24-707.

[68] Kalnes, T, Marker, T, & Shonnard, D. R. Green diesel: a second generation biofuel. International Journal of Chemical Reactor Engineering (2007). Article A48.

[69] Wijffels, R. H, & Barbosa, M. J. An outlook on microalgal biofuels. Science (2010). , 329(5993), 796-799.

[70] Chisti, Y. Biodiesel from microalgae. Biotechnology Advances (2007). , 25(3), 294-306.

[71] Lestari, S, Mäki-arvela, P, Beltramini, J, Lu, G. Q, & Murzin, D. Y. Transforming triglycerides and fatty acids into biofuels. Chem Sus Chem (2009). , 2(12), 1109-1119.

[72] Radakovits, R, Jinkerson, R. E, Darzins, A, & Posewitz, M. C. Genetic engineering of algae for enhanced biofuel production. Eukaryotic Cell (2010). doi:10.1128/EC. 00364-09.PMCID: PMC2863401, 9(4), 486-501.

[73] Wisconsin Grasslands Bioenergy NetworkBioenergy conversion technologies. http://www.wgbn.wisc.edu/conversion/bioenergy-conversion-technologiesaccessed 27 September (2012).

[74] U.S. Department of Energy, Office of Energy Efficiency and Renewable Energy. National Renewable Energy Laboratory, Biomass Research. http://www.nrel.gov/biomass/biorefinery.html (accessed 14 October 2012)

[75] VirentInc., Technology, BioForming. http://www.virent.comaccessed 14 October (2012).

[76] Connor,E. The integrated forest biorefinery: the pathway to our bio-future. Thermochem Recovery International, Inc. http://www.tri-inc.net/pdfs/Forest%20biorefinery%20white%20paper%20jan%2007%20without%20%20econ%202%20col.pdf (accessed 14 October 2012

Section 3

Environment

Chapter 10

Environmental Considerations About the Life Cycle of Biofuels

Estelvina Rodríguez Portillo,
José Ricardo Duarte Ojeda and
Sully Ojeda de Duarte

Additional information is available at the end of the chapter

http://dx.doi.org/10.5772/52481

1. Introduction

Environmental concerns, along with the goals of energy diversification and rural development, accompanied by policies that drive the use of liquid biofuels for transport in several countries, notably in Europe. Since they started these policies, the perception of the environmental benefits of biofuels has changed significantly. At present there is controversy regarding the benefits and possible negative consequences in terms of environmental impact.

The questions are given based on two negative consequences directly: the occupation of natural areas to expand cultivation and, on the other side, energy efficiency in addition to the possible influence on food prices. The latter reinforces the transformation of natural areas to agricultural land.

This chapter discusses environmental considerations in the life cycle of one of the most widely used biofuel today, the biodiesel obtained from sunflower oil. The environmental impacts along the life cycle approach are described and quantified as well, based on the structure proposed by the ISO: 14040, results of empirical research, conducted in the southern part of eastern Paraguay. To do this, we performed field inventory and assessment of impact by weighting and standardization, supported by LCD SIMAPRO 7.0 program. By unifying features protection areas, evaluations were obtained by assigning a weight range of 40% importance to human health, 40% for ecosystem quality and 20% for resources. Topics to be covered in this chapter include descriptions of the agricultural phase, production of sunflower oil and the industrial stage of biodiesel production. Finally, the proposed units in each category were assessed according to the characterization factors and corresponding formulae and graphs were generated impacts in relation to the three protected areas: damage to human health, ecosystem quality and resources, being this the identification of potential impacts.

2. Life cycle assessment

ISO: 14040, 97 defines Life Cycle Assessment (LCA) as "the collection and evaluation of the inputs and outputs and potential environmental impacts of product system throughout its life cycle. "It is a methodology promoted by the United Nations Program for Environment, driven at the World Summit on Sustainable Development in Johannesburg and standardized in the series 14040, within those for ISO 14000 environmental own efforts. This analysis includes four main stages (Figure 1) comprising the definition and scope, inventory analysis, impact assessment and interpretation of results.

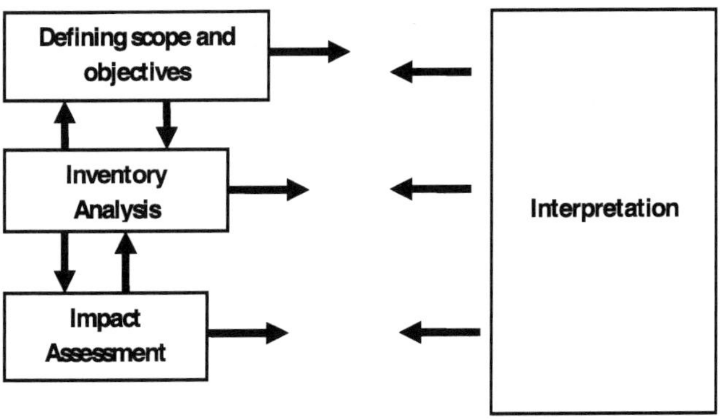

Figure 1. Structure of the life cycle

2.1. Definition and scope

This stage of the process/service/activity begins by defining the global benefits of the research, in which the purpose of the research, the product, the potential market, the importance or significance of the research (limitations of the system), the functional unit, the necessary data and the analysis to be carried out are established.

2.2. Inventory analysis (Life Cycle Inventory LCI)

The analysis for the life cycle for biodiesel consists of processes and systems connected by their common end: the making of the product. Part of this process is the analysis of the inventory that refers to a quantified list of all the incoming and outgoing flows of the system during its working life, and which are extracted from the environment or produced in it by

http://dx.doi.org/10.5772/52481

calculating the energy requirements and materials of the system, and the energetic efficiency of its components, as well as the emissions in each of the processes and systems.

2.3. Impact assessment (Life Cycle Impact Assessment – LCIA)

An evaluation of the classification and assessment of the inventory is carried out according to the inventory analysis lists, and its results are related to evident environmental effects.

2.4. Results interpretation

All results of the preceding part are assessed together in a rational way and according to the previously defined objectives of the research so as to establish conclusions and recommendations for the decisions to be made.

The LCA method is a dynamic one and the four stages in which it is carried out are all connected, as shown in figure 1. Thus, data may be modified or improved as soon as new results are achieved. Social aspects are not typically taken into account by the LCA and it is therefore necessary to use other techniques which provide us with the social problems panorama on this matter (Rajagopal and Zilberman, 2007).

The inventory or environmental balance was divided into the following subsystems: agricultural phase, industrial phase of the oil and industrial phase of the biodiesel, as well as the different means of transport for distribution and use.

3. Content of the agricultural phase – Sunflower crop

This section describes the agricultural process of growing sunflower and quantified inputs and energy expenditure. The raw material is the sunflower species *Elianthusannus*, short cycle (three months or so) from November to January. The technology used is the double disk tillage (Marchetti et al., 2007).

The flows of energy during the lifespan of biodiesel are divided directly and indirectly. A flow of direct energy corresponds to the energy consumed in the way of fossil fuels; firewood, electricity, water steam, and diesel used in; farming operations such as plowing before sowing, defense, and harvest of the crops (Donato, 2007). Fuel used in the transportation of the harvested grain from the field to the stocking plant, industry and drying, denominated in this case short-haul freight (adding short distance of 30 km and 150 km), according to the Márgenes Agropecuarios, 2009. Fuel used in the process of drying which is carried out in a stocking plant whose objective is the reduction of the humidity percentage of the harvested grain, using firewood and water steam for this effect. Electrical energy used in the industrial process for the process of transesterification and maintenance of buildings, which comes from hydro electrical energy (used in the country).

Figure 2. Scheme of the Agricultural Phase Process.

Indirect contributions of energy include the quantities of the agricultural supplies (seeds, fertilizers, and agrochemicals) in order to produce a determined amount of grains per hectare, which is taken as raw material to the process of transformation into biofuels, and the raw materials used in the elaboration of oil and then biodiesel. The outflows of solid, liquid, and gas effluents in each phase, multiplying the amounts by their corresponding energy coefficient. Consistency and balance analysis were carried out to verify the homogeneity of the values (Anderson and Valenzuela, 2007).

http://dx.doi.org/10.5772/52481

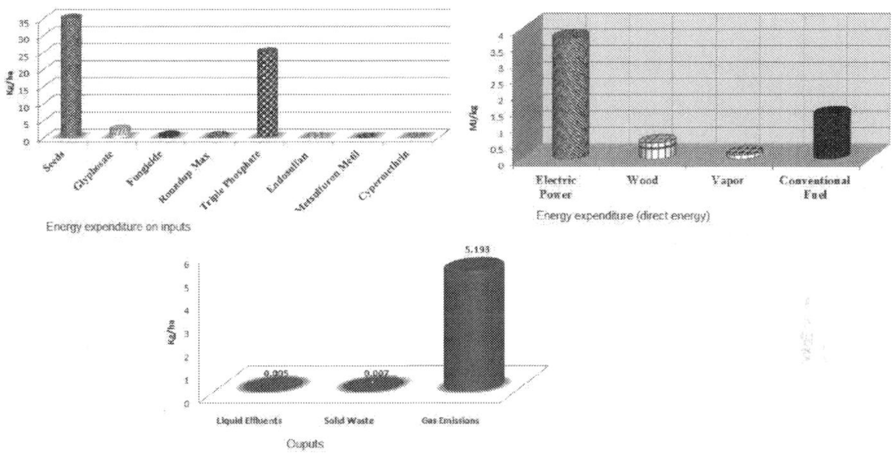

Figure 3. Energy expenditures and outputs.

4. Sunflower oil collection

The technology used in developing the oil is a mixed system, mechanical (pressing) and chemical (solvent extraction).It describes the followed process and the main elements involved in input and energy expenditure.

Once sunflower has been harvested, it has to be cleaned, dried and stored under the adequate conditions. During the storing process seed are rolled in cylinder mills to be moisture-conditioned. Oil is extracted by pressing, and then solvent extraction is done using hexane. Oil is separated from the solvent by using the distillation of micelle and hexane is eliminated from the flour and this is now dried and toast. Finally, crude oil is extracted; refined and sunflower pellet is processed.

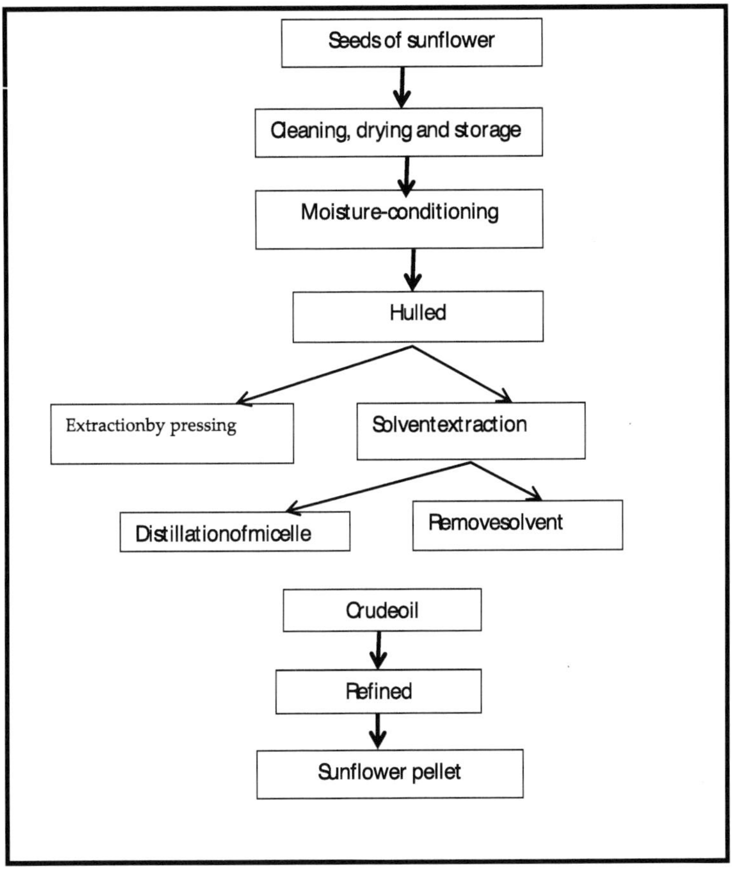

Figure 4. Biodiesel scheme of production.

http://dx.doi.org/10.5772/52481

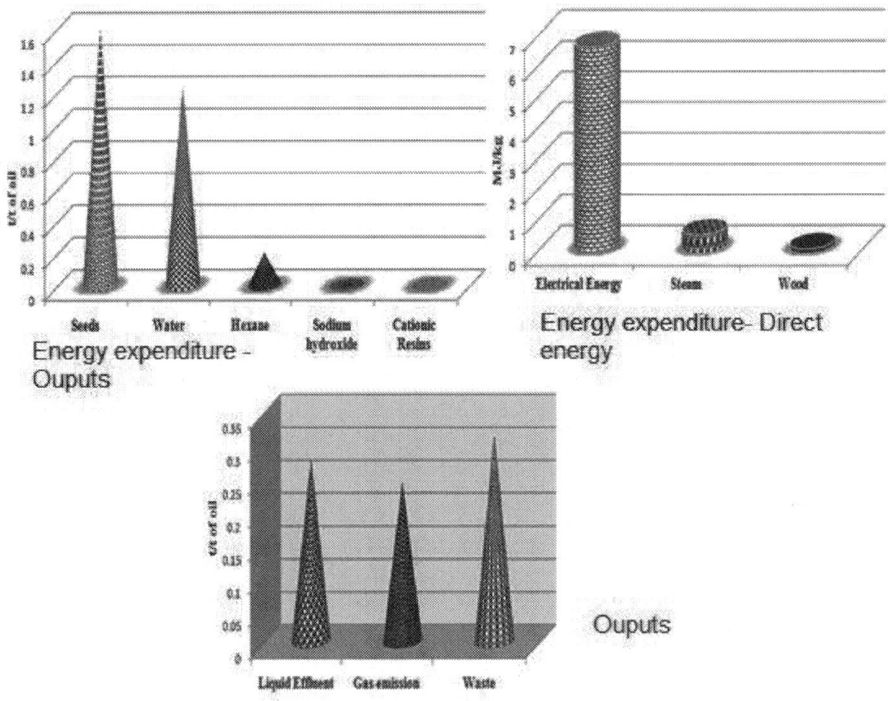

Figure 5. Balance analysis.

5. Biodiesel production

The biodiesel is made from sunflower oil transesterification in a continuous system using mainly methanol and potassium hydroxide as a catalyst obtained as major by-product glycerin. This process involves the use of oils or fats containing free acidity, and in its first phase the free fatty acids are also transformed into methyl ester.

Alternative fuel is produced from vegetable oils, turning triglycerides into methyl or etil esters. This is commonly achieved trhough a method known as transesterification, in which reactions from the three acid chains (ester chains) of each triglyceride molecule are produced by an alcohol and, as a result, these chains are separated from the glycerin molecule.

This separation needs temperature and a powerful base catalytic, such as a hydroxide, for the reaction to be complete. Methanol is generally used for this purpose, although many other alcohols, like ethanol, propane or butane, can be used. Glycerin is obtained as a by-side product and this can be used in other industrial matters, making it a positive factor from the economic point of view (Ma and Hanna, 1999).

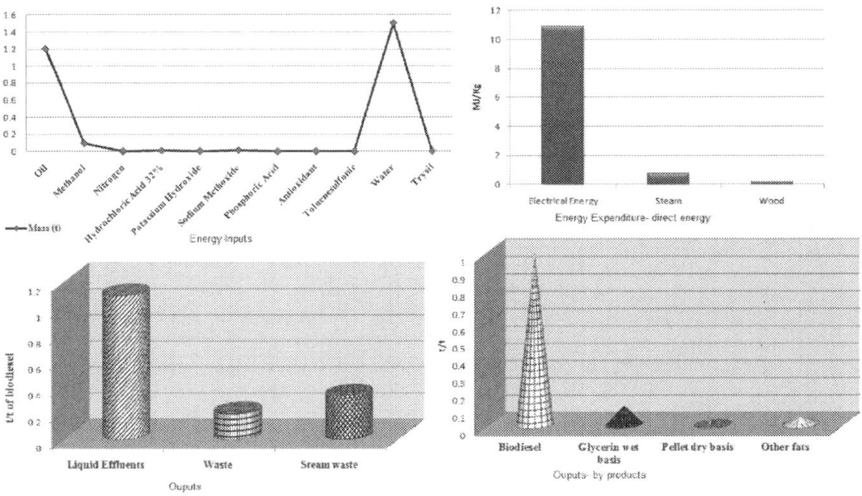

Figure 6. Biodiesel energy balance.

The total of energy consumption for the production of biodiesel was obtained by adding the industrial and farming stages and the different transport of the intermediate products, In figure 6, the total energy expense compared by phases can be observed; in the same figure, it can be observed that the highest direct expense is presented during the stage of industrial transesterification of biodiesel because of the high use of electricity in this phase. However, the largest inflow in supplies and indirect energy expense comes in obtaining the oil because of the higher amount of energy contents in the materials involved in this process.

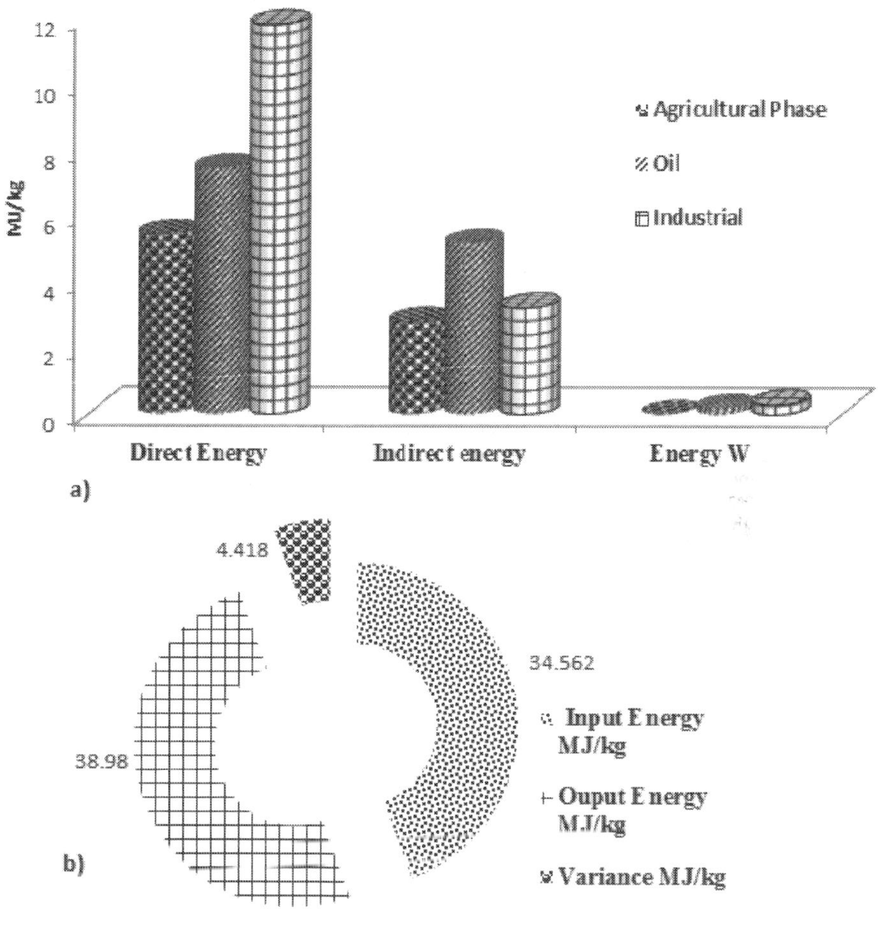

Figure 7. Total energy expense of the cycle, in direct energy consumption, indirect energy, and emission outflow energy, and final energy balance.

6. Identification of impacts on the life cycle

This section gives an overview of the environmental impact of the system under study which results indicate that it is strictly linked to the agricultural phase, although the indicators evaluated the impact is minor compared to other countries such as Argentina and the U.S. The use of fossil fuels, pesticides and phosphate fertilizers are important influencing factors in the Life Cycle Assessment. There is a need to diversify the raw material, for obtaining biodiesel to avoid overuse of soil and other environmental considerations that are practice-based agricultural monoculture. A biodiesel involvement of second and third generation would therefore be relevant.

6.1. Environmental burden – Flow of materials

Environmental impact per mass and energy units. This was determined through the flow of the mass vector, for which the following was taken into account: each flow in the process ($Kg.s^{-1}$) is associated to an eco vector v, which elements are expressed in mass (Kg of the polluting agent per Kg of the product) or in energy ($Kj.\ Kg^{-1}$).

6.2. Energy balance

Coefficient of measurable energetic flow that goes in and out of the biodiesel Life Cycle to determine the quantity of renewable energy that is accumulated in each gram of biodiesel per unit of not renewable energy wasted during its life cycle. This was carried out through the Input/output Ratio connection, from the vector flow for energy (Harding *et al, 2008)*

To assess the impact of the Life Cycle the SIMAPRO LCD software, version 7.0, was used. This is a model of an environmental mechanism which implies three main factors: the geographical characteristics, the elements to be used and the applied technology. An analysis of the inventory was carried out in the field in order to do so, and the following aspects were taken into account:

Energy income: (water steam, electricity, fuels)

Products – by-side products

Raw material income (sunflower seeds, oil, methanol)

Auxiliary income (acids, fertilizers, pesticides, water-soil use)

Waste (urban solids, industrial solids, dangerous solid)

Emissions to water, air and soil (gas, liquid effluents, sunflowers pellet, pesticides containers)

Assignment method: the assignment methods used were the quantification (*mass quantity, using scales),* the estimation (from the estequiometric connections) and calculation (heating power of the used energy).

The proposed units for each category are valued according to the description factors and the corresponding formulae, and impact graphics of the three protection areas pointed out by ISO 14040 were generated: human health damages, resources and ecosystem quality. The description enhances the analysis of the impact of biodiesel upon the affected environmental factors through descriptive factors for each component. Categories such as Never: *with factor=0 with results=0* are omitted.

http://dx.doi.org/10.5772/52481

The next graph shows all the effect categories. Since these are expressed in different units, they are shown in a percentage scale. They indicate the relative contribution for each phase of the product (11 phases).

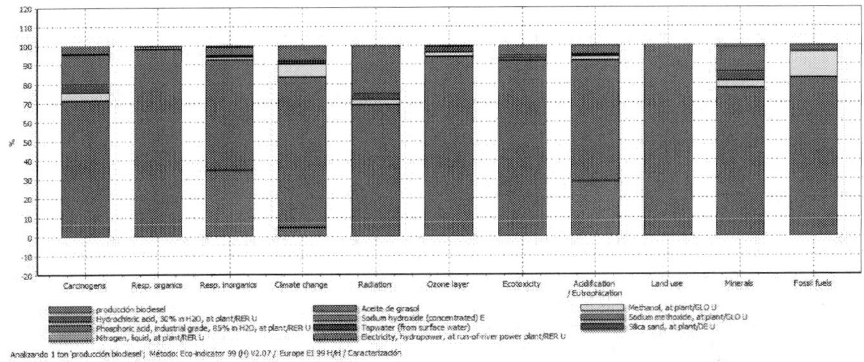

Figure 8. Results obtained from the description of impacts in the Life Cycle of Biodiesel.

Sunflower oil is the most relevant since it is present in all the impact categories, which indicates that the raw material is fundamental in all the soil, ecotoxicity and ozone layer impact cycles, among many others.

6.3. Impact valuation using normalization and weighting

Assessments were obtained by weighting and by unifying the characteristics in the protection areas. The most important areas were human health, with 40% of importance; the quality of the ecosystem, with 40%, and 20% for the resources. The results are shown in the following figure:

The quality of the ecosystem is highlighted as the most affected area in the graph, and the sunflower oil, which is the raw material, is the most important factor. This can be observed in the three areas, but the human health and the resources are less affected. However, when dividing the same factors by a normal value (normalization), there are not significant differences between one and the other form of valuation.

In the implicit ranking of this classification, it is evident that the primary sector determines massive intervention processes and transformation of elements, and natural processes. The secondary sector, through its emissions, effluents and waste production, generates an impact that is, at first sight, less evident, but it may be more dangerous, for there are new factors which are dangerous for the human health or the environment. Finally, a big deal of the tertiary sector generate moderate impacts, with transport as the exception (Pérez, 2007).

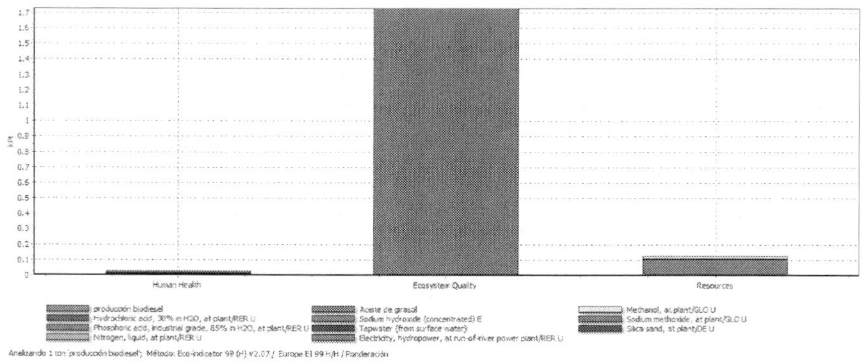

Figure 9. Results obtained from the assessment of impacts of the Life Cycle of Biodiesel (Weighting)

Both assessments show the contributions of the negative impact in the quality of the ecosystem mainly connected to soil erosion. Thus, the use of the land to be used for raw material production should be the first thing to take into account when implementing biodiesel. The seriousness of these impacts depends on the crops expansion and the environmental goods and services that provide the affected ecosystems (Blanco and Azqueta, 2007).

6.4. Damage assessment

Damage assessment results are shown in graph 26, and consist of the relative contribution of the stages of the Life Cycle in negative impacts on the environment and affect the three protection areas issued previously.

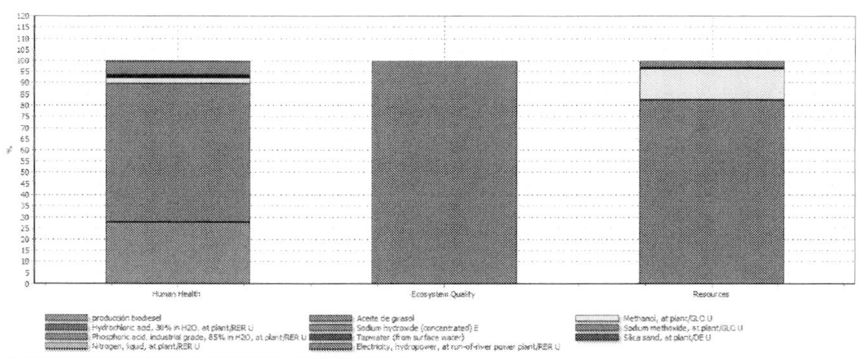

Figure 10. Environmental Damage Assessments in the Life Cycle of biodiesel.

Sunflower oil (integrated in the agricultural phase) has a 100% effect in the quality of the ecosystem, a 60% effect in human health and a similar percentage in the use of resources. Most of the environmental negative impacts start in the agricultural production of the oil, linked to the intensive use of agrochemicals, soil erosion and overexploitation of the resources. Such situations are derived from specific local conditions and are therefore not general.

This may be due to the huge quantity of soil (natural resources) that is used to grow sunflowers and to the agricultural practices which include the use of pesticides that may affect the state of the environmental conditions of nearby ecosystems mainly in terms of biological quality; for the use of pesticides is linked to soil, water and biota pollution in bio-accumulation processes, and in high concentration it may provoke great damage in human health (Timmer, 2002).

6.5. Singular rate

A singular indicator of the impact of the Life Cycle was obtained when the description, weighting and year evaluation were integrated (figure 11). In this indicator there are two processes to be considered: sunflower oil and imported methanol.

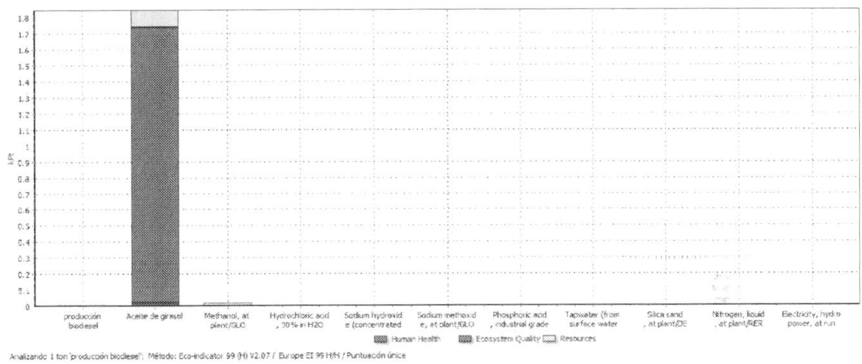

Figure 11. Indicator of the Life Cycle of Biodiesel.

The first one is highly relevant (EPA, 2002), because this would be the indicator of the Life Cycle in which almost all of the impacts affect, being the quality of the ecosystem the most affected protection area by the production and use of sunflower biodiesel in Paraguay.

7. Conclusion

The energy balance is positive; indicating that under the considerations made in this study, the energy system is efficient. This is due to the relationship between energy generated and consumed is greater than one, implying that the system meets the energy function. The environmental impacts evaluated in the biodiesel life cycle phase is related to agricultural ecosystem quality being the most affected area of protection. It affects 100% of the impact factors which influence the use of fossil fuels, pesticides and phosphate fertilizers. Besides the lack of treatment of discharges and waste in the industrial phase, high use of natural resources (land) is another important factor to consider but biofuels may represent a valid alternative energy if there are any mitigating impacts taking into account environmental considerations mentioned.

Author details

Estelvina Rodríguez Portillo[1], José Ricardo Duarte Ojeda[1] and Sully Ojeda de Duarte[2]

1 Universidad Autónoma del Estado de México. Toluca, Estado de México, México

2 Universidad Católica Nuestra Señora de la Asunción, Paraguay

References

[1] Rajagopal, D. y Zilberman, D.. Review of environmental, economic and policy aspects of biofuels.World. Bank Policy Research Working Paper No. 4341. Washington, DC, World Bank; 2007.

[2] Marchetti, J. *et al.* Possible methods for biodiesel production. Renewable and Sustainable Energy Reviews 11, 2007; 1300–1311.

[3] Donato, L. B. Estimación del consumo potencial de gasoil para las tareas agrícolas, transporte y secado de granos en el sector agropecuario, IX Congreso Argentino de Ingeniería Rural y I del MERCOSUR - CADIR 2007. Edited in CD. September 19 to 22; 2007

[4] Anderson, K. and Valenzuela, E. The World Trade Organization's Doha Cotton Initiative: a tale of two issues. The World Economy, 2007; 30(8): 1281–1304.

[5] Ma, F. and Hanna M.A. Biodiesel production: a review. Bioresour Technol 1999;70:1-15.

[6] Harding, *et al.* A life – cycle comparison between inorganic ando biological catálisis for the production of biodiesel. Journal of Cleaner Production. Volume 16.- 3-2008; Pag. 1327-1436.

[7] Pérez C Análisis general de la cadena de valor del biodiesel en El Salvador, Servicio Holandés de Cooperación al Desarrollo SNV y el Consejo Empresarial Salvadoreño para el Desarrollo Sostenible (CEDES); 2007.

[8] Blanco M and Azqueta D Can the Environmental Benefits of Biomass Support Agriculture? Energy Policy; 2007, Volume 36, Pag. 357-466

[9] Timmer, C.P.. Agriculture and economic development .En B.L. Gardner y G.C. Rausser, eds. Handbook of agricultural economics, Vol. 2A. Amsterdam, North-Holland; 2002.

[10] EPA, Environmental Protection Agency. A Comprehensive Analysis of Biodiesel Impacts on Exhaust Emissions. Draft Technical Report. United Status. 126 p.; 2002.

Chapter 11

Environmental Assessment of a Forest Derived "Drop-in" Biofuel

Anthony Halog and Nana Awuah Bortsie-Aryee

Additional information is available at the end of the chapter

http://dx.doi.org/10.5772/52436

1. Introduction

The United States Department of Energy (DOE) has been championing efforts to ensure that the next generation of biofuels will be regarded as "drop-in" biofuels. According to the former DOE Undersecretary Kristina Johnson, "drop-in" biofuels can be defined as fuels produced from various biomass feedstocks which are compatible with the over $9 trillion energy refinery and gas station infrastructure currently in use in the United States [1]. The U.S. National Advanced Biofuel Consortium (NABC) considers it as infrastructure-compatible – they can either be used directly or blended with their petroleum-derived fuels. The European Commission (EC) defined it on the basis of quality specifications (standards) developed by the American Society for Testing and Materials (ASTM International) [2]. Thus, drop-in biofuels in any category (jet fuel, diesel, gasoline) should meet their respective ASTM D standards in order to be classified as such. Jet drop-in biofuels in this scenario are defined by the EC as fuels which meet the ASTM D 1655 and can be used either alone or blended to a certain percentage volume with a conventional fuel before its use (the final blend should have similar properties that meet the standards of ASTM).

Biofuel production involves extracting biomass materials from the environment, which are then transported to processing sites where the biomass is converted into biofuels. From these processing sites, the produced biofuel is then conveyed to end users through distribution points. However, it is critical to identify the effects of these activities and processes at each stage of the life cycle of biofuels with relevance to sustainable development – economic feasibility, environmental soundness and societal acceptability. A Life Cycle Assessment (LCA), which looks at describing the environmental profile of the whole supply chain of drop-in biofuels, has been looked at in this study.

Environmental LCA is a valuable life cycle assessment method used by scientists and researchers in order to assess the environmental aspects and potential impacts associated with a product, process or activity [3]. This assessment method, according to [3], involves four major steps - goal and scope definition, inventory analysis, impact assessment and the interpretation phase. LCA can be used to identify opportunities for improving the environmental performance of processes and activities, informing decision makers in industry, government and non-governmental organizations in order to aid them in strategic decision making and selection of relevant environmental performance indicators; and for marketing purposes (implementing an eco-labeling scheme, making an environmental claim, or producing an environmental product declaration (EPD)). There are two major types of LCA – the attributional LCA which uses average data for each unit process, and the consequential LCA which relies on marginal data for its analysis [4]. Additionally, attributional LCA analysis defines the status quo whilst the consequential LCA measures the impacts through changes in the physical flows.

LCA studies concerning drop-in biofuels are few considering that it is a relatively new form of advanced biofuel which aim to utilize existing infrastructures. However some studies conducted [5-7] emphasize the importance of this type of biofuel in reducing costs associated with replacing existing infrastructures with newly built ones which will be specifically designed for this biofuel type.

2. Problem statement

Meeting the energy needs of the world is necessary for continuous economic growth, enhancing social and even environmental benefits. It has become very clear that conventional biofuels won't be widely accepted if they cannot perform in the same way as conventional fuels. It is for these reasons that nations continue to search for new forms of primary sources of energy. Although this is critically important, an in-depth understanding of the effect of such activities on ecosystems is needed to help make better decisions which will drive the bio-energy revolution positively forward. This study forms part of such investigations.

3. Application area

The biofuel supply chain, according to [8 - 9], has been categorized into 5 phases – feedstock production, feedstock logistics, conversion technology, transportation and end-use of the fuel. For this research in particular, the life cycle stages considered are:

- Biomass production (forest raw materials)
- Biomass transportation
- Biomass conversion (chosen technological process)
- Fuel distribution

http://dx.doi.org/10.5772/52436

- Vehicle fuel use

The study examined the environmental impact of activities across the life cycle by considering the following impact categories across the phases/stages of the supply chain: climate change, eutrophication potential, acidification potential and most importantly, as a result of current discussions, land use change.

4. Research course

4.1. Goal and scope of study

This study of the environmental profile of "drop-in" biofuels production steadily examines the effects of the Thermal Deoxgenation (TDO) process to produce bio-crude from forest-based biomass, along its supply chain. This study aims to answer the following research questions:

- What are the environmental impacts at every stage of drop-in biofuel supply chain?
- What conclusions can be drawn from an LCA study of a relatively new form of advanced biofuel system?

This study is relevant to all stakeholders in the field of energy and environmental impact studies. The diagram below shows the supply chain that has been adapted from the biofuels supply chain described by the 2008 National Biofuel Action Plan [8].

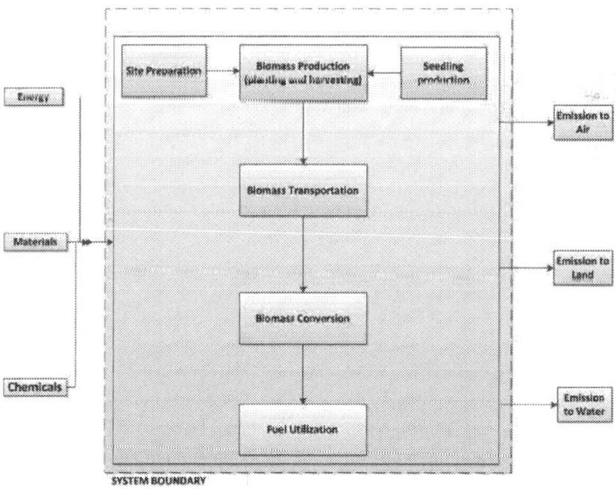

Figure 1. System boundary and process flow diagram for the research study

	Product Specific	Site specific	General	1	2	3	4	5	Comments and Major Literature consulted
Biomass Production									
seedling production	X	X					X		Values obtained from Neupane et al (2011)
site preparation	X	X					X		Values obtained from Neupane et al (2011)
stand preparation	X	X					X		Values obtained from Neupane et al (2011)
harvesting and processing	X	X					X		Values obtained from Neupane et al (2011)
			X		X				FAO (2010)
Biomass Transportation									
Fuel amount used for transportation			X		X				Simapro
Emissions			X		X				Simapro
Biomass Conversion TDO Process									
Electricity used	X				X	X			Data from UMaineEng/ Clayton Wheeler, Author
$Ca(OH)_2$ used	X			X	X				Data from UMaineEng/ Clayton Wheeler, Author
TDO Oil produced	X			X	X				Data from UMaineEng/ Clayton Wheeler, Author
Heat produced	X					X		X	Data from UMaineEng/ Clayton Wheeler, Author
H_2SO_4 Produced	X				X				Data from UMaineEng/ Clayton Wheeler, Author
Emissions produced (CO_2, NOX, CH_4)		X			X				US DOE
Fuel Distribution									
Bio- gasoline to be transported	X				X		X	X	Author
Fuel for distribution vehicles			X				X		Pradhan et al (2009), NREL
Emissions (CO_2, NOX, CH_4)	X		X		X				Date from EIA, EPA
Fuel Use -Biogasoline use									
Emissions (CO_2, OX, CH_4)			X		X				EPA

1. Measurement 2. Computations made by author 3. Data obtained from technology 4. Data obtained from similar technology 5. Approximation

Table 1. Data quality scoring system developed for the study

4.2. Functional unit

The functional unit used for all inputs and outputs used in the study is 1 liter of the TDO process drop-in bio-oil.

4.3. Method used

4.3.1. Data quality requirements

Data (e.g. fossil energy consumptions, carbon dioxide emissions) are analyzed carefully in order to map out the energy component and its associated emissions during the life cycle phases of the bio-oil. The raw material usage component which includes the use of land (the impact to be examined in that respect is land use change impacts) as well as Fertilizers, pesticides and other chemicals known for their use at any stage are accounted for by categorizing them into certain impact categories like acidification potential and eutrophication potential. The study follows procedures as stated in [10] and compiles data from existing databases available in Simapro, National Renewable Energy Laboratory (NREL), Environmental Protection Agency (EPA) as well as from literature in determining the effects with relevance to resource depletion, climate change and water use. A data quality scoring system is provided in Table 1 left.

4.3.2. Life cycle inventory analysis

In order to collect data in accordance with ISO 14040 guidelines, data representativeness, accuracy and consistency were considered. In most cases, specific data for the northeastern region and specifically for the state of Maine were used. For the processes elaborated on in subsequent parts of this study, the economic and environmental input and output flows have been calculated..

4.3.3. Allocation

It is understandable from the principles of LCA that inputs (resource consumptions), outputs and related environmental impacts can be allocated based on different basis (e.g. energy or mass basis). In this study, allocation was done with respect to mass basis of the amount of levulinic acid and formic acid (forming the bio-oil). This was necessary in calculating the amount of energy required in producing the TDO process oil only whilst considering other by-products like char, water and carbon dioxide.

4.3.4. Inputs and outputs

The table below shows all the inputs and outputs that were considered for the supply chain of drop-in biofuels.

Process (stage)	Inputs	Outputs
Biomass production	Nitrogen, phosphorus, potash for seedling production Water for seedling production Electricity for seedling production Fuel for site preparation, stand preparation and harvesting Lubricant for harvesting and processing	Harvesting of raw biomass Emissions
Biomass transportation	Fuel	Emissions –CO2, NOx, CH4, SO2, VOC, Volatile organic compounds
Biomass Conversion process (TDO)	Electricity Ca(OH)2	TDO Oil Heat H2SO4 Emissions-CO2 Steam
Fuel Distribution	Bio-gasoline from TDO oil Fuel for transporting bio-gasoline	Bio-gasoline Emissions-CO2, NOx, CH4
Fuel use	Bio-gasoline	Emissions- CO2, NOx, CH4

Table 2. Inputs and Outputs associated with each stage of the supply chain

5. Status and results

5.1. Biomass production

The biomass production process assumes 100% biomass supply from willow (tree diameter of 6 inches at 4 ½ feet just like Aspen [11] which is typically a hardwood (lignocellulosic biomass). The willow tree (S. alba) has a cell wall composition of 49% cellulose, 27% hemicell-lose and 23% lignin [12, 13]. This stage of the supply chain was comprised of seedling production, site preparation, stand preparation, harvesting and processing. It was assumed that fertilizers were applied to aid the growth of the biomass. Data used for the early stage of biomass production in the state of Maine was obtained from Neupane et al.'s [14] work conducted on assessment of woodchips for bioethanol production.

The amount of harvested biomass was estimated to be 0.014102 tonnes based on the functional unit of 1 liter of bio-gasoline to be used by vehicles. Details of the 0.014102 tonnes include 18% of the described willow tree biomass on dry basis (a single tree). Nitrogen, phosphorus and potash applied were calculated to be 0.00818g, 0.0129g and 0.00818g, respectively. Water use amounted to 0.9 liters whilst electricity use, fuel use and lubricant use in the biomass production process were 0.0028 kWh, 0.936 liters and 0.04779 liters, respectively.

http://dx.doi.org/10.5772/52436

Total carbon dioxide equivalent emissions accounted for in this life cycle phase was 0.772kg.

5.2. Biomass transportation

The second stage of the biofuel supply chain involves the transportation of the biomass produced from the production site to the processing site (TDO processing point). The assumption of single unit trucks each weighing 60 tons [15] and making one trip per day was made for this phase of the life cycle. The choice of 72 kilometers in the project was based on previous work done by Neupane et al. [14]. From the calculations done, the resulting outputs of the unit process were emissions associated with the transportation CO2 (173g CO_2eq), NO_x (1.237g), CH_4 (0.0042g), dinitrogen monoxide (0.0062g), sulfur oxides (0.00327g) and VOC (0.0085g). The data used for the calculation of the emissions were derived from the Ecoinvent V2.1 database found in SIMAPRO.

5.3. Biomass conversion – TDO process

This process of biomass conversion to TDO which is unique to the University of Maine was used in the biofuel processing phase of the biofuel supply chain. This process produces a drop-in biofuel, which has been found to have boiling points similar to that of jet fuel, diesel and gasoline. Although further refining is needed in order to meet biofuel emission standards, all other technical properties make the new fuel attractive for use in existing fuel infrastructure without much further processing.

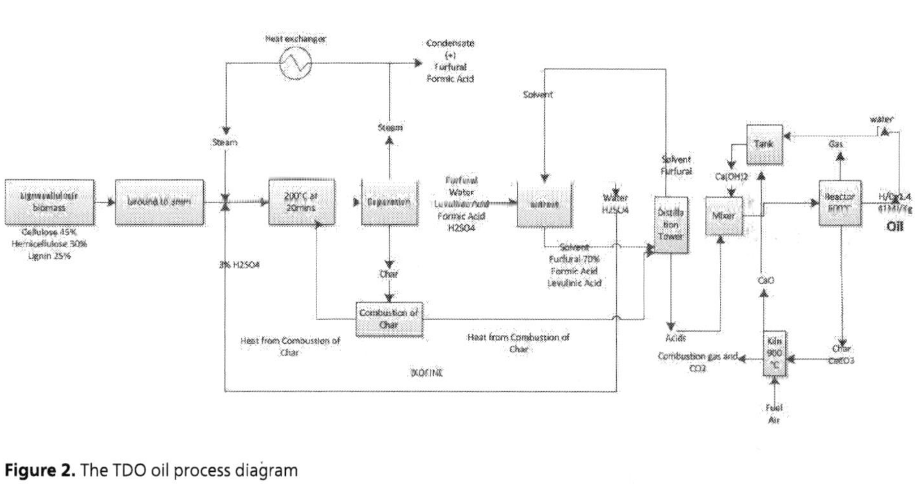

Figure 2. The TDO oil process diagram

This type of drop-in oil was produced through the thermal de-oxygenation process (TDO). This process starts with the conversion of cellulose to organic acids. The acids are combined with calcium hydroxide to form a calcium salt. The salt is heated to a higher temperature whilst being stirred. The resulting reaction from this stirring leads to the formation of a dark amber colored oil. The reaction also removes the oxygen from the oil. This key step (i.e. removal of oxygen) distinguishes the TDO process from other biofuel processes. The oxygen is removed as both CO_2 and water without the need for any outside hydrogen supply. Therefore most of the energy in the original cellulose is contained in the new oil. On mass basis, 13% of the initial lignocellulosic raw material is recovered as TDO bio-oil output. The oil has less than 1% percent oxygenates and has an energy value of 41MJ/Kg. The diagram below shows the TDO process[1] in detail.

From lab scale experiment, we found that to produce a liter of the final product of bio-oil (i.e. bio-gasoline), 0.09kWh of electricity was consumed over 24 hours. 0.077kWh of heat is also expended. Sulfuric acid which is produced in the system and recycled for usage is also needed as an input to treat 76% [13] of the raw material (Holocellulose). Outputs of this stage of the supply chain include the TDO oil (1 liter) – also the functional unit, Ca $(OH)_2$, NO_x(0.00246g), CH_4 (0.0056g) and steam (0.23[2]tonnes). This specific process did not produce carbon dioxide emission from the electricity usage because the primary source of energy for the electricity used is from the TDO biofuel. Most importantly, the biomass used for the TDO biofuel production is assumed to meet all sustainability criteria. As a result of this, there are no emission factors associated with it [16]. Secondly, since the carbon dioxide produced by the burning of the biofuel is offset by the biomass regeneration, it is technically justifiable to put the emisssion factor to zero.

5.4. Fuel distribution

This phase of the supply chain involves the transportation of the produced TDO oil to the distribution point where it will be used by vehicles or other usage facilities. It was assumed from average data gathered in the United States that 47% [17] of crude oil is converted into gasoline. Thus, from the 0.57 US gallon of TDO oil, 1 liter of bio-gasoline product is derived. With regard to the transportation fuel needed to convey the derived bio-gasoline product, it was assumed from literature that 0.08 gallon of fuel was needed to transport 1 gallon of biofuel (i.e. biodiesel) [18]. At the end, the total energy of the fuel used in transportation was estimated to be 0.183 kWh. Outputs of this stage included the transported bio-gasoline, and emissions – CO_2 (19.23g), NO_x (0.861g) and CH_4 (2.14g).

5.5. Fuel use

In the final stage of the supply chain, the major assumption was the choice of an EPA Tier 2 vehicle using an average of 9.6 liters of gasoline per 100km [19]. Through the use of 1 liter[3] of bio-gasoline, emissions produced included CO_2(2230g), NO_x (22.9g) and CH_4 (112.4g) [20]. Further analysis on the gathered data revealed the transportation stage as the most crucial stage due to the high energy usage and also the high amount of carbon dioxide associated with the energy usage. The following tables show some calculated linked flows normalized to the functional unit and the flows passing the system boundary.

1 TDO process description provided by Dr. Clayton Wheeler. University of Maine Biological and Chemical Engineering Department

2 Steam amount produced estimated by Dr Clayton Wheeler to be 200000lbs/hr

3 Calculated from Energy Information Administration, Documentation for Emissions of Greenhouse Gases in the U.S. 2005, DOE/EIA-0638 (2005), October 2007, Tables 6-1, 6-4, and 6-5. (Non-biogenic carbon content and gross heat of combustion for motor gasoline and diesel (distillate fuel))

Total Energy Calculated	(kWh/FU)
Electricity	0.082418539
Fuel (gasoline)	2.76
Heat	0.069743855

Table 3. Total Energy Calculated

Total Water calculated	
Water (tonnes/FU)	0.120283142
Steam (tonnes/FU)	0.212063842

Table 4. Total Water Calculated

Charts were drawn to show graphically the contribution of each stage in the life cycle of the TDO drop-in biofuel. It is important to note in Figure 3, transportation stage contributes a major chunk of the carbon emissions associated with the supply chain.

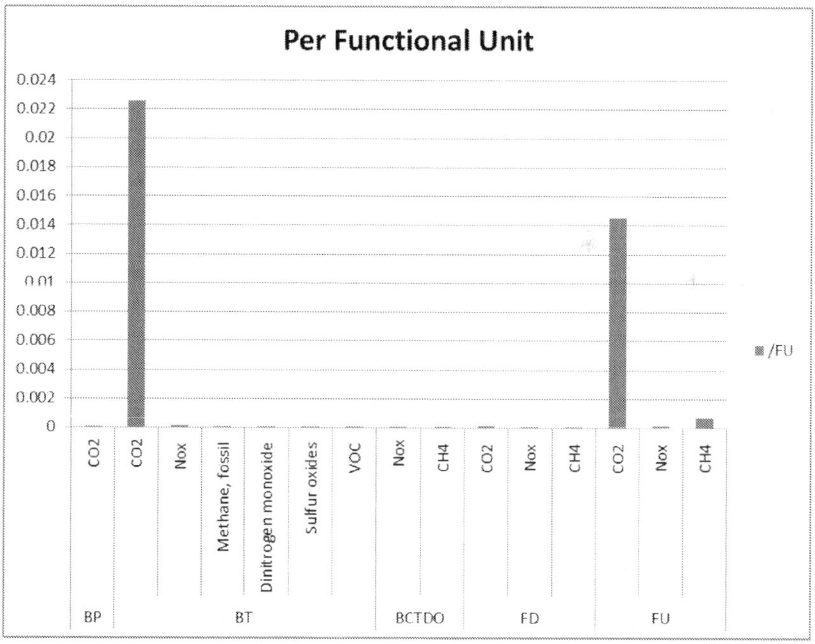

Figure 3. Chart showing life cycle emissions normalized per functional unit; BP=Biomass Production, BT= Biomass Transportation, BCTDO=Biomass conversion TDO Process, FD=Fuel Distribution, FU=Fuel Use

5.6. Life cycle impact assessment

According to the operational guidelines in the ISO standards of LCA, the impact assessment phase of LCA aims to make sense of the data obtained from the inventory analysis phase of the procedure. To interpret the environmental impacts and societal preferences, the following baseline impact categories were chosen in line with the CML method (refer to [21]): climate change, photo-oxidant formation, acidification, eutrophication potential.

Climate change: As stated by [22], this is expressed as

$$\sum_i GWP_{a,i} \times m_i$$

This indicator result is expressed in kilograms (Kg) of the substance of reference, CO_2. *GWP a,i* is the global warming potential for substance *i* over a period of years. *m* is the quantity of substance that *i* emitted over those years. *GWP* over a period of 100 years is used.

Photo-oxidant formation: According to [23], photo-oxidant formation is measured as

$$\sum_i POCP_i \times m_i$$

This indicator result is expressed in kg-ethylene equivalent. $POCP_i$ is the photochemical ozone creation potential for substance *i*. m_i on the other hand is the quantity of substance *i* emitted. In this case study, NO_2 was assumed to be 15% of the total NO_x emissions. This assumption was based on the paper by [24].

Acidification Potential: This potential impact category is expressed in kg-SO2 equivalent. The formula for calculating the potential value is

$$\sum_i AP_i \times m_i$$

AP_i is the acidification potential for substance *i* emitted to the air. m_i is the emission of substance *i* to the air [25].

Eutrophication Potential: The eutrophication potential is expressed as $\sum_i EP_i \times m_i$

This indicator unit is kg PO_4^{3-}. EP_i is the eutrophication potential for substance *i* emitted to air, water or soil whilst m_i is the emission of substance *i* to air, water or soil.

Impact category	Quantity	Unit	Normalization Factors	Normalized Values
Climate Change (GWP 100)	0.156335657	Kg CO2 equivalent	6.83E+03	2.29E-05
Photo-oxidant formation	0.000119795	Kg ethylene equivalent	8.04	1.49E-05
Acidification Potential	0.000320742	Kg SO2 -equivalent	5.29E+01	6.06E-06
Eutrophication Potential	4.3899E-05	kg PO43- equivalent	2.28E+01	1.93E-06

Table 5. Characterization of Chosen Impact Categories

The annual extent of the world's baseline impact categories (mid-1995) was used in normalization of the characterized results.

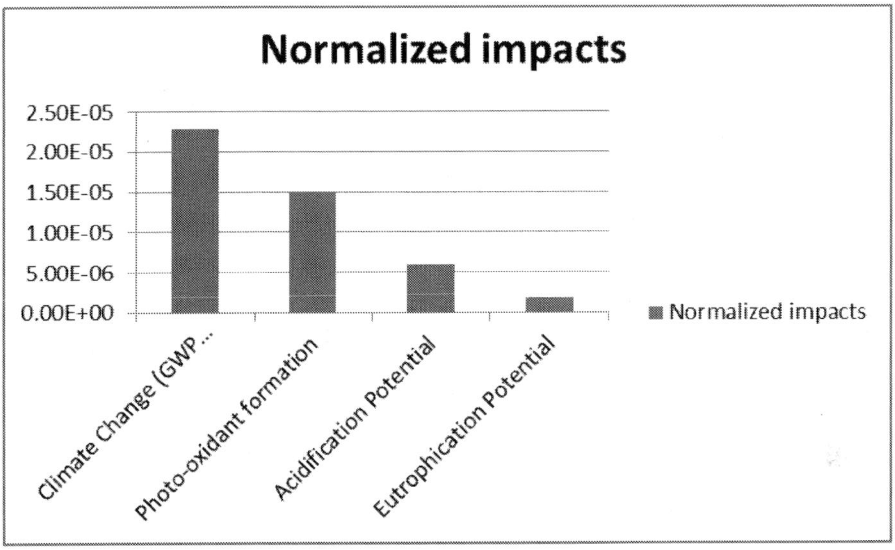

Figure 4. Normalized results showing the contribution of each impact to the environment relative to the functional unit of 1 liter of TDO oil

6. Discussion

6.1. Interpretation

Results obtained from the environmental profiling of the supply chain of the University of Maine TDO drop-in biofuel estimate the potential effects of the activities identified. Based on initial experimentation and inventory analysis, it was determined that the transportation phase of the bio-oil supply chain played a major role in the release of emissions to the environment. This was due to the high amount of fossil energy supplied externally that was assumed to be consumed in the transportation phase. The conversion step which was also observed to be energy intensive will have produced far bigger emission results if there was a reliance on electricity from the grid instead of using internally generated electricity with the TDO biofuel as its primary source.

Based on the inventory analysis conducted and after careful analysis, four major impact assessment categories were chosen: climate change (global warming potential), photo-oxidant formation, acidification potential and eutrophication potential. Many assumptions were also made in order to arrive at the results obtained. Such assumptions included the non-reuse of Ca $(OH)_2$ and sulfuric acid, which did not factor in the calculation of the impact assessment categories chosen.

The global warming potential (GWP) (100) impact assessment investigation revealed that in order to produce 1 liter of this new novel biofuel, 0.15kg of CO2 equivalent are emitted. The normalized results of this with respect to the worldwide normalization factor showed a value of 2.29E-6 for GWP. The highest contributor to this indicator of the amount of heat trapped in the atmosphere by the greenhouse gases emitted is the transportation sector (biomass transportation and fuel distribution), which as mentioned in earlier paragraphs involves the use of a very significant amount of fossil based energy in the form of electricity (supplied externally). It is important to note that transportation phase in the life cycle plays a significant role in the emission of large amounts of carbon dioxide which is also reflected in the calculation of the global warming potential.

With respect to photo-oxidant formation, 0.000119 kg of ethylene equivalent was released relative to the functional unit. Assumptions made from literature estimated the NO_2 composition of NO_X to be 15%. Based on this assumption, NO released along the supply chain accounted for close to 96% of the photo-oxidant formation potential.

Acidification potential (AP) and Eutrophication potentials (EP) are environmental effects that were important in understanding the environmental profile of the scaled-up TDO oil supply chain. The resultant AP and EP were estimated to be 0.00032 kg SO_2 equivalent and 4.38E-5 kg PO4 3- equivalent respectively. The normalized impacts in terms of global factors were calculated to be 6.06E-6 and 1.93E-05 respectively. The major contributing factor to the acidification potential was the NO component of the NO_x. This was also the case in the eutrophication potential. The phase which contributed the most to these impact assessment categories was the conversion phase. This was also associated with the external electricity supply for the conversion process.

7. Conclusion

This study initially evaluated the environmental life cycle impacts across the supply chain of a new drop-in biofuel, developed by the University of Maine (UMaine) College of Engineering, which is still at the bench scale. The study made use of primary data (available in UMaine Chemical Engineering Department) for the biomass conversion stage and utilized Maine's regional data as well as generic data developed in the United States. The study had a few limitations in terms of data quality and uncertainty. Important issues which are very relevant for LCA practices in the field of biofuel development and which have not been addressed in this paper were water and land use as well as the impacts on ecosystem goods and services. Land use change as a result of the activities involved in the growth and harvesting of willow is measurable in the sense that land use change (increase of land competition) can be estimated to be 35.9 m^2.yr (average age of willow in Northeastern America is 55 years to full maturity).

In the case of water use, it is clear from the inventory analysis that the biomass production phase requires a significant amount of water intake to enable the biomass to grow to maturity. A water intake of 0.120 tonnes shows the importance of finding a suitable source of water to use in the first phase of the life cycle and not relying on potable water for the growth of the biomass.

It is important to note that even though the global warming potential associated with carbon dioxide is very high, it should be understood that growing biomass actually helps reduce the atmospheric carbon dioxide in conjunction with photosynthesis. This actually leads to the conclusion that, compared to the conventional oil produced from fossil fuels, the UMaine TDO oil could be better. This is because, internal physiological processes of the biomass during the biomass production phase reduce the carbon dioxide emitted during the complete life cycle phase by using it to further the growth of the trees.

This study is an attributional LCA study, which can be further improved by a thorough life cycle sustainability assessment study [26] which takes into account the effects on the economic and social well-being of the stakeholders and the state of Maine in general. Such a study also advances that LCA is important in making decisions that will affect the long term sustainability of rural communities.

Acknowledgement

Financial support for this research was given by the United States Department of Agriculture (USDA) through the Agriculture and Food Research Initiative (AFRI). Special assistance was also provided by Dr. Clayton Wheeler of the University of Maine Forest Bioproduct Research Institute.

Author details

Anthony Halog[1,2*] and Nana Awuah Bortsie-Aryee[1]

*Address all correspondence to: anthony.halog@gmail.com

1 School of Forest Resources, University of Maine, Orono, USA

2 School of Geography, Planning and Environmental Management, University of Queensland, Brisbane, Australia

References

[1] Johnson K. "Drop-in" Biofuels: Excepts from speech at the 4th Annual Cellulosic Biofuel Summit 2009. http://vtbio.org/VTBIO_News_files/Drop-In%20Biofuels.pdf. (accessed 08 December 2011).

[2] EC (European Commission). Technical Paper- 2 Million tons per year: A performing biofuels supply chain for EU Aviation. European Advanced Biofuel Flight path 2011. http://ec.europa.eu/energy/renewables/biofuels/doc/20110622_biofuels_flight_path_technical_paper.pdf (accessed 14 October 2011).

[3] ISO (International Organization for Standardization). Environmental Management – Life Cycle Assessment – Principles and Framework 1997.

[4] Ekvall T, Andrae A. Attributional and consequential environmental assessment of the Shift to lead-free solders. The International Journal of Life Cycle Assessment 2006; 11 (5), 344 – 353.

[5] Wu M, Wang M, Liu J, Huo H. Life-Cycle Assessment of Corn-Based Butanol as a Potential Transportation Fuel. Argonne National Laboratory 2007.

[6] Shonnard DR, Williams L, Kalnes TN. Camelina-Derived Jet Fuel and Diesel: Sustainable Advanced Biofuels. American Institute of Chemical Engineering. Environment Program Sustainable Energy 2010; 29: 382–392. doi: 10.1002/ep.10461

[7] Pfromm HP, Amanor-Boadu P, Nelson R, Vadlani P, Madl Ronald. Bio-butanol vs. bio-ethanol: A technical and economic assessment for corn and switchgrass fermented by yeast or Clostridium acetobutylicum. Elsevier 2010.

[8] Biomass Research and Development Board. National Biofuels Action Plan. Federal Interagency Biomass Research and Development Initiative. US Department of Agriculture (USDA) and US Department of Energy (DOE) 2008. October 2008.

[9] National Academy of Sciences. Expanding Biofuel Production: Sustainability and the Transition to Advanced Biofuels. Summary of a Workshop (2010). Science and Technology for Sustainability Program (STS) 2010. United States.

[10] ISO (International Organization for standardization). Environmental Management – Life Cycle Assessment – Principles and Framework 2006.

[11] Philip J. Units of Measure and Conversion Factors for Forest Products. Bulletin #7103. University of Maine Cooperative Extension 2004.

[12] Möller R, Toonen M, van Beilen J, Salentijn E, Clayton D. Crop Platforms for Cell Wall Biorefining: Lignocellulose Feedstock. Outputs of the EPOBIO Project. April 2007. United Kingdom

[13] Fengel D, Wegener G. Wood - Chemistry, Ultrastructure, Reactions. Walter de Gruyter; 1989 Berlin, New York.

[14] Neupane B, Dhungel S, Halog A. Attributional Life Cycle Assessment of Woodchip for Bioethanol Production. Journal of Cleaner Production 2011; 19: 733-741.

[15] Gonzalez-Garcia S, Berg S, Feijoo G, Moreira MT. Comparative environmental assessment of wood transport models. A case study of a Swedish pulp mill. Science of the Total Environment 2009;407 3530-3539. Elsevier.

[16] EU Mayors. Technical annex to the SEAP template instructions manual. The Emission Factors. Covenant of Mayors. www.eumayors.eu/IMG/pdf/technical_annex_en.pdf (accessed 1 August 2012).

[17] Newton. Oil to Gasoline. electronic community for Science, Math, and Computer Science K-12 Educators, sponsored and operated by Argonne National Laboratory's Educational Programs. Available at: http://www.newton.dep.anl.gov/askasci/gen99/gen99675.htm. (accessed 12 November 2011).

[18] Pradhan A, Shrestha DS, McAloon A, Yee W, Haas M, Duffield JA, Shapouri H. Energy Life Cycle Assessment of Soybean Biodiesel. U.S. Department of Agriculture, Agricultural Economic 2009 Report Number 845.

[19] Cheah L, Evans C, Bandivadekar A, Heywood J. Factor of Two: Halving the Fuel Consumption of New US Automobiles by 2035. Laboratory for Energy and the Environment. Massachusetts Institute of Technology 2007. Publication No. LFEE 2007-04 R.

[20] Energy Information Administration. Documentation for Emissions of Greenhouse Gases in the United States 2005, DOE/EIA-0638. October 2007, Tables 6-1, 6-4, and 6-5. (Non-biogenic carbon content and gross heat of combustion for motor gasoline and diesel (distillate fuel)).

[21] Guinée BJ, Gorrée M, Heijungs R, Huppes G, Kleijn R, de Koning A, van Oers L, Sleeswijk AW, Suh S, de Haes HAU. Life Cycle Assessment: An Operational Guide to the ISO Standards. Kluwer Academic Publishers; 2002. Dordrecht / Boston / London.

[22] Houghton JT, MeiraFilho LG, Callander BA, Harris N, Kattenberg A, Maskell K .Climate Change 1995: The Science of Climate Change. Cambridge University Press; 1996. Cambridge.

[23] Huijbregts MAJ, Thissen U, Guinée BJ, Jager T, Van de Meent D, Ragas AMJ, Sleeswijk AW, Reijnders L. Priority Assessment of Toxic Substances in Life Cycle Assessment I: Calculation of Toxicity Potentials for 181 Substances with the Nested Multi-Media Fate, Exposure and Effects Model USES-LCA. Chemosphere 2000; 41(4): 541-573.

[24] Soltic P, Weilenmann M. NO2/NO Emissions of Gasoline Passenger Cars and Light-Duty Trucks with Euro-2 Emission Standard. Atmospheric Environment 2003; 37: 5207–5216.

[25] Heijungs R, Guinée BJ, Huppes G, Lankreijer RM, de Haes HAU, Sleeswijk AW, Ansems AMM, Eggels PG, van Duin R, de Goede HP. Environmental Life Cycle Assessment of Products. Guide and Backgrounds. CML, Leiden University; 1992. Leiden.

[26] Halog A., Manik Y. Advancing Integrated Systems Modelling Framework for Life Cycle Sustainability Assessment. Sustainability. 2011; 3(2):469-499. (available at http://www.mdpi.com/2071-1050/3/2/469)

[27] Bortsie-Aryee Life Cycle Assessment – Study of Umaine TDO Lignocellulosic Drop-in Biofuel. Paper submitted as FTY 582 Requirement. University of Maine, Orono; 2011.

[28] Dickerson K., Rubin J. Maine Bioproducts Business Pathways. Margaret Chase Smith Policy Centre, http://www.forestbioproducts.umaine.edu/library/Maine-Bioproducts-Business-Pathways/203/. Forest Bioproducts Research Initiative, School of Economics, University of Maine, OronoME ; 2008.

[29] Food and Agriculture Organisation.Impact of the Global Forest Industry on Atmospheric Greenhouse Gases. FAO Forestry Paper 2008;159. Rome. Italy.

[30] Hetland J, Teimuraz G. Security of Natural Gas Supply through Transit Countries. NATO Science Series II.Mathematics, Physics and Chemistry 2003; 149.

[31] NABC (National Advanced Biofuel Consortium.Biofuels for Advancing America.About.http://www.nabcprojects.org/about.html 2010. (accessed 14 October 2011).

[32] Pitcher JA, Mcknight JS. Black Willow. http://forestry.about.com/library/silvics/blsilsalnig.htm. (accessed 18 November 2011).

[33] United States Department of Energy. Updated State level Greenhouse Gas Emission coefficients for electricity generation (1998-2000) 2002.

[34] United States Environmental Protection Agency. Inventory of U.S. Greenhouse Gas Emissions and Sinks:1990-2009. 2011; EPA 430-R-11-005.

Chapter 12

Evaluation of Gaseous Emission in the Use of Biofuels in Brazil

Marcelo Silva Sthel, Aline Martins Rocha, Juliana Rocha Tavares, Geórgia Amaral Mothé, Flavio Couto, Maria Priscila Pessanha de Castro, Victor Habez Perez, Marcelo Gomes da Silva and Helion Vargas

Additional information is available at the end of the chapter

http://dx.doi.org/10.5772/52773

1. Introduction

In the seventies, the human society was faced with an important issue: the environmental problems. Is it possible to continue the exploitation of natural resources indefinitely? What to do with industrial waste and toxic gases emitted by industries and vehicles that use predominantly fossil fuels as energy source? Since the beginning of environmental awareness to the present day, many events have occurred, mobilizing international public opinion. In 1992, the International Convention on Climate Change has occurred and in 1997, the Kyoto Protocol aimed at reducing emission of greenhouse gases.

The Intergovernmental Panel on Climate Change (IPCC) report [1, 2] concluded that the use of fossil fuels, intensified after the Industrial Revolution, is closely related with the increasing of the average temperature of the earth, due to increased emissions of greenhouse gases (GHGs).

A possible alternative to minimize this problem is the use of renewable energy in large scale in order to reduce greenhouse gas emissions [3]. The IPCC report suggests that, in the transport sector, biofuels can play an important role in reducing emissions of greenhouse gases, depending on their production pathway.

It is projected that biofuels used as additives / substitutes for petrol and diesel will increase its participation to 3% of the total energy demand for transport in 2030. This value can increase to about 5-10%, depending on future oil and carbon prices, improvements in vehicle efficiency and the success of technology in the use of cellulosic biomass [4, 5].

Policies aimed at solving issues such as traffic congestion, air quality and energy security are closely associated with emission reductions in the transportation sector.

Currently, the transport matrix in Brazil is predominantly by road, with intensive use of fossil fuel sources (diesel and gasoline). Thus, the government created two programs that encourage the use of biofuels in the transport sector: the Biodiesel National Program and the Ethanol National Program. Currently, all diesel fuel sold in Brazil has 5% of biodiesel (B5), whereas the gasoline used in the vehicles has 20% of ethanol. Recently, with the increased use of flex-fuel vehicles, which may be fueled with gasoline and ethanol in varying percentages, the use of biofuels becomes more prominent.

Brazil has a great ability to produce bioenergy, once it is a tropical country (insolation), with extensive arable lands, regular rainfall, water resources and rich biodiversity, enabling the success of ethanol and biodiesel programs.

The use of biofuels has generated controversy in the scientific, technical and policy areas around the world [6-8], particularly in regard to emissions of carbon dioxide, oil prices, fuel and food security, destruction of forests, soil erosion, impacts on water resources, energy balance and energy security.

In relation to the greenhouse effect, not only the carbon dioxide emissions from burning fuels must be considered, but the hole production chain. It is necessary to consider the use of nitrogen fertilizers, the land use and nitrous oxide emissions [9-11]. In the case of ethanol, we must consider the emissions generated by agricultural machines, sugarcane transportation, ethanol transportation to consumption centers, generation of carbon dioxide during the fermentation of ethanol and the energy source that powers the ethanol industry. Similar concern should be considered in the production cycle of biodiesel, produced predominantly from soybean in Brazil. All these criteria need to be evaluated in the sustainability studies of biofuels production chain.

This chapter presents some environmental aspects of Brazilian studies biofuels (biodiesel and ethanol) evaluating pollutant emissions from gasoline and ethanol vehicles exhaust and also from bench engines powered by biodiesel and diesel. Furthermore, we evaluated the emissions of greenhouse gases in sugarcane soils.

For this study, we used techniques that ensure high selectivity to identify the gaseous species in a mixture of gases, high sensitivity for detecting low concentrations of gases, good temporal resolution and the possibility of online measurements. The photoacoustic technique has all the characteristics to be used in these measurements.

Photoacoustic spectroscopy is used for the detection of several gases in the concentration range of parts per billion by volume (ppbv) and sub-ppbv [12-21]. This methodology is based on the generation and detection of pressure waves (sound) inside a resonant cell, where the gas samples are placed. These samples are exposed to the incidence of modulated radiation, absorbing it at determined wavelengths. The resonant absorption of radiation generates a modulated heating in the sample. Therefore, a pressure wave or a sound signal is produced and detected by highly sensitive microphones, inside the cell. These microphones convert the sound signal into an electric signal, which is filtered and detected by a lock-in amplifier. We will use this technique to detect the gas ethylene (C_2H_4) in parts per million by volume (ppmv) and ppbv concentrations.

As a complementary technique, electrochemical sensors [22, 23] and infrared analyzer, URAS [24] were used. By combining these two techniques, gases such as CO, NO, NO_2, CO_2, N_2O, could be detected in ppmv and ppbv concentrations.

2. Ethanol in Brazil

2.1. Historical aspects

In Brazil, the first sugar production centers were established in the mid-sixteenth century, with seedlings of cane sugar from the Madeira Island. In the same century, the first mills were built in northeastern of Brazil, which soon became a major producer of spices. The climate was a factor in the success of the sugarcane planting. [25]. During the colonial period, Brazilian sugar production has changed a few times, keeping intact some characteristic features: grown on monoculture plantations and using compulsory labor for its development.

The first ethanol powered vehicle in Brazil was produced by Peugeot and brought by the family of Alberto Santos Dumont. This idea fascinated the Brazilian President, Rodrigues Alves. In 1919, a decree of the Pernambuco government officiated ethanol as a "National Fuel". [26]

By 1930, Eduardo Sabino de Queiroz published some articles on the use of ethanol as a fuel in vehicles. In this period, the President Getulio Vargas became interested in such articles, and in February 1930, he promulgated the Decree No. 19117, which established the addition of 5% alcohol in each liter of imported gasoline in Brazil. At September 1931, another decree was intended to support research on the ethanol fueled engine, charging a tax per liter of imported gasoline, to finance research. [26]. Despite the incentives given to ethanol in the 30's, due to external pressures and disaffection of subsequent governments, ethanol lost its strategic importance in the Brazilian energy matrix.

After the first oil shock in 1973, the sudden rise in the price of the oil barrel has created a major crisis in Brazil. To reduce oil imports (90% of the gasoline consumed was imported) and to offer an alternative market for sugar, ethanol again figured as a strategic option. The National Alcohol Program (Proalcohol) was thus created, through Decree 76.593/75. [27]

In the first phase of the program, the Brazilian government established a mandatory blend of 22% of ethanol in gasoline, low-interest loans and guarantees for construction of new units for ethanol production. These actions produced an increase in more than 500% of ethanol production between 1975 and 1979. [28]

The second phase started in the second oil shock in 1979, when the Brazilian government encouraged the manufacture of vehicles that run exclusively on hydrous ethanol. To encourage the purchase of such vehicles, various tax incentives were offered, ethanol price was fixed below gasoline price (65% in 1980 and 59% in 1982) and taxes on the sale of ethanol cars were significantly reduced.

In 1984, ethanol vehicles accounted for 94.5% of national production. The alcohol industry has replaced since 1976, over 1.44 billion barrels of oil [28]. Brazil was the first country to use ethanol on a large scale. The program was economically and technologically successful. [29]

In the early 90´s, with the fall in gasoline prices, the price of ethanol fuel, initially set at 64.5% of the price of gasoline reached to achieve a percentage of 80%, losing some of its competitive advantages. The Brazilian government has given incentives tax to the so-called "popular cars", that run on gasoline, which came to exert strong competition to the alcohol-fueled cars.

Alongside this, there was an increase in international prices of sugar and more profitable export market, which resulted in the decrease of ethanol production, generating a distrust of regular supply of this fuel. In 1990, the country was forced to import ethanol and methanol used in blending with gasoline. [28]

These factors contributed to the sale of alcohol-fueled cars in Brazil were close to zero in the mid-90s, dismantling the main program of biomass fuel in the world. Only taxis and rental cars continued to be manufactured with alcohol engine, although the mandatory blend to gasoline has been maintained. [28, 30]

Nowadays, the advent of flex-fuel vehicles allows the vehicle to be fueled with alcohol or gasoline, eliminating the uncertainty about the irregularity in supply. Initially developed in the United States and Europe, this technology was introduced in Brazil in 2003. The first Brazilian flex vehicle was the "Gol 1.6 full flex", released by Volkswagen. Then, flex vehicles technology was spread rapidly for nearly all automakers. Currently, there are in the Brazilian market more than 16 million flex-fuel vehicles of sixty different models produced by ten manufacturers.

There are also studies for ethanol use in heavy-duty engines. They have been introduced experimentally in a few numbers of buses (E95) in the city of São Paulo. Brazilian ethanol is considerate an ally to reduce GHGs emissions. Estimates based on analysis of the ethanol life cycle show that when produced from sugar cane can reduce GHG emissions in up to 90% [31]. Energy balance is also excellent for Brazilian ethanol, once for every unit of fossil energy used in the production of ethanol, 9.3 units of renewable energy are generated [32].

In Brazil, the first flex motorcycle was developed by Honda engineers, being considered as an important project by the Honda Research and Development Center in Japan [33].

In Botucatu (SP), Neiva Aeronautic Industry, manufacturer of aircraft components for agricultural and regional Embraer jets, started, on October 2004, the production of Ipanema, aircraft with the certification of the Aerospace Technical Center (CTA) and 100% powered by hydrated ethanol. The Ipanema is the first series production aircraft in the world to leave the factory certified to fly on this type of fuel. [34]

2.2. The Brazilian industry of sugarcane

The Report of the Food and Agriculture Organization (FAO) [35] estimates that in 2010, 23.8 million hectares of sugar cane were grown in over 90 countries worldwide, resulting in a world production of 1.69 billion tons of sugarcane. The report also shows that Brazil is the world largest producer of sugarcane, followed by India, China, Thailand, Pakistan and Mexico. Brazil has accumulated experiences with vehicles powered by biofuels over the past 30 years [4, 34, 35], including pure ethanol and ethanol blended with gasoline. The Brazilian sugar industry has experienced huge and technological developments. [36] Brazil is the world's largest producer of sugarcane, producing about 490 million ton per year in an area of 7.8 million hectares, which is 2.3% of the arable land in the country. Brazil also has the lowest production cost among the main contenders in the international market, and leads the knowledge of sugarcane biotechnology, with Australia and South Africa.

The Brazilian sugarcane industry has some peculiarities that differentiate it from other sugarcane industries around the globe. First, most industries produces a very high proportion of sugarcane that it processes. Only one-third of the raw material processed is acquired from third parties [29]. Another important factor is the diversity of commercial products that are made from the juice of sugar cane and from the solid and liquid waste grinding, beyond the co-generation of electricity by burning bagasse. Most of ethanol and sugar production comes from plants with attached distilleries, capable of designating a portion of the juice of sugar cane for sugar production and partly for the manufacture of the alcohol.

Another important aspect is the distribution of production units of cane sugar in the whole national territory. The provision of a large portion of land in the North-South provides a wide variety of microclimates that allow the production in economic scale of the majority of commercial crops in use worldwide. As a result of the distribution of production units and combining state of the harvest periods, the country maintains with different intensities, the sugar and alcohol production by virtually every month of the year [29]. The successful Brazilian ethanol program is now supported by 2 basic factors. The mandatory use of ethanol blend in gasoline, and expanding market for flex-fuel. The gasoline sold in Brazil has 20% to 25% of anhydrous ethanol, and approximately 90% of the new cars sold use flex fuel engines.

2.3. Environmental aspects: Pollutant emissions

Air pollution is a growing concern for all human society. Phenomena such as acid rain, photochemical smog, depletion of the ozone layer and global warming are directly linked to this type of pollution, with serious consequences to human health and climate change worldwide. The transportation sector is considered a major factor in the generation of air pollution, generating a large variety and quantity of polluting gases such as CO, CO_2, NO, NO_2, N_2O, C_2H_4, VOCs (Volatile Organic Compounds). According to data from the World Automotive Industry, the global fleet of motor vehicles reached in 2008 for the first time, the mark of 1 billion units. Considering that this number continues to grow significantly for many years, the transport will become increasingly relevant in greenhouse gas emissions.

Nowadays, in Brazil, there is a boom in sales of cars and motorcycles, which use various fuels such as gasoline, ethanol, diesel, biodiesel and natural gas. The environmental problem tends to rise with the new perspectives of the Brazilian economy.

The use of biofuels is surrounded by many international, scientific, technical and political controversies [4-6, 36, 37] about carbon dioxide emission, oil prices, fuel and food security, forest destruction, soil erosion, impacts on water resources, energy balance and energy security. Under the particular aspect of the greenhouse effect, it's necessary to consider not only the carbon dioxide content in the fuel, but also the entire chain of production. For the specific case of ethanol, we have to consider the industries of nitrogenous fertilizer, land use and emissions of nitrous oxides [9-11, 37, 38], agricultural machinery, cane transportation from the land to the plant, ethanol transportation to the consume centers, fermentation, generation of carbon dioxide at the plant (boilers) and, finally, we have to consider the source of energy that supplies the plant. All these criteria must be considered in order to ensure the sustainability of biofuels production chain.

3. Biodiesel in Brazil

3.1. Historical aspects

In Brazil, Count Francisco Matarazzo pioneered the use of biofuels. In the sixties, Matarazzo Industries sought to produce oil from the coffee beans. Ethanol from sugar cane was used to wash the coffee in order to remove impurities, unfit for human consumption. The reaction between alcohol and coffee oil resulted in the release of glycerol and ethyl ester, a product that is now called biodiesel [39].

In 1980, the project "PRODIESEL" was released at the Federal University of Ceará, by teacher Expedito Parente. The project was stalled due to disinterest of funding agencies, according to teacher Parente, responsible for the first patent for a process of Brazilian biodiesel in 1980 (PI - 8007957, requested the INPI - National Institute of Intellectual Property), produced from the mixture of castor oil and methanol [40].

The National Biodiesel Production and Use (PNPB) was established by the Decree of 23 December 2003. The production and consumption of biodiesel in Brazil were determined by Laws no. 11097 and 11116 in 2005. An important aspect of the law is assigned to the National Petroleum, Natural Gas and Biofuels Agency (ANP) the competence of the regulator of the biodiesel industry, as well as oil, natural gas and its derivatives.

The main thrust of the program is to implement a model of sustainable energy from the production and use of biodiesel produced from various oil sources, promoting social inclusion, ensuring competitive pricing, product quality and supply. According to PNPB, from January 2005, the addition of 2% biodiesel to diesel (B2) becomes mandatory. From January 2010, the addition of 5% biodiesel to diesel (B5) becomes mandatory.

http://dx.doi.org/10.5772/52773

3.2. Environmental and economic aspects

In 2011, Brazil produced 2.7 billion liters of biodiesel, moving 2.89 billion dollars. Brazil was the third largest producer, after the USA and Germany. In 2014, the Brazilian government provides for the addition of 10% biodiesel to diesel (B10) [41, 42]. Brazil is a tropical country, with a vast territory, intense sunlight, high biodiversity in oil plants, abundant water resources, regular rainfall and advanced agricultural technology, which generates economic advantages in the production of biodiesel.

The use of biodiesel provides excellent lubrication, maximizing engine life and low risk of explosion, facilitating transport and storage. The engines in trucks, tractors or machines generally do not require changes when using percentages to 20% of biodiesel added to diesel.

The use of biodiesel reduces dependence on petroleum and provides environmental benefits and life quality, improving public health in cities, due to the reduction of emissions of air pollutants, such as particulates, hydrocarbons (aromatic) and sulfur compounds. Moreover, unlike fossil fuels, the CO_2 released by burning biodiesel is recycled by absorption during the growth of oilseeds (photosynthesis) [43, 44]. Thus, the production of biodiesel is part of a cyclical process that assists in reducing the greenhouse effect, because there is a balance between the mass of carbon fixed and released into the atmosphere. Once it can be produced from vegetable oils used in cooking, biodiesel also helps reduce the problem of disposal of waste oils.

Biodiesel also provides interesting socioeconomic advantages, because it acts as a regulatory element of the vegetable oil markets, uses agricultural and industrial waste, reducing material sent to landfills and also reducing air pollution in large urban centers. Biodiesel production creates jobs, contributes to the establishment of man in the field and does not require major technological changes in engines [45, 46].

3.3. Ethanol and biodiesel: Gaseous emissions

Among the anthropogenic activities responsible for emissions, transport stands out due to large-scale increase in the number of motor vehicles that circulate everywhere on the planet. In this context, two noteworthy fuels: gasoline, fossil fuel hegemonic entire world's fleet of light vehicles and ethanol, which has gained importance with the advent of flex vehicles, especially in Brazil.

Due to the use of biodiesel (B5) in the Brazilian fuels matrix, studies of pollutant emissions from the combustion of biofuels blended with conventional diesel are required. We present an evaluation of emissions of an engine bench, with different mixtures of biodiesel and diesel.

For this, TEMPEST Electrochemical Analyzer to detect carbon monoxide (CO), the nitrogen oxides NO_x (NO+ NO_2) in the range of ppmv and the infrared analyzer (URAS), to detect carbon dioxide (CO_2). We also carried out the calibration of a photoacoustic spectrometer coupled to a CO_2 laser, which allows us to perform the detection of ethylene (VOCs) in the range of ppbv [23, 47-49].

4. Methodology

Photoacoustic technique presents a set of fundamental requirements to gas detection, such as high selectivity, multicomponent detection, high sensitivity, large dynamic range and good resolution time. This technique has been widely used in the detection of different gases in the concentration range of ppbv and sub-ppbv.

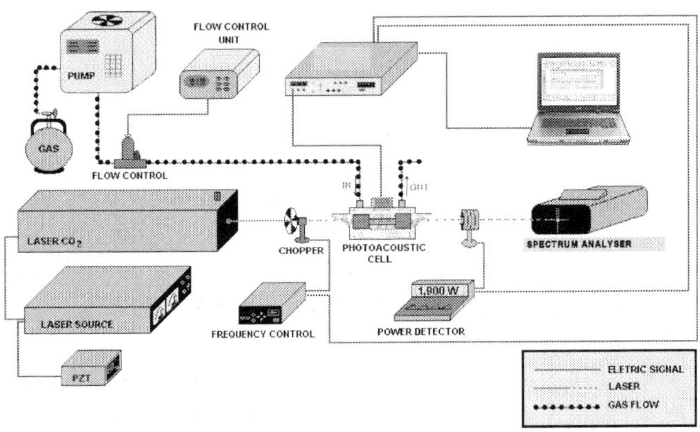

Figure 1. Scheme of the photoacoustic experimental setup

This methodology consists on the generation and detection of pressure waves (sound) inside a resonant cell, where the gas samples are placed. These samples are exposed to the incidence of modulated radiation, absorbing it at determined wavelengths. The resonant absorption of radiation generates rovibrational excitation in the gaseous molecules and a subsequent relaxation by collisions (heat). The modulate heating occurs in a constant volume, generating a pressure wave (sound), which can be detected by highly sensitive microphones inside the cell. These microphones convert this sound signal into an electric signal, which is filtered and detected. The Photoacoustic signal has a linear response with the concentration of the analyzed gas. The photoacoustic spectrometer (Figure 1) consists on a CO_2 laser (Lasertech Group Inc., - LTG, Model LTG150 626G; 1.9W power) as a source of infrared radiation, a chopper (New Focus, 3605) for mechanical modulation of the beam, a photoacoustic cell developed by the group of teacher M. Sigrist (Swiss Federal Institute of Technology (ETH) – Zurich), flow meters which control the entry of gaseous samples, a spectrum analyzer, a power meter, a lock-in amplifier (Stanford SR850) and a computer for data acquisition. The quality factor (Q = 24.7), the coupling constant (C = 40.2 $VcmW^{-1}$ and the resonance frequency (ν = 2.4kHz) of the photoacoustic cell were experimentally obtained. From these parameters we could determine photoacoustic signal of monocomponent sample by the relationship

$$S(\lambda) = CP(\lambda)Nc\sigma(\lambda) \quad (1)$$

where P is the power emitted by the laser, N is the density of molecules (≈2.5×10^{19} molecules/cm^3 to pressure 1013 hPa and temperature 20°C), c is the mole fraction of gas absorber and σ is the cross-section of the gas absorber. In multicomponent samples, it is possible to determine the concentration of different gas species and the photoacoustic signal for different wavelengths (λ_i = 1, 2, 3,...), based on the absorption spectrum of each component to be analyzed.

4.1. Photoacoustic cell calibration and sensitivity measurements

To know the cell performance it is required to perform the calibration of the spectrometer in order to determine the limit of detection of the photoacoustic method. Calibration was performed gradually diluting a sample of ethylene in nitrogen gas, initially at a concentration of 1.4 ppmv (certified by White Martins). The reference signal obtained in the initial condition of maximum concentration can be used to adjust the wavelength of the laser to the absorption maximum of the line. The measurements were carried out in ethylene emission line 10P14 (λ =10.53 μm) of carbon dioxide laser, where the ethylene has its highest absorption. In this same region, there are also considerable absorption intensities from water and carbon dioxide. Therefore, in order to avoid interference of chemical species, chemical filters of potassium hydroxide and calcium chloride, respectively, were used.

The electrochemical techniques are powerful tools in the detection of gaseous species, especially because of some advantageous features, such as high sensitivity of the determinations, portability, easy of automation, the possibility of miniaturization and low cost. Currently, a wide variety of electrochemical sensors are being used to detect gas with numerous applications, among which we highlight the environmental application [22, 50].

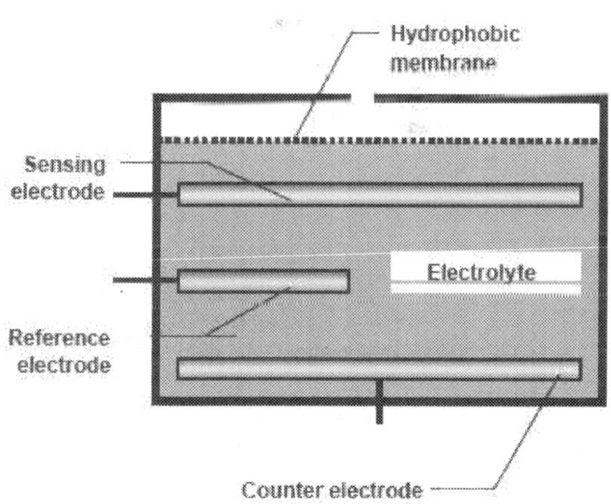

Figure 2. Detection scheme of an electrochemical sensor.

The electrochemical sensors (Figure 2) are composed of a sensing electrode, a counter electrode, a reference electrode and a reagent electrolyte inserted between the electrodes. Furthermore, a barrier permeable to gas, also known as hydrophobic membrane, must cover the sensor to avoid the entry of undesirable gases and water, and to control the amount of gaseous molecules that reach the electrode surface.

When the gas enters the sensor, it reacts with the electrodes and a process of oxidation-reduction occurs. As the electrodes are connected to a resistor, an electric current is generated between the cathode and the anode. The current generated is proportional to the gas concentration [50].

The Gas Analyzer URAS (Figure 3) uses a photoacoustic detection scheme capable of detecting a specific gas out of a multicomponent gas mixture avoiding cross interferences. [51] The selectivity of this instrument is achieved by comparing the direct absorption in a sample cell to that in a reference cell. After passing through the sample cell, each attenuated light beam enters a second detection cell filled only with the gas of interest as the detection cells are interconnected by a membrane connected to a capacitor. Since the dual beam is modulated, the difference in acoustic energy reflects the difference in absorption and thus the concentration difference between the sample cell and the reference cell. The species under investigation enables the wavelength to be selected in such a way that all wavelengths at which absorption occurs are simultaneously active. When there is no spectral overlap from other gases, additional absorptions in the sample cell do not contribute to the acoustic signal and the light passes the detection cell not attenuated. The photoacoustic instrument used in this work has been developed for gas detection concentrations in the ppmv range.

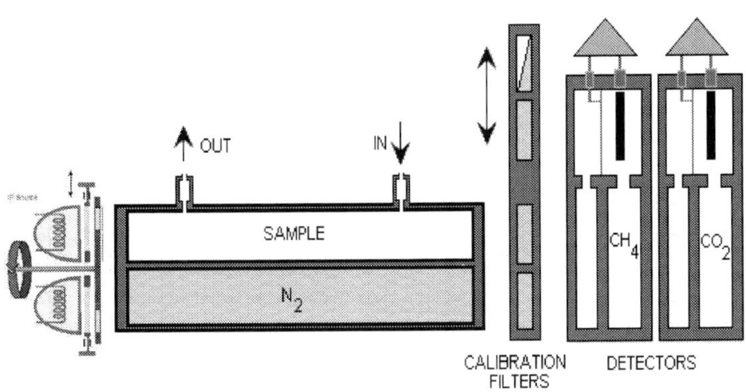

Figure 3. Detection scheme of Infrared Analyzer URAS

5. Experimental procedure

The gas samples were collected from the exhaust of ethanol powered vehicles [52] and were stored in previously evacuated metallic canisters (SUMMA Andersen Instruments). The canisters are made of stainless steel and the samples were taken to the laboratory and coupled to our photoacoustic cell inlet. The gas sample was then pulled into this cell by a mechanical pump (Ambient Volatile Canister Sample AVOCS). Filters were used to remove the particulate matter larger than 2 μm. This collection was performed in two sequences in each vehicle: the first one with the bus engine turned on and without acceleration (*i.e.*, 1000 rpm of rotation speed) and the second one, with the bus engine turned on and with acceleration (*i.e.*, 3000 rpm of rotation speed). The gas samples were analyzed by a photoacoustic method at a pressure of 1 atm for detection of ethylene gas (C_2H_4). In conventional absorption spectroscopy, the absorption of the radiation power transmitted through the sample is measured. On the contrary, in photoacoustic spectroscopy, the absorbed power is determined directly via its heat and hence the sound produced in the sample.

The Electrochemical Analyzer Tempest promosed the detection of CO, NO_x (NO+NO_2) from the exhaust of ethanol powered vehicles. The measurements were performed directly in the exhaust of these vehicles due to the portability of this equipment.

Other samples were collected from the exhaust of diesel engines (TOYAMA Model TD70F6.7 HP) and were stored in previously evacuated metallic canisters, similar procedure with the exhaust of ethanol powered vehicles. The sample collecting procedure was performed in two sequences: the first one with the bus engine turned on and at low rotation (3000 rpm) and the second one, with the engine turned on and at high rotation (9000 rpm). This procedure was adopted in the collection of samples from the diesel-biodiesel blends B5, B10, B15, B20 and B25 [53]. The measurements started in B5 because it is the diesel Brazilian standard. In this experiment, we used the soybean biodiesel and the gas samples were analyzed by the photoacoustic method at a pressure of 1 atm, for detection of ethylene gas (C_2H_4). The samples are also taken to Infrared Analyzer URAS was employed to detect CO_2 emissions. The Electrochemical Analyzer Tempest promised the detection of CO, NO_x (NO+NO_2) from the exhaust of diesel engines. The measurements were performed directly in the exhaust of these vehicles due to the portability of this equipment.

6. Results and comments

6.1. Ethanol engine

Figure 4 shows the emission of CO from the exhaust of ethanol powered. The measurements were performed in seven different kinds of vehicles manufactured in different years [51]. CO concentrations in the range 88-19240 ppmV were obtained. The CO emission was higher for vehicles under low engine speed. Probably, this mode of operation the combustion efficiency is smaller, so that the gases have been produced by incomplete combustion. We must also consider some variables such as manufacture year of the vehicle, make, model, power, maintenance, combustion temperature and rotation speed.

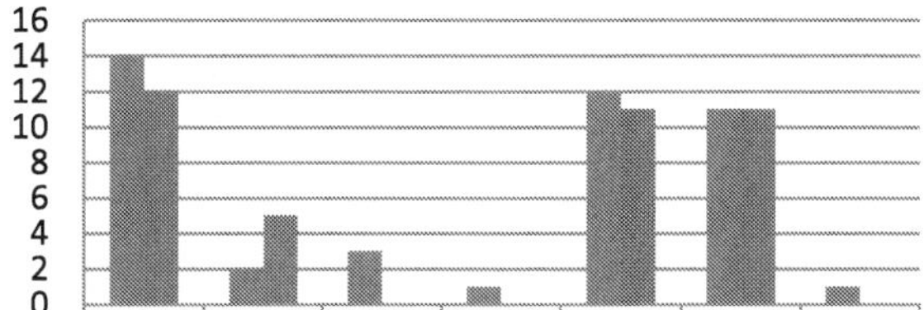

Vehicles	Astra 2006	Vectra 2007	Meriva 2007	Vectra 2007	Fiesta 2007	Focus 2007	Ford Ka 2008
NOx (ppmv) not accelerated	14	2	3	1	12	11	1
NOx (ppmv) accelerated	12	5	0	0	11	11	0

Figure 4. Emission of carbon monoxide by vehicles using ethanol engine

For the same group of ethanol vehicles, we carried out the measurement of nitrogen oxides (NO_x). In this case, the presence of nitrogen oxides is not related to the fuel, since the presence of nitrogen compounds cannot be observed in ethanol composition. The emission of NO_x is only due to oxidation of air nitrogen. We can notice (figure 5) a large fluctuation in the values of NO_x emitted by cars, which are not directly related to the model or year of manufacture of vehicles. This differential pattern is justified, since the temperature in the combustion chamber is the predominant factor in greenhouse gas emissions.

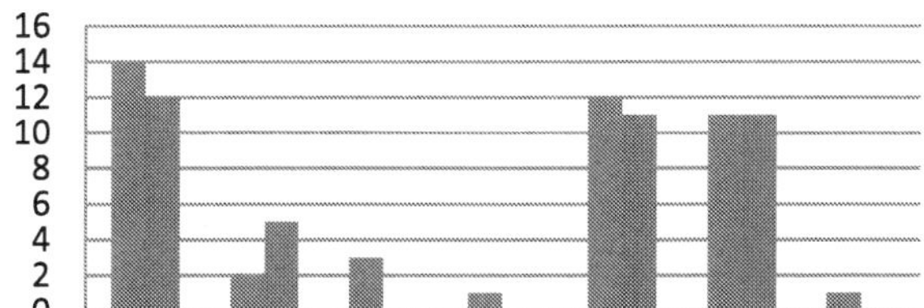

Vehicles	Astra 2006	Vectra 2007	Meriva 2007	Vectra 2007	Fiesta 2007	Focus 2007	Ford Ka 2008
NOx (ppmv) not accelerated	14	2	3	1	12	11	1
NOx (ppmv) accelerated	12	5	0	0	11	11	0

Figure 5. Emission of nitrogen oxides by vehicles using ethanol engine

http://dx.doi.org/10.5772/52773

Collection of gaseous samples was performed in another group of vehicles powered by ethanol, to identify the greenhouse gas CO, NOx and C_2H_4. CO concentrations in the range 350 to 7666 ppmv, NO_x concentrations in the range 1 to 464 ppmv [53] and C_2H_4 concentrations in the range 1 to 1198 ppmv were obtained.

6.2. Diesel-biodiesel engine

In the analysis of gas emissions from diesel engine exhaust, using binary mixtures diesel-biodiesel (B5, B10, B15, B20 and B25), CO_2 concentrations in the range of 1.95% and 3% were obtained in all mixtures tested. CO_2 emissions were higher in the accelerated mode of engine, and a reduction in the emission of this gas could not be observed with increasing percentages of biodiesel [52].

Moreover, CO concentrations in the range from 1400 to 2397 ppmv were obtained. A reduction in CO was not observed with the increase in the proportion of biodiesel. NO_x concentrations in the range 184-346 ppmv were also obtained. We observed an increase in the concentrations of NO_x to increase the proportion of biodiesel, especially for B15 and B25.

In the evaluation of C_2H_4 gas emissions, concentrations ranging from 67 to 123 ppmv were obtained in all mixtures evaluated. Emissions C_2H_4 were higher in an accelerated engine. A decrease in emissions of C_2H_4 with increasing percentages of biodiesel could not be observed [49].

7. Gaseous emissions in soils of sugarcane production

The use of fertilizer nitrogen-based synthetic has grown substantially in recent years, mainly in Brazil, which is the main agricultural activities in growing crops such as soybeans, corn, rice and beans, and large-scale cultivation of sugarcane sugar. In order to define the cycle of sustainability of ethanol production, we performed preliminary studies of the gas emissions of carbon dioxide (CO_2) and nitrous oxide (N_2O) from soils of sugar cane production in the region of Campos dos Goytacazes, RJ. [54]

The analyzes for the flow of nitrous oxide (N_2O) and carbon dioxide (CO_2) emitted from soils cultivated with sugarcane fertilized (urea and ammonium sulphate) and not fertilized, were made by the method known as "static dome." This method is the storage of gases within a dome made of PVC, glass, metal or acrylic, and may vary in size depending on the crop to be studied [55]. This method of static dome is the most commonly used and reported in the literature [56-58], because the gas concentrations vary linearly with time [59]. The gas samples are collected from the tanks and taken to be analyzed in the laboratory. The methods for determining gases concentration were previously described in methodology. CO_2 concentration was determined by Infrared Analyzer (URAS), N_2O concentration was determined by a photoacoustic spectrometer, similar to that described in Figure 1, but using a semiconductor quantum cascade laser (QCL) [20, 60-70] with an emission band ranging from 7.71 μm to 7.88 μm as the excitation source and a resonant differential photoacoustic (PA) cell as detector. The laser was fed applying pulsed current (26.2 mA) with a repetition rate of 400 kHz and a pulse duration of 50ns (duty cycle of 2%). With this methodology, it was possible to detect the emission of CO_2 gas from 500 to 1700 ppmv for the not fertilized area and concentrations from 600 to 1800 ppmv for the fertilized area. The N_2O emissions were from 84 to 321 ppbv for the not fertilized area and from 384 to 2066 ppbv for the fertilized area. The concentration value of the background region (380 ppbv) must be added to the N_2O concentrations.

8. Emission gases by the use of biofuels: Environmental problems

Air pollution generated by gases emitted by the use of biofuels (ethanol, diesel-biodiesel) and emissions from soils cultivated as sugar cane. It can produce serious environmental problems as produce a large amount of air pollutants [71], such as nitrogen oxide (NO_x), generators of acid rain and harmful to human health [72-76], carbon monoxide (CO), another pollutant emitted by ethanol and diesel-biodiesel engines, is not considered a direct greenhouse gas, but it is able to influence the production of methane and tropospheric ozone, which are important greenhouse gases [77-84]. Ethanol and diesel-biodiesel engines also produce volatile organic compounds (VOCs), such as ethylene [84-87].These chemical species are precursor for the generation of the tropospheric ozone [88-90], which is present in photochemical smog and directly affects human health. Tropospheric ozone can also trigger serious respiratory problems and cardiovascular effects [92-97]. Besides, it is a powerful greenhouse gas, whose formation is greatly potentiated by the incidence of sun radiation and the presence of nitrogen oxides (NO_x) [90, 91]. Other species, such as nitrous oxide (N_2O) and carbon dioxide (CO_2) are also present in the emissions studied, being considered important greenhouse gases.

9. Conclusion

The techniques presented, as photoacoustic spectroscopy, electrochemical sensors, and infrared analyzer (URAS) were sensitive and selective for the detection of gaseous pollutants in the range of ppmv and ppbv. Gases such as CO, NOx, CO_2, N_2O, C_2H_4 could be identified in our samples. Such gases are causing serious environmental problems, such as acid rain, global warming and the generation of tropospheric ozone, in addition to severe damage to human health. These gases are produced in the use of biofuels in the transport sector (engine exhaust) and in the cycle of sugar cane production (emissions from soil). We need to develop careful research in gas detection to identify the real contribution of the production chain of biofuels and biomass in general. Research should be encouraged in order to evaluate the real impacts on the use of biofuels. Thus, the use of new techniques, more accurate and selective, and the development of new methodologies for the identification of gases in trace level are essential.

http://dx.doi.org/10.5772/52773

Author details

Marcelo Silva Sthel[1*], Aline Martins Rocha[1], Juliana Rocha Tavares[1], Geórgia Amaral Mothé[1], Flavio Couto[1], Maria Priscila Pessanha de Castro[1], Victor Habez Perez[2], Marcelo Gomes da Silva[1] and Helion Vargas[1]

*Address all correspondence to: sthel@uenf.br

1 Laboratory for Physical Sciences (LCFIS) of the Center for Science and Technology (CCT), Universidade Estadual do Norte Fluminense. Campos dos Goytacazes,, Brazil

2 Laboratory of Food Technology (LTA) of the Center for Agricultural Science and Technology (CCTA), Universidade Estadual do Norte Fluminense. Campos dos Goytacazes,, Brazil

References

[1] IPCC, Report of Working Group I, Climate Change 2007: The Physical Science Basis, Paris, February, 2007.

[2] IPCC, Report of Working Group III, Climate Change 2007: Mitigation of Climate Change, Bangkok, May, 2007.

[3] Meinshausen M, Meinshausen N, Hare W, Raper SCB, Frieler K, Knutti R, Frame DJ, Allen MR. Greenhouse – Gas Emission Targets for Limiting Global Warning to 2 ° C. Nature 2009; 458, 1158-1162. DOI:10.1038/nature08017.

[4] Goldemberg J. The promise of clean energy. Energy Policy 2006; 34, 2185-219. DOI: 10.1016/j.enpol.2005.03.009.

[5] Goldemberg J. Biomass and Energy, Quim. Nova 2009; 32 (3) 582-587. DOI:10.1590/S0100-40422009000300004.

[6] Goldemberg J, Coelho ST, Guardabassi P. The sustainability of ethanol production from sugarcane. Energy Policy 2008; 36, 2086-2097. DOI:10.1016/j.enpol.2008.02.028

[7] Macedo IC, Seabra JEA, Silva JEAR. Greenhouse gases emissions in the production and use of ethanol from sugarcane in Brazil :The 2005/2006 averages and e prediction for 2020. Biomass and Bionergy 2008; 32 (7) 582-59. DOI:10.1016/j.biombioe.2007.12.006.

[8] The International Council for Science (ICSU), Biofuels: Environmental consequence & implications of Changing Land Use, Executive Sumary, Rapid Assessment on Biofuels and the Environment: Overview and Key Findings, Cornell University, September (2008).

[9] Crutzen PJ, Mosier AR, Smith KA, Winiwarter W. N_2O release from agro-biofuel production negates global warming reduction by replacing fossil fuels. Atmos. Chem. Phys. Discuss. 2007; 7, 11191-11205. DOI: 10.5194/acpd-7-11191-2007.

[10] Galloway JN, Townsend AR, Erisman JW, Bekunda M, Cai Z, Freney JR,. Martinelli LA, Seitzinger SP, Sutton MA. Transformation of the nitrogen cycle: recent trends, questions, and potential solutions. Science 2008; 30 (16) 889-89. DOI: 10.1126/science.1136674

[11] Searchinger T, Heimlich R, Houghton RA, Dong F, Elobeid A, Fabiosa J, Tokgoz S, Hayes D, Yu T. Use of U. S. croplands for biofuels increases greenhouse gases through emissions from land-use change. Science 2008; 319 (5867) 1238-1240. DOI: 10.1126/science.1151861

[12] Berrou A, Raybaut M, Godard A, Lefebvre M. High-resolution photoacoustic and direct absorption spectroscopy of main greenhouse gases by use of a pulsed entangled cavity doubly resonant OPO. Appl. Phys. B 2010; 98 (1) 217-230. DOI: 10.1007/s00340-009-3710-x

[13] Thomas S. Photoacoustic Spectroscopy for process analysis Anal. Bioanal. Chem. 2006; 384 (5) 1071-1086. DOI: 10.1007/s00216-005-3281-6

[14] Sigrist MW, Bartlome R, Marinov D, Rey JM, Vogler DE, Wächter H. Trace gas monitoring with infrared laser-based detection schemes. Appl. Phys. B 2008; 90 (2) 289-300. DOI: 10.1007/s00340-007-2875-4

[15] Angelmahr M, Miklos A, Hess P. Photoacoustic Spectroscopy of formaldehyde with tunable laser radiation at the part per billion level. Appl. Phys. B 2006; 85 (2-3) 285-288. DOI: 10.1007/s00340-006-2295-x

[16] Schramm DU, Sthel MS, da Silva MG, Carneiro LO, Souza AP, Vargas, H. Diesel engines gas emissions monitored by Photoacoustic Spectroscopy. Rev. Sci. Instrum. 2003; 74 (1) 513-515. DOI:10.1063/1.1520313

[17] Miklos A, Hess P, Bozoki P. Application of acoustic resonators in photoacoustic trace gas analysis and metrology. Rev. Sci. Instrum. 2001; 72 (4) 1937-1955. DOI: 10.1063/1.1353198

[18] Sthel MS, Schramm DU, Faria RTJr, Castro MPP, Carneiro LO, Ribeiro WS, Vargas H. Photoacoustic Spectroscopy-based analysis of gas samples in a bus station. J. Phys. IV 2005; 125 (1) 881-883. DOI: 10.1051/jp4:2005125204

[19] Sigrist MW. Air Monitoring by Spectroscopic Techniques. Chemical Analysis; John Wiley & Sons, Inc.; New York, NY, USA, 1994; Volume 127. ISBN-10: 0471558753 | ISBN-13: 978-0471558750

[20] Filho MB, Silva MG, Sthel MS, Schramm DU, Vargas H, Miklos A, Hess P. Ammonia detection by using quantum cascade laser Photoacoustic Spectroscopy. Appl. Opt. 2006, 45 (20) 4966-4971. DOI: 10.1364/AO.45.004966

[21] Thony M, Sigrist MW. New developments in CO_2-laser photoacoustic monitoring of trace gases. Inf. Phys. Technol. 1995; 36 (2) 585–615. DOI: 10.1016/1350-4495(94)00046-N

[22] Janata J, Josowics M. Conducting polymers in electronic chemical sensors. Nat. Mat. 2003; 2, 19-24. DOI: 10.1038/nmat768

[23] Bard AJ, Falkner LR. Electrochemical Methods: Fundamentals and Applications; John Wiley & Sons, Inc.: New York, NY, USA, 2001. ISBN: 978-0-471-04372-0

[24] Hekkert S L. Photoacoustic Spectroscopy in Trace Gas monitoring. Encyclopedia of Analytical Chemistry, John Wiley & Sons, Chichester 2000

[25] Paranhos P. O açúcar no norte-fluminense. Histórica – Revista eletrônica. http://www.historica.arquivoestado.sp.gov.br/materias/anteriores/edicao08/materia02/. (accessed 25 may 2012).

[26] Netto JN. A saga do álcool: fatos e verdades sobre os 100 anos de história do álcool combustível em nosso país. Novo Século, Osasco, São Paulo, 2007. ISBN-13 9788576791140

[27] DECRETO Nº 76.593, DE 14.11.1975 - DOU 14.11.1975. http://nxt.anp.gov.br/nxt/gateway.dll/leg/decretos/1975/dec%2076.593%20-%201975.xml. (accessed 26 may 2012).

[28] Bastos VD. Etanol, Alcoolquímica e Biorrefinarias. BNDES Setorial, Rio de Janeiro, 2007; 25, 5-38.

[29] Agricultural Outlook 2007-2016. Organisation for Economic Co-operation and Development Food and Agriculture Organization of United Nations - OCDE FAO, 2007. http://www.oecd.org

[30] Macedo IC, Nogueira LAH. Biocombustíveis. Cadernos NAE – Núcleo de Assuntos Estratégicos da Presidência da República nº 2. Secretaria de Comunicação do Governo e Gestão Estratégica, 2004. ISSN 1806-8588

[31] Goldemberg J. Review, The Brazilian biofuels industry, Biotechnology for Biofuels 2008; 1 (1) 1-7. DOI: 10.1186/1754-6834-1-6

[32] Associação Nacional dos Fabricantes de Veículos Automotores. Assessoria de Planejamento Econômico e Estatístico, 2009.

[33] Gordinho MC. Do álcool ao etanol: trajetória única. Editora Terceiro Nome, 2010.

[34] Food and Agriculture Organization of the United Nations (FAO). Crop production. http://faostat.fao.org/site/567/DesktopDefault.aspx?PageID=567#ancorRetrieved 2010-06-17. (accessed 22 may 2012).

[35] Sthel MS, Tavares JR, Lima GR, Mothé G, Schramm DUS, da Silva MG, Castro MPP. Atmospheric pollution: Global warming and a possible use of bio-fuels wide scale. Int. Rev. Chem. Eng. 2010; 1(6) 564-570.

[36] Howarth RW, Bringezu S, Bekunda M, Fraiture C, Maene L, Martinelli L, Sala O. The International Council for Science (ICSU), Biofuels: Environmental consequence & implications of Changing Land Use, Executive Sumary, Rapid Assessment on Biofuels

and the Environment: Overview and Key Findings, Cornell University, September (2008).

[37] Barton L, Kiese R, Gatter D, Butterbach K, Buck R, Hink C, Murphy . Nitrous Oxide emissions from cropped soil in a semi-arid climate. Global Change Biology 2008; 14 (1) 177-192. DOI: 10.1111/j.1365-2486.2007.01474.x

[38] HISTÓRIA E BIODIESEL, disponível em http://www.biodiesel.ind.br/historia.htm, (accessed 3 november 2003)

[39] Benedetti O, Plá JA, Rathmann R, Padula AD. Uma proposta de modelo para avaliar a viabilidade do Biodiesel no Brasil Teor. e Evid. Econ. Passo Fundo Ed. Especial 2006; 14, 81-107

[40] www.celuloseonline.com.br (accessed 6 august 2012).

[41] Indústrias de biodiesel apostam em mistura de 7% já em 2013. www.exame.abril.com.br (accessed 6 august 2012).

[42] Sheehan J, Camobreco V, Duffield J, GraboskI M, ShapourI H. Life Cycle Inventory of Biodiesel and Petroleum Diesel for Use in an Urban Bus. Colorado: National Renewable Energy Laboratory 1998; 286. NREL/SR-580-24089

[43] TÜV, Future Application Technologies – Biodiesel for Vehicles. TÜV Bayern Holding AG, 2002. 53p.

[44] Ramos LP, Domingos AK, Kucek KT, Wilhelm HM. Biodiesel: Um projeto de sustentabilidade econômica e sócio-ambiental para o Brasil. Biotecnologia: Ciência e Desenvolvimento 2003; 31, 28-37. http://www.biotecnologia.com.br

[45] Geller DP, Goodrum JW. Effects of specific fatty acid methyl esters on diesel fuel lubricity. Fuel 2004; 83 (17-18) 2351-2356. DOI: 10.1016/j.fuel.2004.06.004

[46] Schramm DU, Sthel MS, Silva MG, Carneiro LO, Junior AJSA, Vargas H. Application of the Laser Photoacoustic Spectroscopy on the Analysis of Gas Samples Emitted by Diesel Engines. Infr. Phys. Technol. 2003; 44 (4) 263-269. DOI: 10.1016/S1350-4495(03)00125-7

[47] Teodoro CG, Schramm DU, Sthel MS, Lima G, Rocha MV, Tavares J, Vargas H. CO_2 laser photoacoustic detection of ethylene emitted by diesel engines used in urban public transports. Infr. Phys. Technol. 2010; 53 (2) 151-155. DOI: 10.1016/j.infrared.2009.10.009

[48] Rocha A, Sthel M, Castro M, Perez V, Mothe G, Brasil L, Vargas H. Detection of Ethylene Emitted bv Diesel-Biodiesel Engine CO_2 Laser Photoacoustic Sensors, Applied Physics Research 2012; 4, 3 DOI: 10.5539/apr.v4n3p16

[49] Harren FJM, Cotti G, Oomens J, Hekkert SL. Photoacoustic Spectroscopy in Trace Gas monitoring. Encyclopedia of Analytical Chemistry, John Wiley & Sons, Chichester (2000).

[50] Tavares JR. Avaliação dos Poluentes Emitidos em Veículos Automotivos. Master thesis. Universidade Estadual do Norte Fluminense; 2008.

[51] Rocha AM. Análise das Propriedades Ambientais e Térmicas do Biodiesel Através de Técnicas Fototérmicas. PhD thesis. Universidade Estadual do Norte Fluminense; current.

[52] Tavares JR, Sthel MS, Campos LS, Mila MR, Lima GR, da Silva MG, Vargas H. Evaluation of pollutant Gases Emitted by Ethanol and Gasoline Powered Vehicles, Procedia Environmental Sciences2011; 4,51-60. DOI: 10.1016/j.proenv.2011.03.007

[53] Couto FM. Avaliação das emissões de N_2O e CO_2 em solo cultivado com cana-de-açúcar, através de Técnicas Fototérmicas. Master thesis. Universidade Estadual do Norte Fluminense; 2011.

[54] Christensen S. Nitrous oxide emission from a soil permanent grass: Seasonal and diurnal fluctuation as influenced by manuring and fertilizantion, Soil Biology Biochemistry 1983; 15, 531-536.

[55] Matson PA, Billow C, Hall S, Zachariassen J. Fertilization practices and soil variations control nitrogen oxide emissions from tropical sugar cane, Journal of geophysical Research 1996; 101 (D13) 18533-18545. DOI:10.1029/96JD01536

[56] Khalil MAK, Rasmussen RA. The Global Sources of Nitrous Oxide, J. Geophysical Research 1992; 97 (D13) 14.651- 14660

[57] Alvalá PC. Observações de metano atmosférico no Brasil. (INPE-5969-TDI/573). PhD thesis – Instituto Nacional de Pesquisas Espaciais, São José dos Campos, 1995

[58] Hutchinson GL, Livingston GP, Healy RW, Striegl RG. Chamber measurement of surface-atmosphere trace gas exchange: Numerical evaluation of dependence on soil, interfacial layer, and source/sink properties. Journal of Geophysical Research 2000; 105 (D7) 8865-8875. DOI:10.1029/1999JD901204

[59] Gmachl C, Straub A., Colombelli R, Capasso F, Sivco DL, Sergent AM, Cho AY. Single-mode, tunable distributed feedback, and multiple-wavelength quantum cascade lasers, IEEE J. Quantum Electron 2002; 38 (6) 569–581. DOI: 10.1109/JQE.2002.1005407

[60] Paldus A, Spence TG, Zare RN, Oomens J, Harren FJM, Parker DH, Gmachl C, Capasso F, Sivco DL, Baillargeon JN, Hutchinson AL, Cho AY. Photoacoustic spectroscopy using quantum-cascade lasers. Opt. Lett. 1999, 24 (3)178–180. DOI:10.1364/OL.24.000178

[61] Rocha M, Sthel M, Lima G, Silva M, Schramm DU, Mikos A, Vargas H. A sulfer hexafluoiride sensor using quantum cascade and CO_2 laser- based photoacoustic spectroscopy. Sensor 2010, 10 (10) 9359-9368. DOI: 10.3390/s101009359

[62] Kosterev AA, Malinovsky AL, Tittel FK, Gmachl C, Capasso F, Sivco DL, Baillargeon JN, Hutchinson AL, Cho AY. Cavity ring down spectroscopic detection of nitric oxide

with a continuous-wave quantum-cascade laser. Appl. Opt. 2001; 40 (30) 5522–5529. DOI:10.1364/AO.40.005522

[63] Hofstetter D, Beck M, Faist J, Nagele M, Sigrist MW. Photoacoustic spectroscopy with quantum cascade distributed feedback lasers. Opt. Lett. 2001; 26 (12), 887–889. DOI: 10.1364/OL.26.000887

[64] Rocha M, Sthel M, Silva M, Paiva LB, Pinheiro FW, Mikos A, Vargas H. Quantum-cascade laser photoacoustic detection of methane emitted from natural gas powered engines. Applied Physics B 2012; 106 (3) 701-706. DOI: 10.1007/s00340-011-4800-0

[65] Webster CR, Flesch GJ, Scott DC, Swanson JE, May RD, Woodward WS, Gmachl C, Capasso F, Sivco DL, Baillargeon JN, Hutchinson AN, Cho AY. Quantum cascade laser measurements of stratospheric methane and nitrous oxide. Appl. Opt .2001; 40 (3) 321–326. DOI:10.1364/AO.40.000321

[66] Kosterev AA, Tittel FK. Chemical sensors based on quantum cascade lasers, IEEE J. Quantum Electron. 2002; 38 (6) 582– 591. DOI: 10.1109/JQE.2002.1005408

[67] Nelson DD, Shorter JH, McManus JB, Zahniser MS. Sub-part-per-billion detection of nitric oxide in air using a thermoelectrically cooled mid-IR-quantum cascade laser spectrometer. Appl. Phys. B 2002; 75 (2-3) 343–350. DOI:10.1007/s00340-002-0979-4

[68] Da Silva MG, Vargas H, Miklós A, Hess P. Photoacoustic detection of ozone using quantum cascade laser. Appl. Phys. B 2004; 78 (6)1513–1516. DOI: 10.1007/s00340-004-1513-7

[69] Lima JP, Vargas H, Miklos A, Angelmahr M, Hess P. Photoacoustic detection of NO_2, and N_2O using quantum cascade lasers. Appl. Phys B 2006; 85 (2-3) 279-284. DOI: 10.1007/s00340-006-2357-0

[70] Mothé G, Castro M, Sthel M, Lima G, Brasil L, Campos L, Rocha A, Vargas H. Detection of Greenhouse Gas Precursors from Diesel Engines Using Electrochemical and Photoacoustic Sensors, Sensor 2010; 10 (11) 9726-9741. DOI:10.3390/s101109726

[71] McGonigle AJS, Thomson CL, Tsanev VI, Oppenheimer C. A simple technique for measuring power station SO_2 and NO_2 emissions. Atmos. Environ. 2004; 38 (1) 21-25. DOI:10.1016/j.atmosenv.2003.09.048

[72] Xie Z, Du Y, Zeng Y, Li Y, Yan M, Jiao S. Effects of precipitation variation on severe acid rain in southern China. J. Geograp. Sci. 2009; 19 (4) 489-501. DOI: 10.1007/s11442-009-0489-y

[73] Brunekreef B, Sunyer J. Asthma, rhinitis and air pollution: Is traffic to blame? Eur. Res. J. 2003; 21 (6) 913-915. DOI: 10.1183/09031936.03.00014903

[74] Ramadour M, Burel C, Lanteaume A, Vervloet D, Charpin D, Brisse F, Dutau H, Charpin D. Prevalence of asthma and rhinitis in relation to long-term exposure to gaseous air pollutants. Allergy 2000; 55 (12) 1163-1169. DOI: 10.1034/j.1398-9995.2000.00637.x

[75] Lee YL, Shaw CK, Su HJ, Lai SJ, Ko YC, Huang SL, Sung FC, Guo YL. Climate, traffic-related air pollutants and allergic rhinitis prevalence in middle-school children in Taiwan. Eur. Respir. J. 2003; 21 (6) 964-970. DOI:10.1183/09031936.03.00094602

[76] Gao J, Wang T, Ding A, Liu C. Observational study of ozone and carbon monoxide at the summit of mount Tai (1534 m) in central-eastern China. Atmos. Environ. 2005; 39, 4779-4791. DOI:10.1016/j.atmosenv.2005.04.030

[77] Lijie G, Meigen Z, Zhiwei H. Model analysis of seasonal variations in tropospheric ozone and carbon monoxide over East Asia. Adv. Atmos. Sci. 2009; 26 (2) 312-318. DOI: 10.1007/s00376-009-0312-9

[78] Odón RS, Martins LD, Ynoue RY, Andrade MF. The Impact on tropospheric ozone formation on the implementation of a program for mobile emissions control: A case study in São Paulo, Brazil. Environ. Fluid. Mech. 2007; 7 (2) 95-119. DOI: 10.1007/s10652-007-9018-7

[79] Bruh C, Crutzen PJ. Reductions in the anthropogenic emissions of CO and their effect on CH_4. Chemos. Glob. Chan. Sci. 1999, 1 (1-3) 249-254. DOI: 10.1016/S1465-9972(99)00028-8

[80] Latha MK, Badarinath KVS. Correlation between black carbon aerosols, carbon monoxide and tropospheric ozone over a tropical urban site. Atmos. Res. 2004; 71 (4) 265-274. DOI:10.1016/j.atmosres.2004.06.004

[81] Seinfeld JH, Pandis SN. Atmospheric Chemistry and Physics - From Air Pollution to Climate Change, 2nd ed.; Wiley-Interscience: Malden, MA, USA, 2006. ISBN-10: 0471720186, ISBN-13: 978-0471720188

[82] Nassar R, Logan JA, Megretskaia IA, Murray L T, Zhang L, Jones BA. Analysis of tropical tropospheric ozone, carbon monoxide, and water vapor during the 2006 El ninho using Tes observations and the Geos- Chem model. J. Geoph. Res. Atmos. 2009; 114, 23-25. DOI:10.1029/2009JD011760

[83] US Environmental Production Agency. Carbon monoxide health and environmental impacts of CO. Available online: http://www.epa.gov/air/urbanair/co/hlth1.html (Accessed 29 October 2010).

[84] McCulloch MT, Langford N, Duxbury G. Real- time trace- level detection of carbon dioxide and ethylene in car exhaust gases. Appl. Opt. 2005; 44 (14) 2887-2884. DOI: 10.1364/AO.44.002887.

[85] Abeles FB, Heggestad HE. Ethylene: An urban air pollutant. J. Air Pollut. Cont. Assoc. 1973; 23(6) 517-521. ISSN 0002-2470

[86] Weidmann D, Kosterev AA, Roller C, Curtel RF, Fraser MP, Tittel FK. Monitoring of ethylene by a pulsed quantun cascade laser. Appl. Opt. 2004; 43 (16):3329-3334. DOI: 10.1364/AO.43.003329

[87] Baird C. Environmental Chemistry, 2nd ed.; Brookman: Porto Alegre, Brazil, 2002. ISBN 85-363-002-7

[88] Seinfeld JH. Urban air pollution: State of the science. Science 1989; 243 (4892) 745-752. DOI: 10.1126/science.243.4892.745

[89] Lefohn S, Foley JK. Establishing relevant ozone standards to protect vegetation and human health: exposure/dose-response considerations. J. Air Waste Manage. Assoc. 1993; 43 (1) 106-112.

[90] Atkinson R. Atmospheric chemistry of VOCs and NO_x. Atmosph. Environ. 2000; 34 (12) 2063-2101. DOI: 10.1016/S1352-2310(99)00460-4

[91] Wolff GT, Korsog PE. Ozone control strategies based on the ratio of volatile organic compound to nitrogen oxides. Journal of the Air & Waste Management Association 1992; 42 (9)1173-117. DOI:10.1080/10473289.1992.10467064

[92] Brook RD, Franklin B., Cascio W, Hong Y, Howard G, Lipsett M, Luepker R, Mittleman M, Samet J, Smith Jr SC, Tager I. Air pollution and cardiovascular disease. J. Amer. Heart Assoc. 2004; 109, 2655-2671. DOI: 10.1161/01.CIR.0000128587.30041.C8

[93] Maitre A, Bonneterre V, Huillard L, Sabatier P, Gaudemaris R. Impact of urban atmospheric pollution on coronary disease. Eur. Heat J. 2006; 27 (19) 2275-2284. DOI: 10.1093/eurheartj/ehl162

[94] Townsend C, Maynard R. Effects on health of prolonged exposure to low concentrations of carbon monoxide. Occup. Environ. Med. 2002; 59 (10) 708-711. DOI: 10.1136/oem.59.10.708

[95] Favory R, Lancel S, Tissier S, Mathieu D, Decoster B, Nevière R. Myocardial dysfunction and potential cardiac hypoxia in rats Induced by carbon monoxide inhalation. Amer. J. Res. Crit. Car. Med. 2006; 174 (3) 320-325. DOI:10.1164/rccm.200601-117OC

[96] Lindell KW, Ramona OH, Karen JC, Susan C, Gregory E, Terry P, James F, Frank O, Alan H. Carbon monoxide poisoning. New Engl. J. Med. 2009; 360, 1217-1225.

Chapter 13

Biofuels in Brazil in the Context of South America Energy Policy

Luiz Pinguelli, Rosa Alberto Villela and
Christiano Pires de Campos

Additional information is available at the end of the chapter

http://dx.doi.org/10.5772/54419

1. Introduction

This article seeks to provide an overview of the current energy policies, focusing on Brazil and regarding biofuels in particular. It does not claim to put forward an exhaustive analysis of the subject. Hence, it focuses on certain particularly relevant aspects, such as climate change and CO2 emissions abatement, rather than entering into the details of all extant aspects. It is aimed to serve as a policy instrument in order to provide a basis for decision-making and planning actions, contextualizing the discussion within the global framework. In this connection, special attention is paid to the impact of the initial soaring of the oil prices, their subsequent drop with the onset of the economic crisis, which reached Latin America in 2009, as well as the latest hike in the price of oil. This impact has as a counterpoint the discovery of off shore pre-salt oil in Brazilian deep waters, which significantly increases petroleum-producing potential of the country.

Among the different renewable primary energy sources in Brazil, the most relevant are hydroelectricity and biomass, from which biofuels are derived. The former has been the subject of several studies and, therefore, will not be discussed as thoroughly as biofuels in the present text, especially sugar cane ethanol. Both hydro and ethanol have stirred heated debates and controversies internationally. The recent soaring food prices worldwide is attributed by some to the supposed prioritization of biofuels production, which, in addition, is blamed, in the case of Brazil, for contributing further to the deforestation of the Amazon. Al of this, in spite of the continuous decrease of the deforestation rates since 2004[1], the increase of ethanol production [2] and the fact that sugarcane for ethanol plantations in Brazil occupy less than 1.5% of the Brazilian crop area [3].

The present situation resembles the not too distant past experience of the crises provoked by the skyrocketing in the price of crude oil on the international market which, until 1973, varied between US$ 1 and US$ 2 only to soar, for a spell in 1979, to US$ 40, then to plummet in the second half of the 1980s, and left to follow an erratic path throughout the 1990s. In 1999, oil prices fell to US$ 10 but, in 2006, exceeded US$ 70 and, in 2008, reached US$ 140. So, in 9 years, the price of oil increased 14-fold, nearly doubling in the span of two years, but then dropped to below US$ 50, maintaining itself around this value for the rest of 2009, to increase again up to reach about US$ 100 in 2011.

For its part, natural gas has been the cause of disputes, in recent years, between Russia and Europe, between Argentina and Chile, and, lately, between Bolivia and Brazil. An important new factor is the increasing participation of shale gas in North America. In 2000 only 1% of natural gas produced in the USA was shale gas, with its share growing to 20% in 2010 and projected to reach 46% in 2035, due to its low cost. The share of natural gas in the Brazilian energy matrix is not significant, although it shall became more important with the import of LNG by ships as well as with the pre-salt natural gas production.

Where electricity is concerned, there were serious instances of rationing in 2001, lasting for many months in Brazil and California - in both cases due to lack of adequate power sector regulation. Energy deregulation played an important part in the process of economic liberalization in the course of financial globalization, which is at the root of the global crisis, which first hit the USA in 2008, and worsening in 2009, it spilled over and reached South America, in particular, Brazil.

The energy crisis has been further aggravated by the overlap of an environmental crisis and a financial one, as a result of climate change, due to the intensification of global warming from greenhouse gas (GHG) emissions, such as carbon dioxide from the burning of fossil fuels. Global warming has become a major global political problem, because it bears on society choices which must not be left to business alone to make. The Nobel Peace Prize of 2008 awarded to the Intergovernmental Panel on Climate Change (IPCC) followed the release, in 2007, of its Fourth Assessment Report which caused great concern around the world.

The repercussions of the high international oil prices on the world economy have been significant, although today's share of oil in the world economy is less than at the time of the 70s oil crises. At a global level, this share in the cost of products is generally half of what it was at that time.

Some particular factors contributed to this strong variation in the oil market:

a. The forecasted decline in world output, despite major discoveries in the Brazilian pre-salt, and increased oil consumption, especially in developing countries, led by China

b. The global geopolitical instability, especially in the Middle East oil producing regions, and the strong dependence of OECD countries on oil imports. To a lesser extent, this instability is felt in South America as in the case of the political tensions between the USA and Venezuela.

c. The global economic crisis which first erupted in the USA in 2008.

http://dx.doi.org/10.5772/54419

d. Environmental pressures, especially due to carbon dioxide emissions from the burning of fossil fuels, which exacerbate the greenhouse effect, thus contributing to global warming.

Finally, with respect to point (d), it is important to point out that the share of renewable primary energy sources is higher in Brazil and in South America as a whole than in other continents, while the use of biofuels in Brazil is widespread and thus, the GHG emissions of the country are mainly from deforestation.

Not considering nuclear energy, released from the fission of uranium and without any chemical combustion, non-renewable fuels, such as coal, oil and natural gas, are responsible for greenhouse gas emissions. Life cycle analysis show that renewable sources such as ethanol from sugar cane and hydroelectricity emit little greenhouse gases.CO2 emitted from biofuels combustion is reabsorbed from the atmosphere during plant growth. However, roughly half of all firewood and charcoal in Brazil comes from deforestation, charcoal mainly used in steel production [4]. The net emissions in the case of alcohol come primarily from the use of diesel for the tractors and trucks on the sugarcane plantations, as well as the production of the synthetic fertilizers and herbicides employed. In the case of hydropower, COPPE's research group carried out measurements at various reservoirs in the country and recorded the carbon dioxide and methane emissions, confirming that those are much smaller than that of the thermoelectric power plants.

2. Survey of energy policy in South America

According to the latest IPCC report [5], there was a 70% worldwide growth in greenhouse gas emissions between 1970 and 2004. Among these, CO2 emissions rose by 80% and represented 77% of the anthropogenic emissions in 2004. The energy sector had the highest growth in emissions between 1970 and 2004 (145%) followed by the transport sector (120%), industry (65%), and land use and deforestation (40%). Table 1 provides the rates of primary energy per capita and CO2 emissions per capita, per energy consumption and per GDP of the South American countries in 2009. It can be seen that countries which have a large share of renewable power, such as Brazil and Paraguay, have better emission indicators than countries such as Argentina and Venezuela, which rely heavily on fossil fuels.

The meeting of the UN Convention on Climate Change in Copenhagen, in 2009, resulted in frustration regarding finding a consensus for more effective commitments to reduce global GHG emissions. However, the commitment to limit to 2° C the rise in global temperature relative to the preindustrial era is encouraging. At the Copenhagen Conference the Brazilian position included this limitation, which entails a major effort to reduce emissions on the part of the rich countries and to keep emissions under control where the developing countries are concerned. One controversial issue refers to the adoption of obligations by developing countries regarding their own emissions. An argument in support of adopting such commitments is the growth of emissions in developing countries, especially China and India. Nevertheless, the per capita CO2 emissions in rich countries are still well above those in developing countries.

Countries	TJ per capita	ton CO2 / capita	ton CO2/ TJ	kg CO2 / US$2000 GDP
Argentina	0.077	4.14	53.6	0.42
Bolivia	0.027	1.31	49.3	1.10
Brazil	0.052	1.74	33.6	0.39
Chile	0.071	3.84	53.9	0.63
Colombia	0.029	1.33	46.4	0.43
Ecuador	0.035	2.09	59.9	1.18
Paraguay	0.031	0.64	20.4	0.45
Peru	0.023	1.32	58.2	0.45
Uruguay	0.051	2.31	45.2	0.26
Venezuela	0.099	5.45	55.2	0.97

Source: [6]

Table 1. Energy Per Capita and CO2 Emissions Indexes from Energy (in Terajoules) Consumption in 2009

It is important to point out that CO_2 emission is not the only indicator to analyze the responsibility among countries. For example, cattle grazing emit a huge amount of CH_4 due to enteric fermentation and N_2O due to manure [7]. Nevertheless, these indicators need to be balanced because they are related to food production, a basic need. In Brazil, an encouraging development has been the creation of the National Climate Change Plan, with its targets for reducing deforestation, responsible for most of Brazil's emissions.

On the other hand, the increased share of fossil fuel use in power generation in Brazil is nothing to cheer about. But the growth of production and consumption of fuel alcohol in cars and the fact that 45% of its primary energy matrix is comprised of renewable sources, including hydroelectric generation and biofuels - as against 13% for the world and 6% for OECD countries - is heartening.

Now, if we consider the different primary energy sources [8], Latin America's share in the world's energy production varies according to the source considered:

- 4.4% of total primary energy
- 9.5% for oil
- 4.9% for natural gas
- 1.4% for coal
- 0.8% for nuclear
- 20.1% for hydroelectricity.

http://dx.doi.org/10.5772/54419

The share of nuclear electricity generation in Latin America represents less than 1% of the world's total, as it is limited to Brazil, Argentina and Mexico. Meanwhile, the share of hydropower exceeds 20%, as Brazil, Venezuela and Peru are among the ten countries with the largest water resources in the world, the first two also being among the top ten producers of hydroelectricity.

Table 2 shows the production, import and export of oil, natural gas, coal and hydroelectric power in the major South American countries. Imports and exports related to oil include oil derivatives in addition to crude oil. With respect to coal, the different types have been computed, as well as coke. The hydroelectricity columns show, in addition to production, the import and export of electrical energy.

Country	Oil (Million toe)			Natural Gas (Million toe)			Coal (million toe)			Hydroelectricity (thousand MWh)		
	Prod	Imp(a)	Exp(a)	Prod	Imp	Exp	Prod	Imp	Exp	Prod	Imp(b)	Exp(b)
Argentina	37.8	1.3	14.9	36.2	1.3	5.4	-	1.0	0.14	34.6	8.0	0.4
Bolivia	2.9	0.2	0.6	9.9	-	8.6	-	-	-	2.5	-	-
Brazil	87.3	28.0	23.4	9.2	7.5	-	2.5	11.3	-	337.4	39.2	0.1
Chile	0.3	14.3	1.7	1.7	5.3	-	0.3	3.9	-	24.8	2.1	-
Colombia	27.4	0.9	16.1	6.1	-	-	38.9	-	34.9	39.8	-	1.7
Ecuador	27.0	2.6	20.6	0.4	-	-	-	-	-	6.8	1.7	-
Paraguay	-	1.1	-	-	-	-	-	-	-	51.2	-	43.8
Peru	5.2	5.9	3.5	1.4	-	-	0.03	0.8	-	19.9	-	-
Uruguay	-	2.3	0.3	-	4.1	-	-	-	-	6.7	1.6	0.8
Venezuela	169.3	-	138.1	23.2	-	-	5. 2	-	5.2	75.0	-	-

(*) Includes crude oil and derivatives; (#) Electricity including hydro and thermal generation

Source: International Energy Agency, 2006

Table 2. Oil, Natural Gas, Coal and Hydroelectricity

According to Table 2, the largest oil producers in South America are Venezuela and Brazil, the latter far behind the former. Brazilian exports (mainly of heavy crude oil) match the imports (of light crude for refining). Argentina, Colombia and Ecuador have a similar production and also export oil.

Argentina is the largest producer of natural gas, followed by Venezuela, Bolivia and Brazil, which is also an importer. The exporters are: Argentina (to Chile) and Bolivia (to Brazil and Argentina). Important consumers of natural gas are: Venezuela, Argentina and Brazil. Coal production is particularly significant in Colombia, also an exporter, while Brazil is the largest producer of hydroelectricity on the continent, followed by Venezuela and Paraguay, the latter also being a major exporter.

To understand the changes in South America, the following aspects should be taken into account:

a. In the years 2000 there has been significant pick-up in economic growth in several countries after a period of stagnation or low growth stretching over quite a few years, under the monetarist policies of economic adjustment under the auspices of the International Monetary Fund and the World Bank with the backing of the rich countries.

b. Social inequality remains high, even if significant improvements are taking place in the social field in some countries. In Brazil, it has been estimated that some 40 million people have been lifted out from the poorer Class D, to the level of Class C income.

The two main energy integration projects in operation between Brazil and South American countries are the bi-national Itaipu power plant with Paraguay, the world's 2nd largest in electricity generation, whose expansion from about 12 GW to 14 GW was completed in 2008, and the import of 30 million m3 per day of natural gas from Bolivia. Both were subject to crises with Bolivia and Paraguay respectively, already settled.

There is an electricity connection between Brazil and Argentina in the South and another one in the North with Venezuela. Furthermore, there is a small connection with Uruguay.

Given the variation in flow without a regulation reservoir to secure the power of these plants, the reservoirs of the hydroelectric plants of the interconnected grid can be used to store water when the flow is high, in order to offset the energy drop during the months with a low flow. However, to avoid very large environment impacts, in the new hydropower plants the flooding area is small. The Santo Antonio and Jirau hydroelectric projects under construction on the Madeira River, near the border with Bolivia, are run of the river, the same being the case of the new Belo Monte hydroelectric power plant.

Brazil is the foremost user not only of liquid biofuels, particularly ethanol in addition to its biodiesel program, but also of solid biomass - firewood and charcoal, widely used in the steel industry. Brazil and Argentina are currently among the world's top five producers of biodiesel, the latter also being a major exporter, mainly to Europe. However, since biodiesel demand in general is still only a fraction of ethanol, most of this article's focus will be on this alcoholic biofuel.

3. Biofuels in Brazil and automotive ethanol

There is an international debate on biofuels, which are being blamed for the high food prices worldwide and which affect the poor most. The Brazilian government has addressed this concern adequately in connection to the production of alcohol from sugarcane. According to [9], the country's sugarcane crop was cultivated over an area of 8.4 million hectares (8.4 Mha) in 2011. 50.3% of all sugarcane was targeted for ethanol production, the remainder used for sugar. On the other hand, soybean, Brazil's most important crop, occupies 25 Mha [10]. According to [11], Brazil has 152 Mha of arable land, of which 62 Mha are currently in use. 177 Mha are pastures So, if one excludes the 440 Mha of virgin forests, there remains 90 Mha left available for expanding agricultural production without deforestation. And these figures do not include the reconversion of degraded pastures. Only a portion of these areas is suitable for sugar cane cultivation and is economically and socially viable for producing biofuels such as ethanol and biodiesel. The latter, to a large extent, comes from soybeans, which, unlike sugarcane, can encourage deforestation in the Amazon, but recently this link cannot be established, because the deforestation rates in Brazil have been decreasing in the last decade [1] in spite of the increase in soy production.

US corn ethanol is subsidized, and, unlike Brazilian sugarcane alcohol, it affects the price of corn, impacting the price of of food and feed. Production of corn ethanol also involves the burning of natural gas. The sugarcane crushed stalk, called bagasse, by contrast, has more than enough energy to meet the plant's heat and electricity demand, even providing surplus power to the grid. Therefore, alcohol produced in Brazil is more efficient energy and environmentally-wise. The capture of CO2 from the air during sugarcane growth roughly balances out the emissions from the production and consumption of alcohol. As a gasoline alternative, it is effective in avoiding the emissions of gases contributing to global warming.

The international biofuels market is poised to increase in the next few years. The US presently consumes twice as much fuel alcohol as Brazil, but its percentage in terms of displacing gasoline is low, around 10%, because of its huge gasoline consumption - 8.74 million barrels per day or roughly 540 billion liters in 2011 [12]. The National Renewable Fuel Standard program (commonly known as RFS) has set an increasing volume of biofuels to be required in the US market [13]. RFS categorizes fuels and caps the so-called "conventional" renewable fuel (corn starch ethanol), so by 2022, 21 billion gallons of the 36 billion gallons (136 billion liters) required must come from cellulosic biofuel or advanced biofuels derived from feedstocks other than corn starch. This categorization of fuels contains specific lifecycle GHG emissions for biofuels relative to lifecycle emissions from fossil fuels and will be further discussed in section 4.

The Energy Independence and Security Act of 2007 (EISA), which established the biofuels mandate, stipulates that indirect, as well as direct, emissions must be accounted for in the lifecycle analysis of any biofuel source, an issue to be explored later on this article. It suffices to say that, currently, the only biofuel currently recognized by the Environmental Protection Agency as being "advanced", from a GHG mitigation standpoint, is Brazilian sugarcane ethanol. If this situation does not change until 2022, at least 19 billion liters will have to be imported from Brazil, for environmental reasons. Considering that Brazilian supply of ethanol in 2011 was 23 billion liters, it will be a major undertaking to supply the projected US demand.

On top of that, in December 2008, the European Parliament approved the Renewable Energy Sources Directive (RED) setting an EU target of 10% for biofuel use in transportation fuels by 2020.

In a nutshell, Brazilian ethanol has generally been regarded the most effective biofuel in terms of mitigating GHG emissions, but, due to the huge markets under consideration, it is unreasonable to expect that Brazil would be able to meet their demands for environmentally appropriate biofuels. Other countries would have to play their part as well.

The issue of biofuels has raised a controversy concerning the competition with food production. In view of the fact that Brazil has plenty of spare land for crop production, as stated above, it should be clear that cultivating sugarcane for fuel alcohol does not interfere substantially with food production, remembering that sugar cane for biofuel occupies 1.3% of the country's agricultural land[3].

In a recent paper, [14] made a comparison between sugar cane and corn for ethanol production, with focus on the present debate about land use dispute for food and energy production. The indicators used to compare the activities are CO2 emissions, energy consumption, co-products from the processes and deforestation. From a methodological standpoint, the study conducted a lifecycle inventory evaluation of sustainability issues, both for developed and developing countries. Brazilian government plans to ensure sustainability are commented. A synthesis of that paper together with other considerations follows below.

There are different biofuels feedstocks, such as forest resources; energy crops; agriculture wastes and urban wastes. Table 3 shows biomass raw materials with the corresponding technologies, products and uses in Brazil, as well as the fossil fuels that they replace. Direct combustion of firewood is important in rural areas for cooking. This does not necessarily entail deforestation as families in rural areas, in general, collect twigs and fallen trees branches. The use of charcoal in the steel industry is important for avoiding GHG emissions. For each ton of pig iron produced, 1.7 tons of CO2 from coke and coal are emitted, while charcoal use in steel production allows, on average, a net capture of 0.9 tons ton of CO2 from the atmosphere, due to tree growth, assuming a planted forest is employed. Thus, if one third of all pig iron were made with charcoal, the steel industry in Brazil could have zero net emission. However, as mentioned, about half of the firewood for charcoal used in pig iron production comes from deforestation, a problem yet to be solved.

The Brazilian Alcohol Program began in 1975 after the first oil shock and its first phase consisted in using ethanol as a gasoline octane booster. After the second oil shock in 1979, a second phase began, with ethanol replacing gasoline in cars, whose Otto cycle engines were adapted for this purpose. Among the historical factors that contributed for the government to deploy the Alcohol Program was the need to reduce the trade balance deficit, affected by crude oil importation. Besides, the Program boosted job generation in sugar cane agro-industry and reduced atmospheric pollution through the elimination of lead as an additive to gasoline, as ethanol has a high octane index [15, 16].

By 1985 more than 90% of new cars sales consisted of ethanol fuelled engines, but in the 1990 decade there was a shortage of ethanol in the country. An ad hoc temporary solution was to adopt a ternary mix composed of ethanol, methanol and gasoline to supply part of the market. The result was a lack of consumer confidence in ethanol, with the consequent reduction of sales of new ethanol fuelled cars to 11% in 1990, 2% in 1995 and 1% in 2000 [17]. The reasons for the ethanol shortage were the fall of crude oil price and lack of continuity in governmental policy for ethanol.

http://dx.doi.org/10.5772/54419

Technology	Biomass Raw Material	Products	Main Use	Fossil Fuels Substitution
Direct Combustion	Firewood Sugar cane bagasse trash and other wastes	Heat	Cooking Industry Electric power	LPG Fuel oil Natural gas
Bioconversion:				
- Fermentation	Sugar cane	Ethanol	Transport	Gasoline*
- Anaerobic digestion	Wastes	Biogas	Potential	Natural gas
Chemical and Thermal:				
- Pyrolysis	Wood	Charcoal	Industry	Coal and fuel oil
-Gasification	Biomass	Synthesis gas	Industry	Natural gas
-Esterification	Vegetable oil and others materials**	Biodiesel	Transport	Diesel
- Cracking	Vegetable oil	Diesel	R&D	Diesel
- Hbio***	Vegetable oil	Diesel	Pilot	Diesel
Hydrolysis	Biomass	Ethanol	R&D	Gasoline*

Obs: (#) Includes urban solid wastes, lixivia from pulp and paper industry, wastes from rice and others; (*) – It can substitute also for diesel oil with some additive; in Brazil gasoline has 25% of ethanol as additive, besides the use of pure ethanol in flex fuel cars; (**) Including animal fat wastes, garbage and micro-algae (R&D); (***) – Petrobras Technology for processing vegetable oil in oil refineries.

Table 3. Uses of Bioenergy in Brazil

Beginning in 2003 there was an ethanol revival due to local production of flex fuel cars. Their engines can work with two different fuels in any proportion, and were first made in the US in the 1980s, but the technology developed by Brazilian engineers is innovative, as it uses sensors that already exist in the car, which match their fuel readings against information stored in the on-board computer to adjust the engine. Early US flex cars used a special sensor to identify the fuel mix and adjust the engine, but it was expensive and not viable for the Brazilian fleet, dominated by low cost, small and midsized compact cars [18].

Figure 1 shows the behavior of ethanol consumption in Brazil, which surpassed 15 billion liters in 1998, going down to 10 billion liters in 2001 and grew again in the years 2000, due to the introduction of flex fuel cars, which boosted the demand for hydrated ethanol.

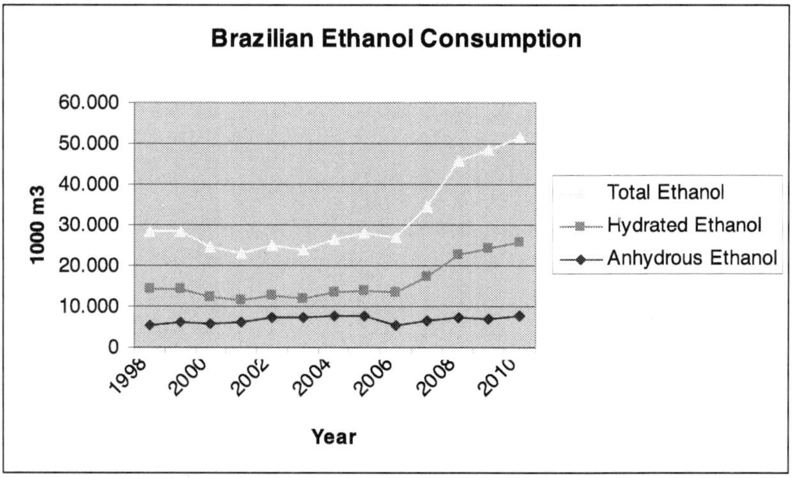

Source: [19,20]

Figure 1. Evolution of ethanol consumption in Brazil (billions liters a year)

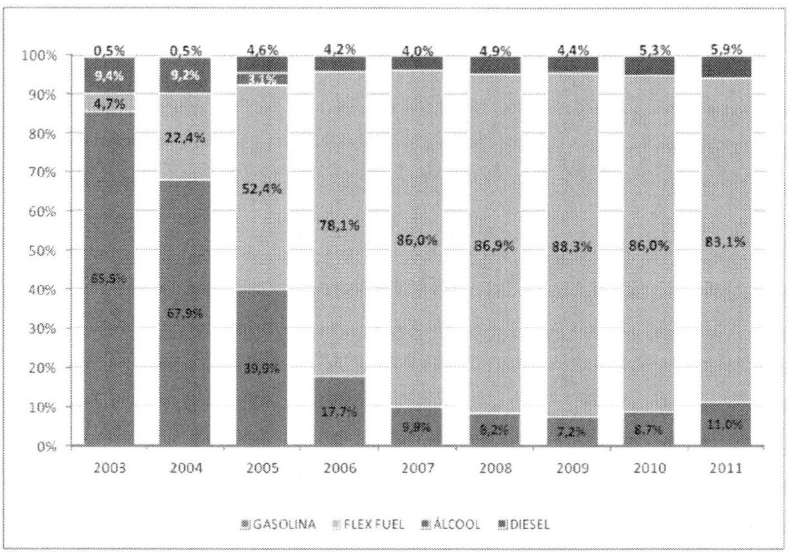

Source: [9]

Figure 2. Sales of Gasoline, Alcohol and Flex Cars in Brazil

http://dx.doi.org/10.5772/54419

Figure 2 shows how this phenomenon was correlated to the quick penetration of flex fuel cars, which currently comprise more than half the total fleet of passenger cars in Brazil The exponential growth of flexible cars after 2003, stimulated by the high gasoline price due to the increase of crude oil price, global warming-related pressure, among others factors. The cost of ethanol in Brazil went down from US$ 20/GJ in 1980 to US$ 6/GJ in 2006, corresponding to US$ 40/barrel of oil [21] following a learning curve. So, while subsides were necessary to start the program they are not needed nowadays.

4. Avoided CO2 emission by ethanol substitution for gasoline

4.1. Life cycle avoided emission: Comparison of sugar cane ethanol with corn ethanol

A problem of ethanol in some OECD countries and China is that it is made from corn. From a global warming standpoint, corn ethanol is less effective than sugar cane ethanol as a substitute for gasoline.

The advantage of biofuels is that when biomass grows up it captures from the atmosphere the CO2 emitted by biofuel combustion in the car engine. However, 1 GJ of fossil fuel is expended to produce 1.3 GJ of ethanol from corn [22]. That is, for each energy unit transformed in heat through corn ethanol combustion, 0.77 energy units is spent producing ethanol from corn, mainly in the natural gas needed for ethanol distillation, as well as the embedded energy in synthetic fertilizers and herbicides.

On other hand, sugar cane has a surplus of biomass enough to generate heat and electricity in the process of ethanol production. For each 1 GJ of fossil fuel consumed in sugar cane and ethanol production there are, on average, 9 GJ of ethanol and this value can reach 11 GJ in the best cases [22].

Therefore, for each unit of energy transformed in heat when sugar cane ethanol is burned in car engines, an average of only 0.11 units of energy from fossil fuel is needed to product it. Besides the bagasse, sugar cane has a significant amount of trash (leaves and top), which is usually burned before harvesting, to allow manual cutting by laborers. However, crop residues are increasingly being recovered, as mechanization, mainly in São Paulo state (responsible for 50% of ethanol produced in Brazil) [23], is becoming more commonplace to harvest cane. To calculate the net avoided emissions, we must subtract, from the gross avoided CO2 emissions due to fossil fuel substitution, the emissions of CO2 from fossil fuels used in sugar cane and ethanol production process, as well as other GHG emitted also for producing cane and ethanol. Therefore, there is the need to express the mass of each non CO2 GHG in terms of equivalent CO2 emission.

In the literature there is a range of values for emissions in sugar cane production[1] and for the avoided CO2, depending on the case study and on methodology. For instance, different papers consider alternatively:

1 It is usual to express the emission in terms of mass of Carbon in the molecule. For instance, the mass of C in CO2 is 12/ (12+2x16) = 12/44 of the CO2 mass and the mass of C in CH4 is 12/ (12+ 4x1) = 12/16 of the CH4 mass.

a. the best case, a particular one or the average among a set of farms and distilleries in a period of time;

b. either only the gross avoided emission of the gasoline replaced with ethanol, or the gross avoided emission of gasoline plus the surplus electricity sold to the grid;

c. either only the direct energy consumption or the life cycle analysis, including emissions in ethanol production and also in gasoline production;

d. hydrated ethanol, anhydrous ethanol or a mix of them in the market.

Sometimes, the assumptions used in published papers are not clear, causing confusion in quotations.

4.2. Numerical results from field research data on sugar cane ethanol

A detailed life cycle analysis was presented in a report supported by the Environment Secretariat of São Paulo State [21]. The data base was composed of three surveys, the first one covering 26 to 31 distilleries, the second one 17 to 22, and the last one a larger set of 98 distilleries throughout the country. From this reference, it is possible to calculate representative values for emissions from sugar cane and ethanol production in percentages of CO2 equivalent (Table 4). The percentage of CO2 that is avoided by the ethanol industry can be found in Table 4.

Emission Source	Emission Type	%
From life cycle (A): equipments, buildings, etc	in cane production	6.6
	in ethanol production	9.5
From fertilizers, herbicides, pesticides etc (B)	in cane production	20.6
From sugar cane burning before harvest (C)	CH4	19.1
	N2O	18.2
From soil (C)	N2O	6.9
From fossil fuel consumption (C)	CO2	19.1
Total	A	16.1
	B	20.6
	C	63.3

Source: Elaborated by using data from [21]

Table 4. GHG Emissions in Sugar Cane Ethanol Production (% of CO2 equivalent)

The results in Table 4 deserve some comments. The lower heat value of ethanol is compensated by the higher compression rate and better efficiency of the engine. In the use of anhydrous ethanol as an additive to gasoline, used in a proportion of 25% in Brazil (E25), 1 liter of ethanol corresponds to 1 liter of gasoline. In the case of hydrated ethanol the proportion is 1.3 liters of ethanol

(E100) to 1 liter of E25, which means 1 liter of ethanol for 0.77 liter of E25 or 0.77 x 0.75 = 0.577 liter of gasoline. The direct emission factor of gasoline is 0.0693 kg CO2/MJ [5], but in a life cycle it becomes 0.0817 kg CO2 / MJ [24]. Instead of bagasse substitution for fuel oil to calculate H′ in Table 5, the emission by electric power generation in Brazilian interconnected grid established for the Clean Development Mechanism can be applied.

More recent data on the average emissions can be obtained from [25], considering the 2005/2006 harvest. Their case study focused on a set of Brazilian distilleries that process 100 Mt of sugar cane per year.

Results from 2002/2003 Harvest	Average	Best Value Scenario
Energy consumption EC (Mcal/t cane)	48.2	45.8
Sugar cane agriculture	11.8	9.5
Ethanol production	60.0	55.3
Total		
Energy production EP (Mcal/t cane)	499.4	565.7
(ethanol + electric energy from bagasse surplus)		
Energy gain (EP/EC)	8.3	10.2
GHG emissions (kg CO2 equiv. / t cane)	19.2	17.7
From fossil fuel consumption	15.3	15.3
Others	34.5	33.0
Total		
Total GHG emission (kg CO2 equiv./ m3 of ethanol) = (A+B+C)	405.8	358.7
Net avoided CO2 (kg CO2 / m3 of ethanol) from gasoline (H) and fuel	2600	2700
oil (H′) used for electric energy: H+H′- (A+B+C)	1700	1900
For anhydrous ethanol		
For hydrated ethanol		
Percentage of avoided CO2	86%	88%
For anhydrous ethanol	81%	84%
For hydrated ethanol		

Source: Elaborated by using data from [21]

Table 5. Energy gain, GHG emissions and percentage of CO2 avoided by the Ethanol industry in Brazil

Using data from this article it is possible to calculate the net avoided CO2 in terms of percentage of fossil fuel CO2 emission. The results are:

a. For sugar cane and ethanol production the total GHG emissions, using GWP [26], are 436 kg CO2 equivalent / m3 of ethanol.

b. The net CO2 avoided emissions are 2323 kg CO2 / m3 of anhydrous ethanol.

c. In [14] the percentage of avoided fossil fuel CO2 emission due to anhydrous ethanol is 84.1 %.

The above results, confirmed by Table 5, conclude that a very high percentage of GHG emission is avoided by sugar cane ethanol substitution for gasoline.

In Brazil about 1 toe of bagasse is consumed to produce 2 m3 of ethanol [19], equivalent to 20,900 MJ/m3. Taking this value as the self consumption of energy in the distillery (from bagasse combustion) and assuming that, instead of bagasse, natural gas is burned, whose emission in life cycle is 0.095 kg CO2/MJ [24], 20,900 x 0.095 = 1985 kg CO2, in a first approximation, the emissions are reduced by 12%.

The above calculation roughly shows the avoided percentage of CO2 when fossil fuel is not necessary to produce ethanol as in corn ethanol. Other factors that make the latter more energy intensive than sugarcane ethanol: better photosynthetic efficiency of sugarcane, producing more biomass; corn produces starch, which must be hydrolyzed (broken down into sugars), before fermentation; corn demands significantly more nitrogen fertilizers than Brazilian sugarcane, which employs biological nitrogen fixation techniques. Besides, Brazilian sugarcane industry is increasingly employing high efficiency steam boilers. All these factors explain why the avoided CO2 from sugar cane ethanol is much higher than that from corn ethanol.

4.3. Potential of GHG emission mitigation through energy efficiency and harvest mechanization

4.3.1. Potential energy improvement from ethanol, bagasse and residues

It is possible to improve the energy balance of sugar cane ethanol by:

a. Increasing sugar cane productivity in tons of cane per hectare;

b. Increasing the amount of ethanol produced from each ton of sugar cane;

c. Obtaining more agricultural residues through harvest mechanization;

d. Improving conversion efficiency of bagasse and trash (sugarcane top and leaves) into heat, mechanical and electric energy.

Ethanol productivity grew up from 2024 liters per hectare in 1975 to 5931 liters per hectare in 2005 [27]. The production of sugarcane in the period 1975-2006 rose from 89 million metric tonnes to 426 million metric tonnes [28]. However, for several reasons – old plantations, poor weather prior and during harvest, sugarcane production and ethanol yield has decreased in 2011, as shown in Table 6.

On average, 55% of sugar cane has been used for producing ethanol in the last five years, but this percentage has decreased to 50.3% in 2011, one of the reasons for the steep decline from 2010 [9]. The best average ethanol yield ever obtained, was 92 liters/t of cane [29], but this performance will take some years to return, due to lack of investments in sugarcane plantation renovation, linked to the Federal Government current gasoline price freeze policy, which has also led to an increase of sugar production instead of ethanol.

The energy available in bagasse and trash is quite significant. Each ton of cane has 280 kg of bagasse with 50% of humidity and 2,130 kcal /kg [19], yielding 596 Mcal / ton of cane. The average value for trash is slightly lower, but the combined energy of bagasse and trash is more than the double of ethanol energy (Table 7) calculated with 92 liter with 0.8 kg/liter and heat value 6,500 kcal/kg [19].

Processed Cane	Ethano	Productivity	
Year	**Mt**	**Mm3**	**liter/t**
2003	359.3	14.5	74.8
2006	426.0	17.7	75.7
2009	622,6	26,2	84.1
2010	627,3	28,0	90,0
2011	565,8	23,0	80,0

Source: [9]

Table 6. Sugar cane and Ethanol Production and Productivity

Mcal/t of cane	
92 liters of ethanol (best value)	478
280 kg of bagasse with 50% of humidity	596
280 kg of trash with 50% of humidity	596

Source: [22]

Table 7. Energy from 1 Metric Ton of Sugar Cane Considering Heat Values

In 2006 bagasse production was 121.0 Mt, from which 71.5 Mt (59.1%) was converted into heat for sugar productiont, 42.0 Mt (34.7%) for ethanol production and 7.5 Mt (6.2%) for electric power, part of it exported to the grid [19]. Sugar cane trash was not computed. So, 94% of bagasse is converted into heat and mechanical work for sugar and ethanol production. If there were a reduction of 20% in this percentage, through efficiency improvement, the energy from bagasse available for electric power would increase by a factor of 4 (24.2+7.5/7.5).

Besides, if 50% of the trash is used, thermal energy for electric generation will increase by a factor (61.5+24.2+7.5 / 7.5) = 12.4. As nowadays only a small part of electric energy from bagasse is sold to the grid, the avoided GHG emission due to electric energy sale to the grid could be multiplied even further.

In this scenario, it would be possible to avoid more CO2 emissions than that from gasoline replaced by ethanol, as can be shown below. The percentage of net avoided CO2 in terms of percentage of fossil fuel CO2 emission is given by the following formula:

$$P' = 1 - \left(A + B + C - H'\right) / H \quad (1)$$

A = emission from fossil fuels to make the equipments and to construct the buildings for cane and ethanol production in an entire life cycle analysis;

B = emission from fossil fuel to produce fertilizers and other materials;

C = emission from fossil fuel in sugar cane production[2] and from soil (N2O), as well as CH4 and N2O emission from cane (trash) burning before harvesting;

H = gross avoided emission of gasoline that is substituted with ethanol, in a life cycle analysis;

H' = gross avoided emission of fossil fuel used for electric generation in the grid, replaced by electric energy sold by the distilleries, using the bagasse (and trash) surplus after their self consumption in ethanol production.

The electric installed capacity using bagasse in 2006 was 2.6 GW [19], 85% of which, 2.2 GW, for self consumption and 15%, only 0.4 GW, sold to the grid. In the same year 8,357 GWh was produced, 1,256 GWh relayed to the grid [BEN, 2007]. In the hypothesis of increasing by a factor 4 the electric power generation from bagasse, as pointed above, the installed capacity could become 10.4 GW and 10.4 – 2.2 = 8.2 GW could be sold to the grid, a twenty-fold increase, expanding generation to 25,120 GWh.

Assuming that it will replace natural gas plants (which has a life cycle emission 0.095 kg CO2/MJ [16]) with 40% conversion efficiency, the avoided CO2 yields 2.13 billion kg of CO2. As in 2006 ethanol production was 17.7 Mm3, the avoided emission would be 1,200 kg CO2/m3 of ethanol. The bagasse computed for electric generation in the above estimate came from ethanol and sugar production as this industry is integrated. Considering that 55% of sugar cane is for ethanol production, H'= 660 kg CO2/m3 of ethanol. Using in formula (1) this figure and the average values from Table 4, the result is P'= 1.10, or 110%.

Therefore, besides compensating the full emission of sugar cane and ethanol production, for each ton of CO2 avoided through ethanol substitution for gasoline, a further 100 kg of CO2 could be avoided due to bagasse surplus conversion into electric power.

4.3.2. Energy and emissions scenario with mechanization

Focusing only ethanol production and consulting again Table 6 and reference [19], it is easy to calculate that bagasse consumption for heat and mechanical work in ethanol production amounts to 42 x 2,130 / (426 x 0.55) = 382 Mcal/ t of cane. Subtracting this value from 596 Mcal (Table 6) there is a 214 Mcal bagasse surplus per ton of cane in ethanol production that

---

2 It includes emissions from diesel oil in tractors, mechanized harvesting and trucks for transportation

can be used for electric energy. By the same token, it can be deduced that 68.2 Mcal of bagasse per ton of cane was used for electric power in 2006.

Considering a substitution of 1 liter of ethanol (average of hydrated and anhydrous) for 0.79 liter of gasoline, with heat value 10,400 kcal / kg and density 0.74 kg / liter, the corresponding energy is 0.79 x 0.74 x 10,400 = 6,080 kcal per liter of ethanol or 75.7 x 6,080 = 460,000 kcal/t of cane. In the case of 92 liters/t of cane the equivalent energy will be 559 Mcal / t of cane.

There are limits for the sugar cane residues recovery because a portion is needed to recycle nutrients as well as protect the soil from erosion and because mechanization cannot be used in more than 50% of the area with present technology, due to declivity. On the other hand, the burning of bagasse and residues could be done with improved thermodynamic efficiency. Table 8 shows a hypothetical scenario regarding the 2006 harvest, based in the above considerations:

	Brazil 2006	**Future Scenario**	**Increase**
Ethanol (energy of displaced gasoline)	460*	559**	21%
Bagasse for electric energy (part to grid)	68	214***	314%
Trash for electric energy (all to grid)	-	298 ****	infinite

* Considering 75.7 liters of ethanol per ton of cane (Brazil's average in 2006)

** With the best value of 92 liters of ethanol per ton of cane

*** It is subtracted self consumption of 377 Mcal / t of cane

**** 50% of total mass

Table 8. Energy (Mcal) from 1 Metric Ton of Sugar Cane

The scenario of Table 7 does not take into account the possible improvement of efficiency in energy transformation that can increase the bagasse surplus, for instance by changing low efficiency steam systems for electric power. In many plants low pressure steam boilers are used. tone can obtain more mechanical and electric energy per ton of cane with higher efficiency systems, decreasing bagasse self consumption. In general the sugarcane sector in Brazil employ boilers and turbines with 22 bars of steam pressure, which can be increased to 60 bars or 80 bars, improving efficiency by at least a factor of 2. Bagasse self consumption in ethanol production is usually divided in the following way: 90% for heat in ethanol distillation, 5% for mechanical work and 5% for electric power. If efficiency is improved in the conversion of heat into mechanical and electric energy, not only the bagasse surplus will be higher, but there will also be more electric power available to the grid per ton of bagasse. Therefore, H' could be even higher.

Harvest mechanization, utilized in such a way to avoid cane burning, can allow not only a higher value of H' due to the use of trash in electric power generation, but also a lower value of C emission in formula 1. 100% of mechanization is not feasible because of the slope in part of the lands where sugar cane is planted. If mechanization is increased in 50% in relation to the case study depicted in Table 3, there could be a reduction of 0.5x 37.3= 18.6% in CO_2 equivalent emission of CH_4 and N_2O from cane burning.

However, 50% more machine-based harvesting will increase emissions from diesel oil in the same proportion. Assuming that half of fossil fuel consumption in Table 3 is diesel oil, the corresponding emissions (0.5x.19.1%=9.5%) will increase by 4.75%. The net result should be an emission reduction of 18.6 – 4.75= 13.85%. Diesel oil can be eliminated by fuelling diesel engines with either biodiesel or ethanol with additive. In this scenario, the higher indirect energy in life cycle of harvesting machines (A in Table 3) is considered negligible.

The problem of increasing harvesting mechanization is the drastic reduction of workers in sugar cane crop. The number of workers in the year 2005 in sugar cane agriculture was 414 thousand, in sugar production 439 thousands and in ethanol industry 128 thousand [30]. However, manual harvesting of sugar cane is a very hazardous activity, often causing stress-related diseases in workers. Also, sugar cane burning is a major source of air pollution, causing respiratory illnesses to local populations.

5. Discussion on land uses, ethanol competition with food and deforestation

5.1. Land uses and deforestation

The issue of food crop displacement due to biofuel competition has been raised recently by several authors, which have concluded that land use change is the main cause of GHG emissions of biofuels in general. Land use change (LUC) is a complex process caused by the interaction of natural and social systems at different temporal and spatial scales. LUC can induce GHG emissions due to oxidation of soil organic carbon and due to burning or decomposition of above-ground biomass.

However, it's important to notice that biofuels account for a very small proportion of global agricultural production; approximately 2%, or around 36 Mha [31] from a total cropland area of around 1,527 Mha [32]. Therefore, the magnitude of GHG emissions due to LUC from global biofuel production is small compared to the total emissions from all LUC: agricultural land expansion for food, feed, fibre, cattle ranching, fuel wood and timber (loggings), and expansion of infrastructure generates the greater part of LUC emissions. With respect to biomass cultivation, LUC can be divided into:

- Direct land Use Change (DLUC) – it occurs when bioenergy feedstock production modifies an existing land use, resulting in a change in above- and below-ground carbon stocks

http://dx.doi.org/10.5772/54419

- Indirect land Use change (ILUC) - occurs when land that was formerly used for the cultivation of food, feed or fiber is now used for biomass production shifting the original land use to an alternative area that might have a high carbon stock, like forests and wetlands. This carbon stock could be reduced if utilized for agricultural purposes. The resulting (indirect) GHG emissions are (at least partly) caused by increasing biomass/biofuel production.

[33] were the first to address the issue of ILUC: they estimated that allocation of 12.8 Mha of corn to produce ethanol in the USA would result in 10.8 Mha of new cropland around the world. The conversion of native ecosystems to cropland would result in indirect emissions potentially twice as large as direct lifecycle emissions, yielding emissions that surpass the gasoline it would replace. As there are no direct measurements that can be made, the authors used a partial economic equilibrium model of the global agricultural sector to assess the indirect emissions. ILUC will be discussed in more detail in the next section.

The problem of ILUC is not new in Brazil. It was exhaustively discussed since the displacement of food crops in São Paulo State by sugar cane was pointed out a long time ago by [34;35]. There was indeed displacement of food crops; however, its dimension is very different in the case of sugar cane in Brazil as opposed to that of corn in US. The US participation in World corn production is higher than the participation of Brazil in global sugar cane production, while the area used for corn production in US is almost 5 times the area for sugar cane production in Brazil (Table 9).

	US corn	Brazil sugar cane
Percentage of World production	38%	21.7%
Crop area	37 Mha	8.4 Mha
Source: [9;36]		

Table 9. Corn in US X Sugar Cane in Brazil

In 2011, the area used for sugar cane production in Brazil was 8.4 million ha, 4.2 million ha each to produce ethanol and sugar. As previously mentioned, Brazil has 152 Mha useful for agriculture, but only 62 Mha are currently cultivated for food, feed and biofuels. Thus, sugar cane for ethanol utilizes only 2.7% of the area useful for agriculture, not taking into account the recovery of degraded pasture lands. Native vegetation area of Brazil comprises 440 Mha, most of it in Amazon rain forest, in the North. The production of sugar cane is concentrated in the Southeast, followed by Center West, South and Northeast regions, but only 0.4% from the North Region (Table 10).

Southeast - 63.7%	Northeast – 11.8%	Center West-16.7%	North - 0.4%	South - 7.3%
Source: [23]				

Table 10. Sugarcane Production in Brazil per Region

In September 2009, to emphasize the Brazilian Federal Government disposition to preserve its main natural resources from sugarcane expansion, EMBRAPA, its agricultural research agency, published the *Agro Ecological Zoning of Sugarcane.* In it, EMBRAPA surveys the entire territory, pointing out the appropriate sites to cultivate sugarcane, from a soil and climate standpoint. The survey explicitly excluded any areas within the Amazon Rainforest, the Pantanal Biome (Wetlands), as well as Indian territories and protected areas according to the National System of Conservation Units. The study identified a total of 64.7 Mha of appropriate land for sugarcane expansion, 37.2 Mha from pasture land. It concludes that those numbers demonstrate the country's capacity to expand sugarcane production without causing deforestation or displacing lands with other crops used for food and feed. Figure 3 shows the resulting map.

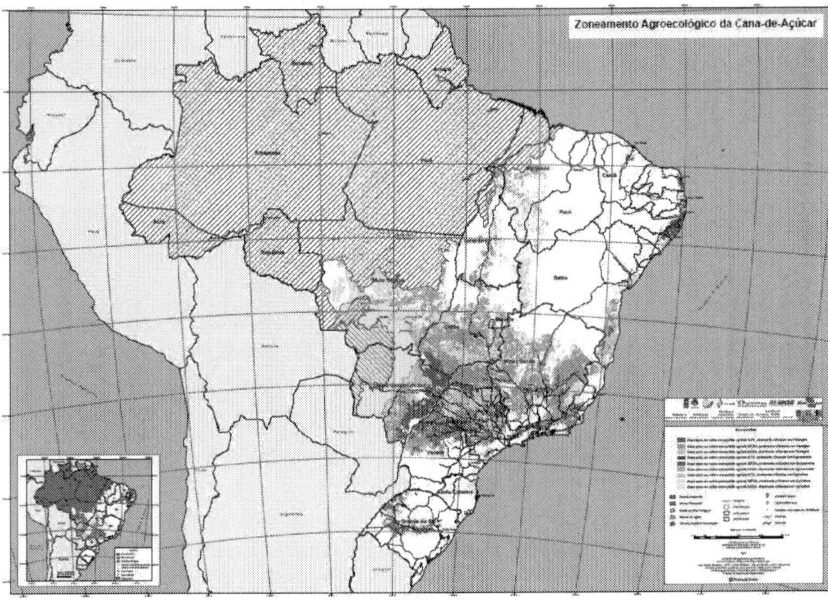

Source: [37]

Figure 3. Brazilian Sugar cane Agroecological Zoning

No correlation between expansion of sugar cane and soy beans crops and deforestation has been established, since Brazil's production of both crops has continuously increased, whereas deforestation rates have been decreasing since 2005 [1].

There are two points to be observed about biofuels and deforestation:

a. The presence of sugar cane in Center-West region (17%) has an impact on the Brazilian Cerrado, although its savanna type vegetation is less dense than rain forest.

http://dx.doi.org/10.5772/54419

b. Soybean, mainly used for feed, but also for biodiesel, is different, because its presence in the North region is high, although Brazilian biodiesel production is only one-tenth of ethanol's. A multi-stakeholder initiative was established in 2006 with a commitment from major buyers not to acquire soybeans from areas in the Amazon that have been deforested after July 2006. The Soy Moratorium was signed by corporate unions which represent more than 90% of Brazil's soybean industry.

Among the goals of the Brazilian Biodiesel Program there is the cultivation of castor, palm, sun flower and other crops by small farmers to supply the raw materials. But, since its inception, soybean oil from large soybean plantations has been the dominant feedstock for biodiesel.

The commonly accepted drivers of deforestation are four, mainly in the rain forest:

a. wood extraction;

b. pastures for cattle grazing;

c. crop plantations;

d. mineral resource exploration.

But in the case of Brazil all these drivers increased in the last decade while the deforestation rates decreased, showing that this mainstream deforestation model does not hold true in Brazil.

Different concepts of deforestation yield different values of deforested areas due to the accounting (or not) of deforested areas where vegetation is growing again. In some studies the deforested areas are only the ones that really changed their land use definitively, but others consider the burning of biomass as deforestation. Nevertheless these differences are mostly qualitative for the Brazilian Legal Amazon. Another issue is the vegetation classification which leads to varying results in terms of carbon emissions for the same area. It depends on the adopted vegetation classification because carbon content can differ a lot. For example, according to HYDE/IVIG database, the Brazilian Legal Amazon land use changes representing agriculture and pasture lands added up to 422,070 km2, in 1990. The natural areas were originally tropical forest, wooded tropical forest and savanna. According to INPE database, the cumulative Brazilian Legal Amazon deforestation until 1990 was 415,000 km2. These numbers show the compatibility of the 2 databases in terms of magnitude but the quality of the information present huge differences. These differences indicate that it is important to adopt a more detailed focus of analysis with new indicators [38].

Deforestation was responsible for about 78% of CO2 emissions of Brazil in 2000/2005 [39]. In 2006, Brazil proposed to the UN Climate Change Convention the creation of an International Fund for Reducing Deforestation. In 2007, the Brazilian Forum on Climate Change[3] presented the Government a formal suggestion for a National Plan of Action, with contributions from universities and research institutions, NGO's and private companies. In December 2008 the National Plan on Climate Change was published.

3 The President is the chairman of the Brazilian Forum on Climate Change. Members include the Minister of Science and Technology, of the Environment, of Foreign Affairs of Energy, members from academia, NGOs and industry.

To achieve its GHG emission targets, NPCC set as one of its main actions a sustained reduction of deforestation rates in all Brazilian biomes, in particular the Amazon Forest. Specifically, it calls for a reduction of 40% in the average deforestation rate by the 2006-2009 period in relation to the average rate of the ten years preceding years (1996-2005). For each of the next two periods of four years, it aims to reach a further 30% reduction, in relation to the previous period.

Since Brazil does not belong to Annex I of the UN Climate Change Convention, it does not have to set a binding emission commitment. But its decision is compatible with the so called road map decided in Bali (in 2007), which called for *"Nationally Appropriate Mitigation Actions by developing country Parties in the context of sustainable development, supported and enabled by technology, financing and capacity building, in a measurable, reportable and verifiable manner"*. NAMAs are to act as a bridge between developed and developing country parties, following the principle of *'common but differentiated responsibilities'*.

5.2. Indirect land use change and GHG emissions

Before [33], biofuel "well-to-wheel" LCAs were mostly attributional, consisting of a linear chain-like series of analytical steps, carrying out no assessments of nonlinear feedback-like effects. As already mentioned, the above authors suggested that when including ILUC, the previously thought GHG-saving corn ethanol turned into a net producer of GHG emissions. This raised serious concerns and new efforts were launched, particularly in the U.S. and Europe, to study biofuels LUC.

The logic that supports ILUC is that biofuel production competes for agricultural resources, which results in an increase in the price of agricultural products, and these price increases cause additional conversions of the world's grasslands and forests to cropland. This additional land conversion results in loss of carbon previously sequestered in grassland and forest ecosystems. These emissions are an indirect result of producing biofuels and should be considered in calculating the GHG implication of adopting biofuels.

The quantification of the net GHG effects of DLUC occurring on a site used for bioenergy feedstock production requires the definition of a reference land use as well as carbon stock data. This data can be uncertain but still allows quantification of emissions with sufficient confidence for guiding policy. ILUC emissions estimation, on the other hand, is highly problematic given the complexities of the economic and social systems that connect biofuel production with land conversion throughout the world.

To make matters worse, ILUC due to the recent expansion of the biofuel industry is hard to assess because this expansion constitutes a very small driver as ILUC effects are not specific to biofuels or bioenergy, but to all incremental land use, so the biofuel impact is likely to be dwarfed by other causes. Besides, as the name implies, ILUC cannot be measured, only modeled, although case studies could offer some useful evidence. The fact is, up to now, there is still no sound and commonly accepted methodology either to calculate or to assign iLUC effects properly. Hence, bioenergy policies worldwide face a dilemma: neglect iLUC effects that in fact exist or take them into account although no sound methodology is available?

Despite these difficulties, the US Environmental Protection Agency, as part of the updated Renewable Fuels Standard (RFS-2), specified that life-cycle GHG are to include "direct emissions and significant indirect emissions such as significant emissions from land use change" to determine if a specific biofuel is eligible to be counted towards the RFS-2 mandate. Biofuels in the RFS were divided into four categories:

- Renewable biofuel – any qualifying renewable fuel, including corn ethanol. Must meet 20% lifecycle GHG threshold
- Advanced biofuel – anything but corn ethanol, can include cellulosic ethanol and biomass-based diesel. Must meet 50% lifecycle GHG threshold
- Cellulosic biofuel – renewable fuel produced from cellulose, hemicellulose, or lignin. BTL, green gasoline also apply. Must meet 50% lifecycle GHG threshold
- Biomass-based diesel – biodiesel, renewable diesel. Must meet 50% lifecycle GHG threshold

Using the FAPRI/CARD model, EPA published in 2009 an analysis of GHG emissions of a set of biofuels: US corn ethanol, Brazilian sugarcane ethanol, soybean biodiesel, among others. Brazilian sugarcane ethanol, after taking ILUC into account, was found to reduce emissions by 26%, which would disqualify it as an advanced biofuel. Corn ethanol did not qualify as a renewable fuel, except when cogeneration was used, nor did soybean biodiesel. There was an immediate backlash from the industry.

ICONE, a Brazilian consultancy, sent a letter to EPA listing shortcomings of the agency's adopted model. It presented BLUM - Brazilian Land Use Model, more sophisticated, with better spatial resolution, and whose calculations resulted in total emission reduction (DLUC + ILUC) of 60%.

On February 2010, EPA concluded that, in fact Brazilian sugarcane ethanol reduced emissions by 61%, qualifying it as an advanced biofuel. The impacts of this decision will be discussed in the following section. EPA also revised the GHG emission reductions of other biofuels: corn ethanol was found to reduce emissions by 21%, qualifying it as a renewable fuel; soybean biodiesel was found to mitigate 57% of emissions, qualifying it as biomass-based diesel.

In the European Union (EU), ILUC calculations were more carefully considered. On April 2009, the EU adopted the Renewable Energy Directive which included a 10% target for the use of renewable energy in road transport by 2020. It set a minimum rate of direct GHG emissions savings – 35% in 2009, rising over time to 50% in 2017. Moreover, the European Commission (EC) was asked to examine the matter of ILUC, including measures to avoid it and report back this issue by the end of 2010. In that context the EC undertook a review of the scientific literature modelling the land use change impacts of biofuels, reviewing over 150 contributions on the topic and reviewing 22 different modelling exercises [40]. Large discrepancies were found in their results, reinforcing the arguments that good estimates of land use indirect impacts are hard to achieve. Comments sent to the EC [41] in regard to the need to improve sugarcane ethanol model assumptions from those studies include:

- Projections on sugarcane and ethanol yield: models project smaller yields as compared to historical trends
- Poor analysis or lack of analysis on pasture intensification, leading to an overestimation of LUC resulting from the expansion of biofuels, as pasture is the largest land user in Brazil and there has been high cropland expansion in this land category in the current decade
- Lack of evidence supporting the criteria used to allocate marginal land demand over native vegetation. The main criterion was historical data, either inaccurate or based on the assumption that additional cropland due to biofuels expansion will determine a frontier advancement similar to what has been observed historically.

The EC published a report in December 2010 setting out four policy options it was considering:

- Take no action for the time being while continuing to monitor.
- Increasing the minimum greenhouse gas threshold for biofuels.
- Introducing additional sustainability requirements for certain biofuels.
- Attributing GHG emissions to biofuels reflecting the estimated ILUC impact

Recognizing the difficulties in establishing a consistent methodology for calculating biofuels ILUC emissions, as of this writing, no decision has yet been made by the EC.

5.3. Potential sugar cane expansion and external markets

The next question concerning possible impacts arising from the expansion of sugar cane production is related to the potential exportation of ethanol to OECD countries.

The European car fleet uses a growing proportion of diesel engines, although there is a non negligible consumption of gasoline either with or without ethanol as additive. For instance, part of Sweden's car fleet uses 80% of ethanol and 20% of gasoline (E80).

The most important ethanol producers, as well as the countries to which Brazil exports can be seen in table 11. Only the USA and Brazil, which together supply 87% of the world's production, use ethanol substitution for gasoline in a large scale. The current percentage of ethanol mixed to gasoline in the US is 10%, due to an EPA blend "wall", which has recently expanded to a 15% threshold.

Until recently, the North-American market, almost all supplied by domestic production, was not open to Brazilian ethanol. However, at the end of 2011, a U$ 0.54/gallon levy on Brazilian ethanol, as well as a US$ 0.45/gallon credit for American producers of corn ethanol was waived, paving the way for a better relationship between these two major players. Also, due to EPA's classification of Brazilian ethanol as an advanced biofuel, the US market, where Otto cycle engines are predominant in the car fleet, has become more attractive than ever. There are other potential importers, as Japan, where Petrobras created a joint venture with Mitsubishi to export ethanol. China, which uses corn to produce ethanol, has also become a major market. Many foreign investors are being attracted to ethanol agro-business in Brazil, including oil majors (BP, Shell) as well as companies researching 2nd generation biofuels, due to the widely held perception of sugarcane comparative advantages over corn, as seen on table 12:

a. Inefficacy of corn ethanol to mitigate global warming (as previously discussed).

b. higher competition of corn ethanol with food agriculture (as previously discussed);

c. lower productivity per hectare and higher cost of corn ethanol (Table 11);

Production in 2011		**Brazilian Export Destination in 2010**	
USA	54.2	South Korea	334
Brazil	21.0	USA	233
China	2.1	Japan	230
Canada	1.8	Netherlands	221
France	1.1	United Kingdom156	
Germany	0.8	Jamaica	107
Spain	0.5	Nigeria	80
TOTAL	86.1 billion liters	TOTAL	1.650 million liters
Source: [42],[43]			

Table 11. Ethanol Supply and Brazilian Exports

	Productivity	**Costs**
Sugar cane ethanol	4000 to 7000 liters / ha	0.19 US$ / liter (S.Paulo)
		0.23 US$ / liter (Northeast Region)
Corn ethanol	3500 to 4700 liters / ha	0.33 US$ / liter
Source: [44;17]		

Table 12. Comparison of Sugar Cane Ethanol with Corn Ethanol

Based on Brazilian ethanol's competitive advantages, it is reasonable to imagine a virtual extreme scenario for future ethanol demand to be supplied through international trade. According to the RFS2 mandate, US biofuels consumption will total 136.5 billion liters/year by 2022, 19 billion from advanced biofuels, for which only Brazilian sugarcane ethanol currently qualifies. The entire American market alone will be equal to 7 times Brazil's current production.

If we consider a technology and productivity freeze, land requirement should increase seven-fold, resulting in 28 Mha for sugar cane ethanol, approximately 25% of the land still available to expand agriculture in Brazil. This percentage is not small taking into account the need of cropland for food and feed to supply the internal market (including biofuels), as well as exports to other countries. If the EU ethanol market expands, due to a more favorable ILUC-wise view of this biofuel over biodiesel, the area needed for sugar cane exports could become too large, especially if we consider the inherent problems of very large monocultures. Second generation technology for ethanol production can change the present prospects.

6. Conclusion

At this point there is no major obstacle from a land use point of view to expand ethanol supply for Brazil's internal market, but it should not attempt to supply all the global ethanol market, if it is decided that biofuels is a valid global warming mitigation scheme. Conversely to aforementioned studies, emissions due to indirect land use change in Brazil cannot beattributed to the increase of biofuel production, because deforestation and ethanol production have presented opposite trends for almost ten years.

Government's scenarios predict a domestic demand of 63 billion liters of ethanol in 2020 [45]. However, due to the recent downfall of ethanol exports from a peak of 5.1 billion liters in 2008, the amount exported in 2020 has been revised to only 6.8 billion liters. Brazil can comfortably supply such quantities as the area needed – 10 Mha - is compatible with the available land for agriculture.

As was pointed out in the present paper, CO2 is dominant among GHG emissions and automotive fleet contributes with 20% of World CO2 emission. It amounted to 890 million light vehicles in 2005 and it consumes half of petroleum products in the World [46]. Besides, the fleet increases 20% per year in China and 3.5% in Brazil, where more than half the cars are flex fuel vehicles running with either gasoline or ethanol. As in Brazil cars use from 25% to 100% of ethanol, CO2 avoided emissions are substantial.

Ethanol per se is not enough to mitigate CO2 emissions at the World level. A deeper change in energy technology, transport, and consumption pattern is needed, including in public transportation, which use more efficient diesel engines. But ethanol can become an important fuel for different technologies, besides Otto cycle engines in cars. With additives, ethanol could feed Diesel engines used in buses, trucks and railway trains. It serves as fuel in hybrid vehicles of electrical propulsion - in which an Otto or Diesel engine is coupled to an electric generator that supplies current to an electrical motor and to accumulate energy in batteries – or in fuel cell vehicles to replace combustion engine based ones.

Sugar cane is the better way to produce bio-ethanol, from both an economic and environmental view point, including GHG mitigation through gasoline replacement. However, ethanol industry in Brazil has to improve, undergo technological changes, some of them concerning efficiency in energy transformation and natural resource use, by applying the best available technologies. The main changes must be, at a first level:

a. Efficiency improvement in the transformation of sugar cane bagasse chemical energy into heat, mechanical and electric energy for self consumption and export to the grid; current participation of bagasse in the Brazilian electric generation matrix is too small and must increase.

b. Utilization of the sugar cane trash, which is burned before harvesting to allow access to manual laborers; the amount of energy that could be converted into electric generation is significant.

c. Item (b) implies the increase of harvesting mechanization in sugar cane agriculture, decreasing the number of workers; however, manual harvesting is known to be hazardous.

d. Job conditions of workers in sugar cane plantation have to improve in some cases, including a social dimension besides the environmental one in clean energy production.

e. Technological improvement in agriculture.

On a second level there are:

a. Gasification of sugar cane bagasse and sugar cane residues;

b. Second generation ethanol production through hydrolysis;

c. Bio-refineries with multiple byproducts or integrated oil & bio-refineries, in an advanced concept.

Gasification could allow either high efficiency conversion in electric energy through combined cycle or could be used to produce liquid fuel from gas. Second generation ethanol consists in an acidic or enzymatic hydrolysis followed by fermentation that converts cellulose from biomass into ethanol.

The commercial use of hydrolysis can reduce sugar cane comparative advantage in relation to other kinds of vegetable to produce ethanol. On the other hand, the entire sugar cane biomass could be used to obtain ethanol, including the hydrolysis of bagasse and residues, as well as allowing fermentation of pentoses from hemicellulose to produce ethanol. [28] predicts a time horizon between 2010 and 2020 for second generation ethanol to become commercial, while gasification will take a little bit longer, 2015 to 2025, in spite of already existing technological uses of wood gasification. In the case of hydrolysis there are prototypes and some recent small scale industrial plants are in construction in the World, but no 2nd generation ethanol production has, to this date, become commercial.

Bio-refineries can produce ethanol together with other chemical byproducts. For instance, biodiesel production needs ethanol or methanol and has glycerol as a byproduct that can be used to produce biogasoline. Since the beginning ethanol production in Brazil was integrated with sugar production in the so-called annex distilleries. A more advanced concept is the integration of bio-refinery with oil refinery.

Finally, biofuels for private cars must not prevent the search for technical and social efficiency in transport, with an emphasis on public transport. Climate Change Policy must be devoted to find realistic solutions for sustainable development with social justice. Elimination of poverty needs more energy per capita in developing countries, but, at same time, it is necessary to change the intensive energy use and consumption pattern of high income and middle classes. It is not possible to radically mitigate global warming without any change in business as usual energy consumption.

Acknowledgements

The authors would like to acknowledge COPPE/UFRJ researcher Rodrigo Ribas contribution to the revision of this article.

Author details

Luiz Pinguelli[1], Rosa Alberto Villela[2] and Christiano Pires de Campos[3]

1 Graduate Energy Planning Department and Director of COPPE/UFRJ, Secretary of the Brazilian Forum on Climate Change, Brasil

2 Graduate Energy Planning Department of COPPE/UFRJ, Brasil

3 Petrobras Research Center, Rio de Janeiro, Brasil

References

[1] INPENational Space Research Institute. Projeto Prodes: Monitoramento da Floresta Amazônica por Satélite, (2012). http://www.obt.inpe.br/prodes/index.php.

[2] MasieroG; Lopes, H- Ethanol and biofuels as alternatives energetic sources: Latin-American e Asian perspectives, Revista Brasileira de Política Internacional, (2008). Print version 0034-7329, 0034-7329.

[3] KohlheppG- Análise da situação da produção de etanol e biodiesel no Brasil. Estud. av. São Paulo, (2010). http://dx.doi.org/10.1590/S0103-40142010000100017Print version 0103-4014, 24(68)

[4] SBS Brazilian Silviculture Society- Fatos e Numeros do Brasil Florestal. (2008).

[5] IPCCIntergovernmental Panel on Climate Change, Fourth Assessment Report, (2007).

[6] International Energy Agency- CO2 Emissions from Fuel Combustion(2011). Edition

http://dx.doi.org/10.5772/54419

[7] Muylaert de AraujoM.S., CAMPOS, C.P., ROSA, L.P., (2007). Historical emissions, by country, of N2O from animal waste manure management and CH4 from enteric fermentation of domestic livestock, Climate Research, b., 34(3), 253-258.

[8] IEA International Energy Agency- Key World Energy Statistics. (2011).

[9] EPE Brazilian Energy Research Agency- Análise da Conjuntura dos Biocombustíveis. (2012).

[10] CONAB National Supply CompanyMinistry of Agriculture- Acompanhamento da Safra de Grãos, (2012).

[11] IBGE Brazilian Institute of Geography and Statistics- National StatisticsGovernment of Brazil, (2007).

[12] EIA Energy Information Administration- How much gasoline does the United States consume?

[13] EPA Environmental Protection Agency- Renewable Fuel Standard.

[14] Rosa, L. P, De Campos, C. P, & Araujo, M. S. (2009). Biofuel contribution to mitigate fossil fuel CO2emissions: Comparing sugar cane ethanol in Brazil with corn ethanol and discussing land use for food production and deforestation. J. Renewable Sustainable Energy 1, 033111 (2009)

[15] Rosa, L. P, & Ribeiro, S. K. Activities Implemented Jointly and the Use of Fuel Alcohol in Brazil for Abating CO_2 Emissions, Energy Policy, Elsevier, Great Britain, nº 2, (1998). , 26, 103.

[16] Rosa, L. P, & Ribeiro, S. K. Avoiding Emissions of Carbon Dioxide through the Use of Fuels Derived from Sugar Cane, Ambio, Royal Swedish Academy of Sciences, nº 6, (1998b). , 27, 465.

[17] Dias de MoraesM. A. F., Reflections on Brazil's Ethanol Industry, in MRE, (2007). , 158.

[18] Joseph JrH.- Flexible Fuel Engine Technology, in MRE, (2007). , 92.

[19] EPE Brazilian Energy Research Agency- National Energy Balance 2008

[20] EPE Brazilian Energy Research Agency- National Energy Balance 2011

[21] Macedo, I. C, & Leal, M. R. L. V. and Ramos da Silva, J. E. A., Balanço das Emissões de Gases do Efeito Estufa na Produção e no Uso do Etanol no Brasil, Environment Secretariat of S. Paulo State, (2004).

[22] Braunbeck, O, Macedo, I. C, & Cortez, A. B. Modernizing Cane Production to Enhance the Biomass Base in Brazil, in Silveira, (2005). , 75.

[23] CONAB National Supply CompanyMinistry of Agriculture- Acompanhamento da Safra de Cana de Açúcar, (2012).

[24] EUCARWell to Wheels, Analysis of Future Automotive Fuels and Powertrains in the European Context, May (2006).

[25] Macedo, I. C, Seabra, J. E, & Silva, J. E. A. R. GHG Emissions in the Production and Use of Ethanol from Sugar Cane in Brazil. The 2005-2006 Averages, Biomass and Bioenergy, 32 (7), (2008). , 582.

[26] IPCCIntergovernmental Panel on Climate Change, Second Assessment Report, (1995).

[27] NastariPlinio, (2005). quoted by Dias de Moraes, 2007

[28] Macedo, I. C. Ethanol: Current Situation and Outlook, in MRE, (2007). , 76.

[29] UNICAUnion of Sugar and Ethanol Industries, Brazil, Report, (2008).

[30] MTEMinistry of Work and Employment of Brazil, (2007).

[31] United Nations Environment ProgrammeAssessing Biofuels, (2009).

[32] United Nations Food and Agriculture OrganizationGlobal land use area change matrix. Forest Resource Assessment, Working Paper 134, Input to the fourth global environmental outlook, (2006).

[33] Searchinger, T, Heimlich, R, Houghton, R. A, Dong, F, Elobeid, A, Fabiosa, J, et al. Use of US croplands for biofuels increases greenhouse gases through emissions from land-use change. Science 319:1238-1240 ((2008).

[34] Homem de MelloF. B.; A Crise Energética e o Setor Agrícola no Brasil, Revista de Economia Rural. S. Paulo, (1980). , 18, 115.

[35] Homem de MeloF. B.; Proálcool, Revista Estudos Econômicos, S. Paulo, (1981). , 11, 83.

[36] USDAUnited States Department of Agriculture. World Agricultural Supply and Demand Statistics, (2012).

[37] EMBRAPA Brazilian Agricultural Reasearch AgencyMinistry of Agriculture. Zoneamento Agroecológico da Cana-de-Açúcar, (2010).

[38] Muylaert de Araujo MS.; Silva, Corbiniano; Campos, C.P.; Land use change sector contribution to the carbon historical emissions and the sustainability-Case study of the Brazilian Legal Amazon, Renew Sustain Energy Rev ((2007c). doi:10.1016/j.rser.2007.10.008.

[39] MCTMinistry of Science and Technology. Brazilian Inventory of GHG Emissions, (2004).

[40] EC European Commission, The Impact of Land Use Change on GHG Emissions from Biofuels and Bioliquids: Literature Review, July 2010

[41] ICONEConsultations on Indirect land Use Change Impacts of Biofuels: Comments by the Institute for International Trade Negotiations, (2010).

[42] REN21 Renewable Energy Network for the 21st CenturyRenewables 2012 Global Status Report. Paris, (2012).

[43] MDIC Ministry of Industry and TradeSecretariat of Foreign Trade, (2012).

[44] Hill, J, Nelson, E, & Tilman, D. Environmental Economic and Energetic Costs and Benefits of Biodiesel and Ethanol, PNAS, (2006). , 103, 11206.

[45] EPEEnergy Planning Enterprise, Plano Decenal de Expansão de Energia 2020, (2011).

[46] SzwarcAlfred; Ethanol and the Control of GHG, in MRE, (2007). , 120.

Chapter 14

Vehicle Emissions: What Will Change with Use of Biofuel?

Lílian Lefol Nani Guarieiro and
Aline Lefol Nani Guarieiro

Additional information is available at the end of the chapter

http://dx.doi.org/10.5772/52513

1. Introduction

In urban areas, large concentrations of chemical compounds are emitted into the atmosphere by industries, vehicles and other human activities. Nearly 3000 different compounds, mostly organic, resulting from human activity have been identified in the atmosphere. This complex mixture of pollutants can have impacts on health and the environment. Thus, the systematic determination of air quality should be, for practical reasons, limited to a restricted number of pollutants, defined in terms of their importance and the human and material resources available to identify and measure them. Generally, pollutants chosen to serve as indicators of air quality are the currently regulated and universally occurring compounds: sulfur dioxide (SO_2), particulate matter (PM), carbon monoxide (CO), ozone (O_3) and nitrogen oxides (NO_x). They are chosen due to their frequency of occurrence and adverse effects on the environment. Thus, the effects of air pollution can be characterized by a deterioration of good quality environmental conditions and the exacerbation of existing problems, which can manifest themselves in health, population welfare, vegetation, fauna, and urban structures. The attention of regulatory authorities and researchers must not only look to the standards of air quality. There are compounds that despite being unregulated deserve attention because of the damage they cause to the environment and, especially, to human health.

The search for alternative fuels to reduce dependence on petroleum and emission of pollutants into the atmosphere has stimulated many scientific studies. The goal is to develop fuels that can be used in existing vehicles without the need for major changes in their engines. A term often used for fuel derived from renewable sources is 'biofuel', which has strong links with the concept of sustainability, whereby the use of natural resources to meet current needs should not compromise the needs of future generations. In this way, the purpose of this chapter is to answer the question about vehicle emissions: what will change with use of biofuel?

2. Emission sources of pollutants to the atmosphere

Air quality in urban atmospheres depends on several related factors: primary pollutant's emissions (emitted directly from sources to the atmosphere), secondary pollutant's emissions (resulting from the chemical reactions occurring in the atmosphere and which involve some primary pollutants) and consumption, geographical and meteorological factors.

Primary pollutants can be emitted by natural and anthropogenic sources. The pollutants emitted from both sources may be in two physical states: adsorbed in the particulate or in the gas phase. In this context, the primary particles emitted by many natural and anthropogenic sources include combustion processes, volcanic eruptions, forest fires, fumes created by certain industrial activities and roadways, the "marine spray" and some biological materials [1]. The pollutants frequently found in the atmosphere are: CO, NO_x, sulfate oxides (SO_x), PM, volatile organic compounds (VOC), O_3 and; some Greenhouse Gases such as carbon dioxide (CO_2), methane (CH_4), nitrous oxide (N_2O), chlorofluorocarbons (CFC) andnitric acid (HNO_3).

In many areas of the developed and developing world, the concentration of tropospheric pollutants has increased to levels significantly affecting various aspects of the environment. Reducing the amount of solar radiation reaching the ground has the potential for important effects on climate by reducing ground temperature, and increasing cloud albedo and stability, which result in global cooling. Increased pollutants in the atmosphere reduce visibility, and have important effects on human health leading to respiratory diseases by inhalation and to rickets due to inadequate sunlight for the production of vitamin D. Reduced solar radiation and changes in atmospheric stability have important effects on atmospheric photochemistry and modeling [2].

In recent decades, there have been concerns with the increase of anthropogenic pollution that can be seen through the initiation of programs aimed at improving air quality in cities around the world. One of the strategic actions to reduce the emission of pollutants in urban environments is the displacement of local industries from urban to non-urban areas [3]. Thus, the improved control of industrial emissions resulted in the current situation, where in large cities vehicles are the main source of emissions of air pollutants, especially CO, hydrocarbons (HC), NO_x, aldehydes and metals [4-7]. Although there are actions and programs that encourage the reduction and control of air pollutants, there are cities where some of these pollutants routinely exceed local air quality standards. Table 1 shows some general pollutants emitted and their sources such as human activity, industrial uses and transportation systems.

http://dx.doi.org/10.5772/52513

Pollutants	Sources
CO	Traffic (especially vehicles without catalytic converters) and Industries.
NO_x	Traffic and General industry (resulting from the combustion of the higher or lower temperatures).
SO_x	Traffic (vehicles using fuel with high sulfur content) and Industry (chemical industry, pulp and paper, refineries and boilers using fuel with high sulfur content, for example, fuel oil).
PM	Traffic, Industries (cement, refineries, steel, paper pulp, chemical industry, inter alia), Construction work and agricultural practices.
Pb (Lead)	Traffic (Leaded fuels) and Industries (manufacturing process from raw materials that integrate Pb).
VOC	Chemical Industry, traffic, Storage of Fuel and petrol stations, Car workshops, Construction Materials, and other activities, involving handling of solvents.
O_3	Is formed at ground level as a result of chemical reactions established between some primary pollutants, such as NO_x, VOC or CO. These reactions occur in the presence of light sunlight, so that higher levels of ozone occur during the emission of primary pollutants in the summer. The sources of primary pollutants that have influence on atmospheric ozone concentration are: Traffic; Industries, Landfills, Paints and solvents (VOC); forests; and Other sources (gas stations, equipment that uses this fuel, etc.).
CO_2	Use of fóssil fuels, deforestation and Change of land use.
CH_4	Production and consumption of energy, Farming and livestock, Landfills and wastewater.
N_2O	Fertilizer use, Production of acids and Burning of biomass and fossil fuels .
CFC	Industry, refrigeration, aerosols, propellants, Expanded foams and Solvents.
HNO_3	Combustion of wood, Fossil fuels, the chemical composition of fertilizersand microbes.

Table 1. Main sources of air pollutants.

The transportation sector can be considered the major source of atmospheric pollutants. In the evaluations of the vehicular emission impacts, measurements have been limited to regulated pollutants such as suspended PM, HC, CO, NO_x and O_3, with this last one being an important secondary product formed by photochemical reactions in the atmosphere. However, some specific compounds, which are not regulated by law to be monitored, have a significant toxic potential. Thus, one can highlight the polycyclic aromatic hydrocarbons (PAHs), Nitro PAHs and the VOC, which result from incomplete combustion, as having a double influence on air quality. These molecules can act as primary toxic pollutants and play a role of precursor in the formation of photochemical oxidant species.

Emission studies of controlled substances have already been done extensively. However, there is a current need for the study of non-regulated emissions. Some studies on unregulated substances, such PAHs, nitro HPAs, carbonyl compounds (CC) and both vapor and particulate of light aromatic hydrocarbons, showed a lower amount of mutagenic compounds being emitted when using biofuels [8]. However, there are some çontradictory results [9], which point to the need for further studies of such substances.

In the last three decades CC, aldehydes and ketones, have received a great deal of attention due to their strong influence on photochemical smog formation and their recognized adverse human health effects. Carbonyl compounds are directly emitted into the atmosphere by combustion sources and also produced from photochemical oxidation of hydrocarbons and other organic compounds.

3. Vehicle emissions

The transportation sector has an active role in rising pollution levels, especially in large urban centers in regions where transport is based on roads, i. e., much of the transportation of goods is done by trucks, and transportation of people is primarily done by bus or car. The gases resulting from complete combustion of fuel used in vehicles are CO_2, H_2O (usually in gaseous state) and nitrogen (N_2). In this reaction, the only product that has concern from environmentalists is the CO_2, due to impacts on the greenhouse effect and global warming. However, as seen before, the reactions that occur in vehicle engines emit other compounds into the atmosphere. Due this fact, the combustion process of vehicles is considered incomplete. Moreover, it is important to note that vehicle emissions are not only those emitted during the combustion process. Emissions of pollutants arising from the use of vehicles can be divided into the following categories [10]:

- Emissions of gaseous and particulates by tailpipe of the vehicle (byproducts released to the atmosphere by combustion exhaust pipe);
- Fuel evaporative emissions (released into the atmosphere through evaporation of the hydrocarbon fuel);
- Emissions of gases from the crankcase of the engine (combustion by products that passing through the piston rings of the engine and the oil vapors lubricant);
- Particulate emissions from wear of tires, brakes and clutch;
- Resuspension of dust and soil;
- Evaporative emissions in the fuel transfer operations fuel (associated with storage and fuel supply).

The main pollutants emitted into the atmosphere by the vehicles are from the process of incomplete combustion in which fuel injected into the cylinder doesn't find the required amount of air for its burning. So, these primary pollutants are emitted directly by the automotive exhaust (CO, NO_x, SO_x, alcohols, CC, HC, PAHs and PM). These pollutants can interact with each other or with the aid of light to form secondary pollutants (O_3, nitrates peroxiacetila - PAN, among others). The latter may be much more harmful to the environment than the primary pollutants.

The PM from engines has three major components: soot formed during combustion, heavy hydrocarbon condensed or absorbed on the soot, and sulfates. Particle size is also an important variation in terms of vehicular emissions, as it has been associated with an increase in health conditions. Ultrafine particles (< 0.01 μm), generated in great amounts mainly by diesel exhaust, have special toxicity due to their ability to penetrate into the cardiovascular system and other organs [11-13].

Even with the technological evolution of vehicle exhaust systems, these emissions remain a serious air pollution problem in many regions. Several reasons can be highlighted [14]:

- Significant increase in size of vehicle fleet and its use;
- High fuel consumption because of lower prices in some countries, the characteristics of the vehicles and driving conditions;
- Malfunction of emission control systems reducing the effectiveness of control;
- Accelerated degradation of components of the car that has a direct impact on increasing the emissions such as design flaw and / or use of inappropriate materials, or also by misuse of the vehicle;
- Lack of care in the maintenance of the vehicles by their owners;
- Lack of preparedness in a considerable number of vehicle repairshops to offer technically appropriate maintenance services;
- Deliberate withdrawal of emissions control devices by the owners of vehicles or inadequate repair services;
- Adulteration of fuel;
- Existence of old vehicles in circulation or vehicles in poor condition, with very high levels of emissions;
- Lack of measures to popularize and encourage the use of public transport, to contain the increasing use of automobiles as a means of individual transportation.

In this context, the factors mentioned above have contributed to overtake the air quality standards in major metropolitan areas. They should be prioritized over the effects of other sources of pollutant emissions such as power plants and industries. Vehicle emissions are an important contributor to the formation of photochemical smog and overall emissions.

Light vehicles (Otto cycle) are becoming more numerous in large urban centers with the main regulated pollutants emitted from these vehicles being: CO, HC and NO_x. On the other hand, diesel vehicles emit the largest amount of the regulated pollutants: PM and NO_x (Table 2).

Among unregulated pollutants, several authors study the emission of individual HC [15,16] and especially those of methane, a gas with a strong greenhouse effect. However, the total contribution of light vehicles (Otto cycle) to global methane emissions is estimated to be very low, not more than 0.3-0.4% of the total methane emissions[17].

Kind of vehicle		Fuel	Emission (%)			
			CO	HC	NOx	PM_{10} [a]
Light vehicles (cars, etc)		Gasoline	46.65	14.47	5.72	-
		Ethanol	8.60	4.13	1.37	-
		Flex (Gasoline/Ethanol)	13.27	6.81	2.46	-
Commercial vehicles		Gasoline	5.42	1.76	0.72	-
		Ethanol	0.78	0.38	0.13	-
		Flex (Gasoline/Ethanol)	0.60	0.30	0.11	-
		Diesel	0.29	0.34	2.84	1.42
Trucks	Light	Diesel	0.16	0.23	1.77	1.35
	Medium		0.81	1.15	8.74	6.55
	Heavy		2.92	3.36	32.00	15.90
Buses	Urban	Diesel	1.87	2.30	19.94	12.01
	Road		0.43	0.53	4.72	2.77
Motorcycles		Gasoline	15.56	12.92	1.15	-
		Flex (Gasoline/Ethanol)	0.04	0.04	0.01	-

[a]Contribution study as a model recipient for inhalable particles (< 10 μm).

Table 2. Estimation of emission sources of air pollution in a Brazilian urban center (São Paulo) in 2010 [10].

Studies have shown that amongst the total gas phase non-methane hydrocarbons emitted for gasoline burning, 75–93% are aromatics species, 6–18% linear and substituted alkanes, 1.2–4.3% alkenes and alkynes, and 0.1–2% CC. The analysis for a diesel engine showed 54–75% of species analyzed are aromatics, 18–31% linear and substituted alkanes, 3–6% alkenes and 2–6.4% CC.In the case of vehicles with diesel engines, the abundance of CC is more significant than Otto cycle engines; and should be related to the composition of the fuel[18].

An important global emission, VOC, is observed at different levels in light vehicles (Otto cycle) compared to diesel vehicles. Generally, depending on vehicle technology and vehicle year, make and model, aromatics compounds are the major species but with a somewhat weaker contribution for diesel cars. Saturated hydrocarbons with weaker percentages which are about 20% for diesel cars are the second most common and 12% for gasoline cars. The CC displayed very low concentration (0.5% for gasoline and 10% for diesel)[18].

4. Profile of vehicular emissions with the use of biofuel

Biofuels are derived from biomass, the name given to the organic matter in an ecosystem or a plant or animal population. Because plants and animals can reproduce continuously, one

http://dx.doi.org/10.5772/52513

can assume that they are renewable sources of energy. Plants, through photosynthesis, convert solar energy they receive into biomass, and animals generate energy by eating organic matter (plants or other animals). There are several types of biofuels that can be produced from biomass, such as alcohol (ethanol and methanol), biodiesel, biokerosene, H_2 and others. The sources for its production can be through animal (for example, tallow fat or chicken), vegetable (e.g., vegetable oils and cane sugar) and biomass materials [19].

All biomass materials can be converted to energy via thermochemical and biological processes. Biomass gasification attracts the reactive and forms stable chemical structures, and consequently the activation energy increases as the conversion level of biomass increases [20]. Biomass gasification can be considered as a form of pyrolysis, which takes place in higher temperatures and produces a mixture of gases with H_2 content ranging 6–6.5% [21]. Hydrogen may be an alternative to gasoline, gas-oil and biofuels for the automotive sector. Hydrogen can be used in internal combustion engines or in fuel cells. However, this chapter will discuss more about biodiesel and ethanol biofuels emission profiles.

Ethanol can be produced from a number of crops including sugarcane, corn (maze), wheat and sugar beet. In general, ethanol is produced through fermentation of sugar derived from corn or cellulosic biomass. Moreover, technically speaking, biodiesel is the alkyl ester from fatty acids, made by the transesterification of oils or fats, from plants or animals, with short chain alcohols such as methanol and ethanol. Glycerine is, consequently, a by-product from biodiesel production [22].

The ethanol obtained from sugar cane is the biofuel with the most energy efficiency: each joule (unit of energy) used in its production allows the return of about seven joules. Brazil developed technologies for producing ethanol and gasoline engines adapted to it, but alcohol is considered by many a luxury fuel, being used only in light vehicles. One great challenge is to develop technologies that enable the use of ethanol as fuel in large vehicles (buses and trucks) and aviation [19].

Over the past 10 years, the number of scientific and technological studies on biofuels has grown exponentially. A refined search done in a scientific database [23] using as keywords: "biodiesel emission" and "ethanol emission", revealed that interest in research on biofuel emissions has increased each year and studies of emissions from burning of ethanol and biodiesel have similar trends, although there was a greater interest for research on ethanol (Figure 1). This fact is justified because ethanol is a biofuel that has been used in the energy matrix since the 70's and biodiesel in the last 10 years. For this reason (Figure 2) the production of ethanol is roughly more than four times the world production of biodiesel.

In this context, there are several reasons for biofuels to be considered a relevant technology for both developing and industrialized countries. They include energy security reasons, environmental concerns, foreign exchange savings, and socioeconomic issues related to the rural sector. The following sections will discuss the results obtained from research evaluating the emission of pollutants when biofuels are used.

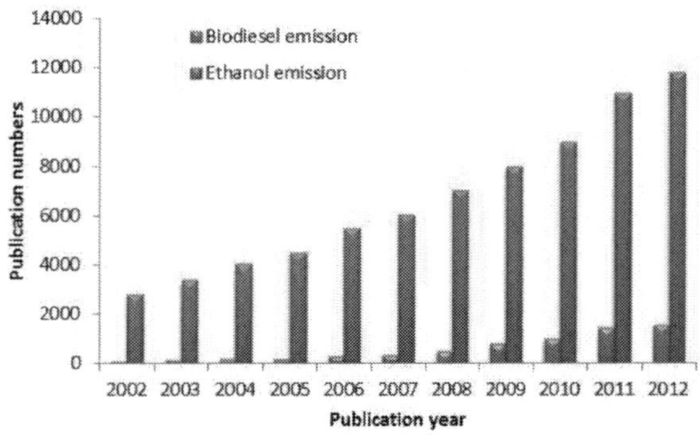

Figure 1. Publication numbers of research about both biodiesel and ethanol emission topics from 2002 to 2012 [23].

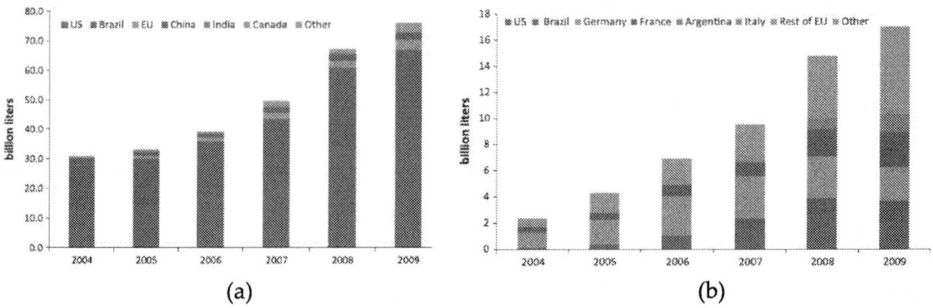

Figure 2. World Ethanol (a) and Biodiesel (b) Production from 2004 to 2009 [24].

4.1. Otto cycle vehicles (gasoline/ethanol)

In general, biofuels are considered climate friendly, even when based on a life-cycle analysis. Ethanol use in gasoline has tremendous potential for a net reduction in atmospheric CO_2 levels. CO_2 is released into the atmosphere when ethanol (like other fuels) is burned in an engine and is also recycled into organic tissues during plant growth [25].

A study was done about the direct vehicle emission impact on the future use of ethanol as a fuel for gasoline cars in Denmark arising from the vehicle specific fuel consumption and emission differences between neat gasoline (E0) and E5/E85 gasoline-ethanol fuel blends derived from emission tests. For vehicles using E5 rather than E0, the average fuel consumption and emission differences are small. For CO, VOC and NO_x the derived average differences are 0.5%, -5% and 7%, respectively. For using E85 rather than E5, the emission differences become even smaller for VOC and NO_x, but greater for CO. The derived average emission differences are in this case 18%, -1% and 5% for CO, VOC and NO_x, respectively [26].

Already, in field studies conducted regarding the use of 10% ethanol additions to gasoline on pollutant formation concluded that PM and CO emissions are significantly reduced. For some of the vehicles tested, CO_2 emissions were also significantly reduced and overall it led to a small deterioration in fuel economy (although this was not significant at 95% confidence level). NO_x emissions were not significantly influenced. However, for some of the vehicles tested, acetaldehyde emissions significantly increased [27].

CO, formed by the incomplete combustion of fuels, is produced most readily from petroleum fuels, which contain no oxygen in their molecular structure. Since ethanol and other "oxygenated" compounds contain oxygen, their combustion in automobile engines is more complete. The result is a substantial reduction in CO emissions. Research shows that reductions range up to 30%, depending on type and age of engine/vehicle, the emission control system used, and the atmospheric conditions in which the vehicle operates [27].

Because of its high octane rating, adding ethanol to gasoline leads to reduction or removal of aromatic HC's (such as benzene), and other hazardous high-octane additives commonly used to replace tetra ethyl lead in gasoline [28]. Adding ethanol to gasoline can potentially increase the volatility of gasoline. However, some studies have identified divergent results about NO_x emissions, showing the ethanol concentration in the fuel increased anywhere from 0% to 20%. So, the ethanol addition can reduce CO and HC, aldehydes and unburned ethanol emissions. NO_X results can vary depending on the operating condition, spark advance timing and other parameters [29].

Adding ethanol to gasoline does emit slightly greater amount of aldehydes during combustion. However, the resulting concentrations are extremely small and are effectively reduced by the three-way catalytic converter in the exhaust systems of all modern vehicles. Generally, benzene and toluene emissions decrease by ethanol addition to gasoline, although this beneficial effect of ethanol was eliminated after the operation of the catalyst. Acetic acid was detected in exhaust gases in some cases only for the base and the 3% ethanol blend fuel [30,31].

There are other toxic emissions (unregulated), which should be considered to ascertain the impact of ethanol blended fuels, such as: acetaldehyde, formaldehyde, propionaldehyde and acrolein, benzene, ethylbenzene, 1-3 butadiene, hexane, toluene, xylene, and fine particulates. Studies indicate a reduction of benzene emission up to 50% with the ethanol-blended fuels. Emissions of 1,3-butadienes were also substantially decreased, with reduction ranging from 24% to 82%. Isolated trends were noted for certain PAHs. There was a decrease in 1-nitrobenzene with use of ethanol in all cases. There was also a general increase in the proportion of heavy PAHs in the particulate phase with ethanol use, and although less pronounced, general decreases in light PAHs in the particulate phase [32].

In summary, it can be said that ethanol produces generally less pollution than gasoline and diesel. Alcohol has a tolerance combustion with excess air, which allows a more complete burn with lower emissions of CO and PM. Moreover, there is an increase in the emission of aldehydes. Under certain conditions (cold start), alcohols are oxidized to aldehydes, especially formaldehyde (in the case of methanol) and acetaldehyde (in the case of ethanol) [33].

4.2. Diesel cycle Vehicles (diesel / biodiesel / ethanol)

Among the biofuels discussed above, we can highlight the use of biodiesel and ethanol in vehicles with diesel engines. In the specific case of biodiesel, this has viscosity close to mineral diesel. These vegetable oil esters contain 10–11% oxygen by weight, which may encourage more combustion than hydrocarbon-based diesel in an engine. Furthermore, biodiesel can form blends with diesel in any ratio, and thus could replace partially, or even totally, diesel in combustion engines that could bring a number of environmental, economic and social advantages. However, biodiesel can be produced from different types of raw material and this can directly influence the final composition of the biofuel and consequently in the emission of pollutants.

Thus, the investigation discovered that biodiesel impacts on emissions varied depending on the type of biodiesel (soybean, rapeseed, or animal fats) and on the type of conventional diesel to which the biodiesel was added. There is one minor exception: emission impacts of biodiesel did not appear to differ by engine model year [34].

The United States Environmental Protection Agency has conducted a comprehensive analysis of the emission impacts of biodiesel using publicly available data. This investigation made use of statistical regression analysis to correlate the concentration of biodiesel in conventional diesel fuel with changes in regulated and unregulated pollutants. The majority of available data was collected on heavy-duty highway engines and this data formed the basis of the analysis. The average effects are shown in Figure 3.

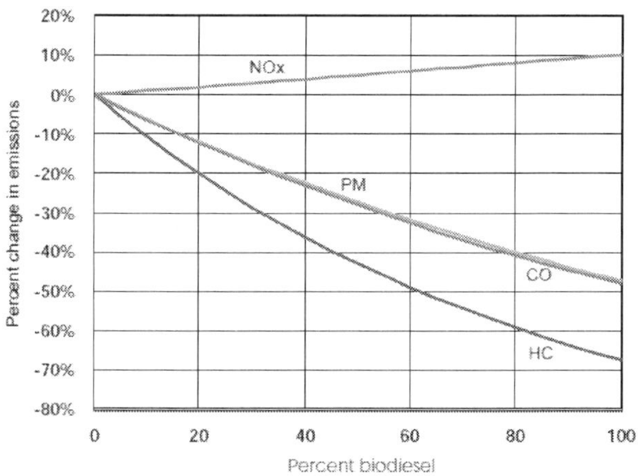

Figure 3. Average emission impacts of biodiesel for heavy-duty highway engines [34].

http://dx.doi.org/10.5772/52513

Increasing the level of biodiesel in the fuel blend increased NO_x while reducing PM. Proportionally, the PM reduction was slightly more than the increase in NO_x, on a percentage basis. The reduction in CO and HC was linear with the addition of biodiesel for the blends tested. These reductions indicate more complete combustion of the fuel. The presence of oxygen in the fuel was thought to promote complete combustion [35,36].

The NO_x forms by oxidation of atmospheric nitrogen at sufficiently high temperatures. Kinetics of NO_x formation is governed by Zeldovich mechanism, and its formation is highly dependent on temperature and availability of oxygen. There are several reported results of slight increase in NO_x emissions for biodiesel [37]. It is quite obvious, that with biodiesel, due to improved combustion, the temperature in the combustion chamber can be expected to be higher and a higher amount of oxygen is also present, leading to formation of a higher quantity of NO_x in biodiesel-fueled engines. However, biodiesel's lower sulfur content allows the use of NO_x control technologies that cannot be otherwise used with conventional diesel.

Biodiesel is free from sulfur, hence less sulfate emissions, and reduced PM is reported in the exhaust. Due to the near absence of sulfur in biodiesel, it helps reduce the problem of acid rain caused by emission of pollutant from fuels burning. The lack of aromatic hydrocarbon (benzene, toluene etc.) in biodiesel reduces unregulated emissions as well as ketone, benzene etc. Breathing particulate matter has been found to be hazardous for human health, especially in terms of respiratory system problems. PM consists of elemental carbon (~31%), sulfates and moisture (~14%), unburnt fuel (~7%), unburnt lubricating oil (~40%) and potential remaining metals and others substances [27].

Regarding environmental concerns, many studies have shown that pure biodiesel, biodiesel/diesel and biodiesel/ethanol/diesel blends may reduce emissions of regulated substances (CO, CO_2, SO_x, HC and PM) [38-40]. However, there is an increasing interestin studying emissions of some unregulated substances, such as carbonyl compounds, PAHs, nitro-PAHs and other toxics that are of concern from both environmental and human health standpoints [22]. Among CC, both formal dehyde and acetaldehyde were the major contributors to the observed total CC levels in diesel and diesel/biofuels blends emissions. Except for acrolein and formaldehyde, all CC showed a clear trend of reduction in emissions when using biodiesel/diesel blends [41].

In general, the addition of higher concentrations of biodiesel to diesel make an improvement in the carbonyl concentration profile at places with high circulation of heavy-duty vehicles, bringing profiles down to levels found at sites less impacted by these kind of vehicles [42].

However, concerning CC emissions, there are some divergences when considering the results obtained using pure diesel and biodiesel blends. Depending upon the author, biodiesel could contribute to increase or decrease in the CC emissions [43-47]. Furthermore, comparing these studies is not straightforward since different authors have used different biofuel sources, engines, and especially, different sampling methodologies or protocols.

Experimental results showed no significant difference in engine function, damage from deposits inside the chamber or the inferior condition of engine oil for 300 h (18,000 km) of engine operation when using biodiesel/diesel blends. The emissions of HC and CO increased with operation time but the emissions of NO_x and PAHs decreased with operation time between 0 and 300 h (18,000 km). The use of biodiesel/diesel blends can reduce the emissions of total PAHs significantly [48]. Other studies on unregulated emissions with use of biodiesel/diesel blend found that, besides reducing PAH emission, there was also a reduction of nitro-PAHs, carbonyl compounds and light aromatic hydrocarbons when this mixture is compared with pure diesel results [44].

PAH concentrations in the samples from a bus station were associated with atmospheric PM, mass size distributions and major ions (fluorite, chloride, bromide, nitrate, phosphate, sulfate, nitrite, oxalate; fumarate, formate, succinate and acetate; lithium, sodium, potassium, magnesium, calcium and ammonium). Results indicate that major ions represented 21.2% particulate matter mass. Nitrate, sulfate, and ammonium, respectively, presented the highest concentration levels, indicating that biodiesel may also be a significant source for these ions, especially nitrate. Dibenzo[a,h]anthraceneandindeno[1,2,3,-cd]pyrene were the main PAH found, and a higher fraction of PAH particles was found in diameters lower than 0.25 μm in a bus station. The fine and ultrafine particles were dominant among the PM evaluated, suggesting that biodiesel decreases the total PAH emission. However, it does increase the fraction of fine and ultrafine particles when compared to diesel [49].

The direct application of ethanol in diesel engines requires changes in the constitution of the engine and the use of additives to improve the ignition. However, diesel/ethanol blends are a more viable alternative and require little or no change in conventional engines. The use of ethanol combined with diesel can significantly reduce the emission of toxic gases and particulate matters when compared to pure diesel. However, there are critical barriers to commercial use of diesel/ethanol blends, as the addition of ethanol to diesel affects properties such as lubricity, viscosity, energy content, cetane number, and, mainly, stability and volatility. The solubility of ethanol in diesel depends, among other factors, on the composition of diesel, the water content in the mixture, and the temperature [50].

Despite the technical problems presented by the use of pure ethanol in diesel cycle vehicles, many studies have been developed using ethanol blended with diesel. Thus, no modification is required in the engines of these vehicles. E-diesels (blends of ethanol in diesel) are currently being used in fleet vehicles in the European Union and the United States. Studies carried out with E-diesel indicated significant reductions of PM, sometimes up to 40%, depending on the test methods and operating conditions. The CO and NO_x emissions were significantly lower when a 20% blend of E-diesel was used in a constant-speed stationary diesel engine, as opposed to diesel fuel. The addition of ethanol to diesel may result in a volumetric reduction in sulphur, by as much as 20%, thus significantly reducing SO_2 emissions [51]. The major drawback in E-diesel is that ethanol is immiscible in diesel over a wide range of temperatures [52].

The diesel/ethanol/biodiesel blends have also emerged as an alternative fuel to reduce emissions in diesel engines. The biodiesel can help the miscibility of ethanol in diesel fuel. Researches have shown that the use of these blends can substantially reduce emissions of CO, HC, and PM[53, 54]. The mixtures (v/v/v) were used in the emission study: diesel/ethanol –

http://dx.doi.org/10.5772/52513

90/10%, diesel/ethanol/soybean biodiesel – 80/15/5%, diesel/ethanol/castor biodiesel – 80/15/5%, diesel/ethanol/residual biodiesel – 80/15/5%, diesel/ethanol/soybean oil – 90/7/3%, and diesel/ethanol/castor oil – 90/7/3%. The diesel/ethanol fuel showed higher reduction of NO_x emission when compared with pure diesel. The combustion efficiencies of the diesel can be enhanced by the addition of the oxygenate fuels, like ethanol and biodiesel/vegetable oil, resulting in a more complete combustion in terms of NO_x emission. In the case of CO_2 decreases were observed. Meanwhile, no differences were observed in CO emission. Among CC studied, formaldehyde, acetaldehyde, acetone, and propionaldehyde showed the highest emission concentrations [50].

There are a great number of previously published studies comparing diesel with biodiesel and ethanol blends. These biofuels have a good energy return because of the simplicity of its manufacturing process, and has significant benefits in emissions as well. It could also play an important role in the energy economy if higher crop productivities are attained.

5. Burning biofuel vs toxicology

Considerable populations are exposed to fuel exhaust in everyday life, whether through their occupation or through the ambient air. People are exposed not only to engine vehicle exhausts but also to exhausts from burning sources such as from other modes of transport (trains and ships) and from power generators.

Increasing environmental concerns over the past two decades have resulted in regulatory action in North America, Europe and elsewhere with successively tighter emission standards for both diesel and gasoline engines. There is a strong interplay between standards and technology – standards drive technology and new technology enables more stringent standards. For diesel engines, this required changes in the fuel such as marked decreases in sulfur content, changes in engine design to burn diesel fuel more efficiently, reductions in emissions through exhaust control technology with some countries investing in the use of biofuels [55].

However, while the amount of particulates and chemicals are reduced with these changes, it is not yet clear how the quantitative and qualitative changes may translate into altered health effects. In addition, existing fuels and vehicles without these modifications will take many years to be replaced, particularly in less developed countries, where regulatory measures are currently less stringent. It is notable that many parts of the developing world lack regulatory standards, and data on the occurrence and impact of diesel exhaust are limited.

Recently in June 12, 2012 after a week-long meeting of international experts, the International Agency for Research on Cancer (IARC), which is part of the World Health Organization (WHO), classified diesel engine exhaust as carcinogenic to humans (Group 1), based on sufficient evidence that exposure is associated with an increased risk for lung cancer. In this context, the biofuels can be an interesting alternative fuel to reduce health impact of petroleum fuel and pollutant emissions into the atmosphere. However, little is known about health impact and effects, and the air quality impacts of biofuels remain unclear. Significant concern exists regarding biofuel's production impacts on food security and nutrition for the poor [56].

The purpose of this section is to describe research that has been done on the toxicity of vehicular emissions, when fossil diesel is replaced by biofuel or when biofuels are added to petroleum fuel, and what happens in the chemical composition, size distribution and toxicity of the compounds emitted and their damages on health. This section will also discuss how the compounds can damage cells and organs and how the chemical composition and physical proprieties can influence the toxicity of pollutants that affect human health.

5.1. Emissions chemical and physical composition

The most commonly found pollutants from burning fuel emissions are regulated by many countries around the world. These pollutants can harm health and the environment, and cause property damage. Among the pollutants, PM and O_3 are the most widespread health threats. United States Environmental Protection Agency (EPA) calls these pollutants "criteria" air pollutants because it regulates them by developing human health-based and/or environmentally-based criteria (science-based guidelines) for setting permissible levels. The main regulated pollutants emitted during burning fuel and their damages to human health are listed in Table 3.

There are many studies reporting the difference in pollutant emission when comparing fuels with biofuel as described above. In general they report a decrease, similarities or increases in emissions using biofuel comparing with fossil fuels. Studies have indicated a decrease release of CO, SO_x, PM from the combustion process of biodiesel and ethanol. On the other hand, they indicated an increase of NO_x from the combustion process [57,58]. These results are important in terms of human health but do not assume the real effects and damages caused because there are some other pollutants emitted that do not fall under regulated pollutants.

Importantly, there are some air pollutants not regulated that can cause damage to human health. The types of components in the gas and PM phases include single aromatic and PAH and your derived (alkylbenzenes, quinones, oxy and nitro- PAH), alkanes, alkenes, CC, metals, inorganic ions (e.g. sulfates, carbonates), among other chemicals. These compounds are related as potential mutagenic and carcinogenic compounds to humans. Some of these compounds, present in fuel and biofuel exhaust, can induce known toxicity in exposed human populations, even causing cellular effects.

An important group of chemical carcinogens is the PAH which are important in health for several reasons: some are known to be potent carcinogens in man; there is strong epidemiological evidence that exposed groups have increased risks of lung, urinary tract, brain and skin cancers. As heavy-duty vehicles are the main contributors to particle emissions, where this kind of compound is present, the large increase in internal combustion vehicles in big cities has intensified atmospheric pollution and consequently the harmful effects on human health. Studies have indicated a considerable decrease in PAH emission when burning biodiesel when compared to those from burning of diesel fuel [49,59-61]. In the same way, good results are found for ethanol emission when compared to gasoline fuel burning emissions [62]. On the other hand, as biodiesel use can increase NO_x emissions, some of the derived PAH are supposedly increased, as Nitro-PAHs. The high emission of NO_2 from burning of biofuel can lead to the nitration of the available PAH forming Nitro-PAH. Nitro-PAH is a potential worse chemical carcinogen than PAH; it is shown to induce mutations in bacterial and mammalian cells, sister chromatid exchanges and chromosomal aberrations in cultured mammalian cells [63]. In addition, there is evidence for carcinogenicity in rats [64]. However, some studies have shown discordant results about PAH emissions when biofuels are used compared with fossil fuels in vehicle engines. Some hypothesis to this result may be due the influence of biodiesel source material being particularly strong in the formation of these pollutants. Both increases and decreases can be observed in PAH, nitrated PAH and oxygenated PAH compounds with the use of biodiesel blends from different origin [62,65].

Another group of chemical carcinogens is the CC. Taking into account important concerns of CC for atmospheric chemistry and their negative impact on human health, the levels of carbonyls and their diurnal variability can be an effective indicator reflecting the status of local air pollution. In this sense, correlations between major aldehydes emitted by vehicles and the level of pollution of these compounds in sites impacted by this source are still relatively scarce [42]. The most observed toxic effects to human health by some CC are irritation of skin, eyes and nasopharyngeal membranes [66]. More seriously, formaldehyde, which is usually the most abundant carbonyl in the air, is also the one of more concern because it is classified as carcinogenic to humans by the IARC [67]. Epidemiological studies suggest a causal relationship between exposure to formaldehyde and occurrence of nasopharyngeal cancer, although this conclusion is based in a small number of observed and expected cases in the studies [67]. Indeed, studies about CC indicated a substantial increase in CC during the biofuel combustion process. Biodiesel emissions show an increase of formaldehyde and acrolein [41] and ethanol emissions are related to a large increase of acetaldehyde and formaldehyde [67]. These CC are shown to be a carcinogen, mugenic and can lead to onset of pulmonary edema, respiratory disturbance and asthma like symptoms [68]. Nonetheless CC is also a contributor to the formation of O_3, and thus,of photochemical smog. However, observations of increased aldehydes released by biofuel combustion needs to be better understood for its contribution to any adverse health effects.

Moreover, several studies indicate there is a decrease in the concentration of transition metals in biofuels emission. Metals are more abundant in petroleum fuel combustion exhaust than biofuels and they have the ability to generate radicals which likely lead to depletion of antioxidants and increases in DNA and protein adducts. However, elemental metal composition analyzed in PM from biodiesel and diesel exhaust was found to have metal bound to the carbon core [69].

Additionally, some fuels have specific compounds that are emitted in their burning. Biodiesel exhaust composition presents a number of methyl ester, cyclic fatty acids and nitro fatty acids. Fatty acids are considered pulmonary irritants and present dual polarity that can play an important role tampering the membrane structures and lead to cell death [69]. This kind of compound can play an important role as a fuel emission marker.

Air pollutants	Damage to Human Health
CO	CO can cause harmful health effects by reducing oxygen delivery to the body's organs (like the heart and brain) and tissues. At extremely high levels, CO can cause death.
NOx	In the group of NOx, NO_2 is the component of greatest interest and the indicator for the larger group of nitrogen oxides. In addition to contributing to the formation of O_3, and fine particle pollution, NO_2 is linked with a number of adverse effects on the respiratory system as airway inflammation in healthy people and increased respiratory symptoms in people with asthma.
SOx	SO_2 is the main component and is linked with a number of adverse effects on the respiratory system such as bronchoconstriction and increased asthma symptoms, also short-term exposure increases visits to emergency departments and hospital admissions for respiratory illnesses.
O_3	Even relatively low levels of ozone can cause health effects. People with lung disease, children, older adults, and people who are active outdoors may be particularly sensitive to ozone. The exposure to ozone can make it more difficult to breathe deeply and vigorously, cause shortness of breath and pain when taking a deep breath, cause coughing and sore or scratchy throat, inflame and damage the airways, aggravate lung diseases such as asthma, emphysema, and chronic bronchitis, increase the frequency of asthma attacks, make the lungs more susceptible to infection and ozone can continue to damage the lungs even when the symptoms have disappeared.
PM	Exposure to such particles can affect both lungs and heart, especially fine particles - containing microscopic solids or liquid droplets that are so small that they can get deep into the lungs and cause serious health problems. Numerous scientific studies have linked particle pollution exposure to a variety of problems, including premature death in people with heart or lung disease, nonfatal heart attacks, irregular heartbeat, aggravated asthma, decreased lung function, and increased respiratory symptoms, such as irritation of the airways, coughing or difficulty breathing.

Table 3. Air pollutants and their damage to human health [34].

The PM emitted during fuel burning present some special characteristics. They are composed of a carbon core and organic compounds are adsorbed in their surface. The physical characteristic, especially the size of PM emitted during fuel burning is directly linked to their potential to cause health hazards. Small particles (PM_{10}) less than 10 micrometers in diameter (Dp) pose the greatest problems, because they can get deep into the lungs, and some may even get into the bloodstream. EPA is concerned about particles that are 10 micrometers in diameter or smaller because those are the particles that generally pass through the throat and nose and enter the lungs. Once inhaled, these particles can affect the heart and lungs and cause serious health effects. EPA groups regulated particle pollution into two categories: "Coarse particles," which are larger than 2.5 micrometers and “fine particles” which are smaller than 10 micrometers in diameter. These particles are deposited into the airways in the head region when inhaled. "Fine particles," are 2.5 micrometers ($PM_{2,5}$) in diameter or smaller and when inhaled they are deposited into lung airways or the tracheobronchial region (Figure 4).

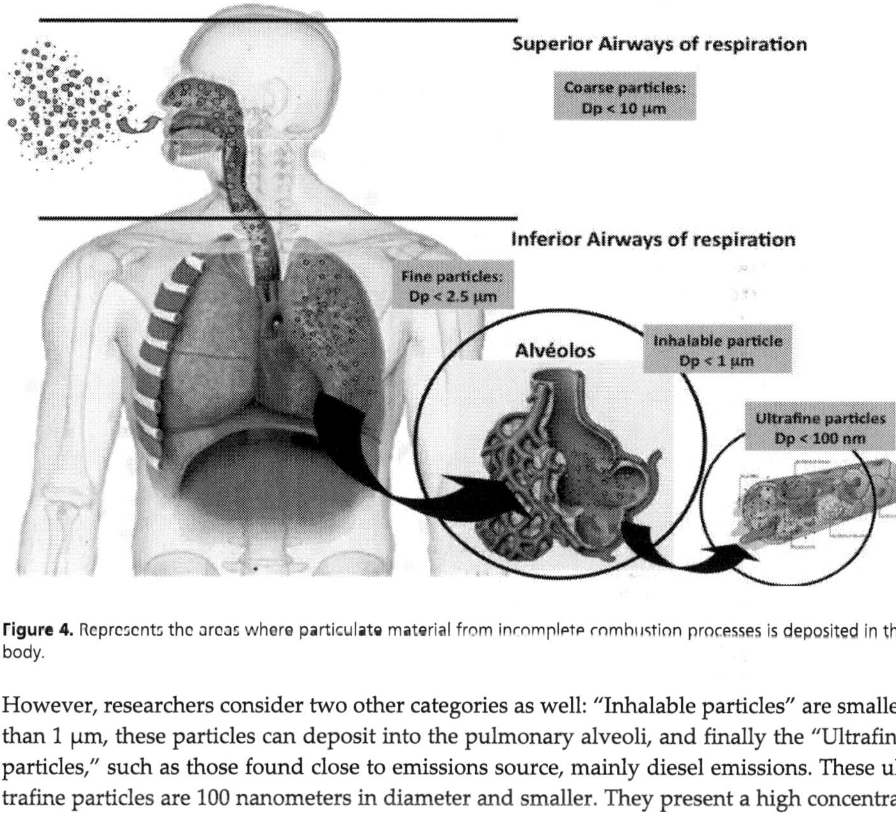

Figure 4. Represents the areas where particulate material from incomplete combustion processes is deposited in the body.

However, researchers consider two other categories as well: "Inhalable particles" are smaller than 1 µm, these particles can deposit into the pulmonary alveoli, and finally the "Ultrafine particles," such as those found close to emissions source, mainly diesel emissions. These ultrafine particles are 100 nanometers in diameter and smaller. They present a high concentration of organic compounds in their composition and can be deposited in the alveolar region and also get into the bloodstream. So, the uses of biofuel can interfere on size distribution of particles emitted as well as chemical composition. Most of the studies published demonstrate a reduction in PM emissions with biodiesel as compared to diesel fuel. This reduction is mainly caused by reduced soot formation and enhanced soot oxidation [57]. In terms of toxicity this is a good gain for the human health. However, in terms of size distribution studies there are worries that the addition of biofuel to petroleum fuel or the use of pure biofuel will change the particle size distribution. These changes have potential implications for the health impacts of PM emissions from biofuel blends.

The addition of ethanol in gasoline fuel changes the particle size distribution, especially in the accumulation mode (30 nm <Dp< 2.5 μm), and decreases the black carbon and total particulate mass concentrations. The molecular weight distribution of the PAH was found to decrease with added ethanol [70, 71]. Generally these changes in the particle size distribution can happen when volatile materials are in excess, leaving insufficient solid area available for adsorption and condensation promoting the nucleation process.

5.2. Emission exposure and toxic effects

The crucial aim of toxicology studies is to identify possible health effects induced by exposure of both the general population and sensitive or susceptible populations, inducing by determination of the exposure threshold, the level needed to induce health effects. The threshold should include not only a concentration but a duration metric, which could be acute or repeated exposures. The strategies to plan and realize the toxic studies should regard that possible health effects may take years of exposure to discern, e. g., lung cancer, fibrosis, emphysema, and mitigation of the exposure and/or effects may be too late for an individual [69].

There are many factors that influence emissions toxicity and the use of biofuel in recent years in some countries has shown a difference in emission contents reflected in the emission toxicity profile. It is important to understand how and what is changing to be able to identify means to improve human and environmental health. In terms of emission from vehicles, the main line of exposure and toxicity effect is on the respiratory system. This is the main point of human contact with the air pollutant and where the first contact and exposure to the pollutant happens. This system is composed of three main regions: the head airways region, lung airways or tracheobronchial region and the pulmonary or alveolar region. Each region differs markedly in structure, airflow patterns, function, retention time, and sensitivity to gases absorption and particle deposition. Inhaled air follows a flow that goes through a sequence of airway branchings as it travels from the trachea to the alveolar surfaces. The first branchings take place in the tracheobronchial region and the remainder in the gas exchange region (Figure 4). In this mechanism the gases are absorbed in the alveolar region and the particles' contents deposited in the lung for varying time durations, depending on their physicochemical properties, their location within the lung, and the type of clearance mechanism involved [72]. Once the pollutant makes contact with the human tissue it starts a series of mechanisms in which the absorption process, the biotransformation and distribution, and lastly the excretion process occur. Thus, the air contents, the pollutants concentration and the physical proprieties are very important since they will determine the acceptable level of toxic exposure and what health effects and damages are caused.

5.2.1. Lung cancer

Lung cancer is a serious health problem and is the main toxic health effect caused by air pollutants. According to WHO, lung cancer accounts for 1.2 million deaths yearly worldwide, exceeding mortality from any other cancer in developed countries. Though the vast majority is caused by tobacco smoking, environmental causes of lung cancer, including air pollution,

http://dx.doi.org/10.5772/52513

have long been a concern as well [73]. WHO recognizes that the exhaust fumes from diesel engines do cause cancer. A panel of experts working from WHO concluded that the exhausts were definitely a cause of lung cancer and may also cause tumors in the bladder [74].

Studies verified that controlled exposures of humans to whole diesel exhaust typically results in lung inflammation as shown with neutrophils entering the lungs [75]; in which these studies are generally 1-2 hr at approximately 100-300 μg/m³ with healthy adults. In these same exposures, several soluble substances which mediate inflammation, e.g., interleukin-8 (IL-8) were shown to be increased by use of lung lavage or inducing sputum production to recover airways secretions. PM from diesel exhaust induced an adjutancy effect using nasal instillations of 300 μg particles in allergic subjects as common biomarkers of allergy (e.g., increased IgE production and histamine release) increased in nasal secretions [76]. Neutrophil influx into the lungs of healthy volunteers exposed to nearly 500 μg/m³ wood smoke for 2 hr was observed [77] suggesting a common outcome from different combusted fuel sources.

Epidemiologic evidence has shown gasoline fuel emissions as a potential lung cancer cause as well [78]. Studies have shown that the gasoline exhaust increased DNA single strand break, promoted lipid peroxidation andoxidative protein damage and decreased activities of superoxide dismutase in lungs and brains. Though, it decreased the activities of glutathione peroxidase in lungs but not in the brain. The present data suggested that gasoline exhaust exposure could cause oxidative damage to lungs and brains of rats. That is to say that gasoline is a toxin to brains of mammals, not only to lungs [79].

However, lung cancer studies on biofuels emissions are limited. In general the findings are for fossil fuels that showed an elevated risk for the development of lung cancers in those with greater exposure compared to workers with lower exposure.

5.2.2. Mutagenicity and genotoxicity assessments

Mutagenicity assays in general can detect the genotoxicity effect of either single chemical and physical agents or heterogeneous mixtures. The Ames mutagenicity test is a short-term *in vitro* assay that has frequently been used to establish mutagenicity [80]. *Salmonella ryphimurium/mammalian* microsome test [81] detects mutagenic proprieties of a wide spectrum of chemicals by reverse mutation of a series of *Salmonella ryphimurim* tester strains. In general, the bacterial stains used to detect frameshift mutagens and base pair substitutions are TA98 and TA 100, respectively [64]. The Ames test is the most frequently used test system worldwide to investigate mutagenicity of complex mixtures like combustion products [82].

Studies about the mutagenicity of biofuel report a wide range of results. Studies about PM exhaust from burning of ethanol or methanol in gasoline blends submitted to the mutagenicity Ames test report that in all the ethanol blended fuel tests, the mass of PM associated to emitted organic compounds from the exhaust was lower than that observed during the control tests using pure gasoline. In the same way, others studies report that in most cases, estimates of the emission of mutagenic combustion products from the exhaust were lower using alcohol blend [83]. However, studies about the influence of ethanol–diesel blended fuels on mutagenic and genotoxic activities of particulate extracts showed higher mutagenicity for E20 (diesel with 20% v.v. of ethanol) compared to E15, E10 and DF (Diesel fuel). Additionally it was found that DF and E20 had a higher genotoxic potential than the other fuel blends [63].

In terms of biodiesel use, studies indicate that biodiesel exhaust is significantly less mutagenic in comparison with diesel fuel [82,84-88]. On the other hand, some studies reported no difference between diesel and biodiesel exhaust or nearly the same mutagenic effects [89, 90]. Moreover, some studies reported increase in the mutagenic effects with the use of biodiesel added to diesel [91-94]. It is important to highlight that the studies found high mutagenicity considering that the biodiesel mutagenicity was generally high or similar compared to diesel and in some other studies were comparing biodiesel to low sulfur diesel.

5.2.3. Oxidative stress assessments

The exposure to air pollutants promotes an event called oxidative stress. It can be defined as a disturbance in the prooxidant-antioxidant balance in favor of the former, leading to potential damage. The hypothesis is that many of the adverse health effects promoted by air pollutants may derivate from oxidative stress, initiated by the formation of reactive oxygen species (ROS) within affected cells [95]. In its simplest form then, oxidative stress is a potentially harmful process that occurs when there is an excess of free radicals, a decrease in antioxidant defenses, or a combination of these events [96].

After inhalation, PM deposited in the lung may stimulate the formation of ROS, such as hydroxyl and superoxide anion radicals. These ROS can be either directly derived from PM or endogenously produced by chemical components of PM, such as transition metals and quinone structures that undergo redox cycling. Furthermore, enhanced ROS formation in the lungs is likely involved in the activation of transcription factors and the induction of cytokines and chemotactic factors. Via these mechanisms, continuous exposure of the lungs to PM-induced ROS formation can cause pulmonary inflammation and eventually cause and/or aggravate impairment of lung development and lung diseases like chronic obstructive pulmonary disease, cystic fibrosis and asthma [80].

The assessment of radical generating capacity has been studied by many methods. Oxygen radicals cannot be detected directly because of their short half-lives, and therefore several alternative methods have been developed. In one method, a molecular probe reacts with the radical species and forms a stable product that can be analyzed with analytical methods, e.g. spectrophotometric analysis of thiobarbituric-acid reactive substances or the dithiothreitol (DTT) assay. Another method is based on biological indicators to assess the formation of ROS by PM, such as the induction of strand breaks in fX174 RF plasmid DNA or the formation of oxidized DNA-bases like 8- oxo-7,8-dihydro-20-deoxyguanosine (8-oxodG). A third method is based on the detection of free radicals by electron spin/paramagnetic resonance (ESR/ EPR) spectroscopy in combination with spin trapping compounds [80].

Studies carried out using Biodiesel and pure plant oil indicated significant reductions of the oxidative potential measured via DTT assay, about 95%, compared to diesel fuel [93]. However, studies measuring oxidative potential of ethanol exhaust are limited.

5.2.4. *Cytotoxicity assessments*

Cytotoxicity assessment is currently essential to evaluate the potential human and environmental health risks associated with chemical exposure, and to limit animal experimentation whenever possible. Several different methods have been used to establish the citotoxicity of air pollutants. Air pollutants have a high potential to damage cells in a concentration- and a time-dependent manner in which the reactive oxygen species play an important role inducing cytotoxicity to the cells [97]. The most important distinctions between these sets are the use of various cell types (lungs cells, macrophages cells, embryonic cells, endothelial cells, fibroclast cell, and others, in which it can be from human or animals) and the time of incubation, varying from 4 to 72h. Also, different fractions and extraction procedures have been used. These differences and variables should be taken into account when results from different studies are compared [80].

Fossil fuels have been studied for many years and a range of researches have shown that the air pollutants promote toxicity and in some case apoptosis (cell death) to animals and human cells [98]. The studies found that biofuel presents a variable cytotoxicity compared to fossil fuels. In general, biodiesel presents an increase in cytotoxicity effects when compared to diesel fuel [82, 93, 99] or no significant differences in cytotoxicity between biodiesel and diesel exhaust [84, 88]. A study carried out using ethanol added to gasoline fuel demonstrated a strong decrease of ethanol exhaust cytotoxicity potential compared to gasoline exhaust [100].

6. Conclusions

Biofuels are promoted in many parts of the world and concern of environmental and social problems have grown due to increased production of this fuels. Production of biofuels promises substantial improvement in air quality through reducing emission from burning of the fuel used in vehicle engines. Some of the developing countries have started biofuel production and utilization as transport fuel in local market. Thus, below are described some important conclusions that we can be done about the use of biofuels by vehicle engines.

Compared to fossil diesel, the emission of regulated and non-regulated compounds from biofuels burning are generally equal or lower. An exception is NOx emission, which is generally higher with use of the biofuels, more specifically of the biodiesel use. The amount of compounds emitted depends considerably of the type engine, its configuration, the load condition and the use of a catalyzer. In most cases, reducing the emission of unwanted compounds requires modification in the standard engines for the use of biodiesel and/or raw vegetable oil and ethanol.

The recent literature demonstrates an increase in research activities on biofuels, especially within recent years. Even with the massive amount of data available, it is still difficult to accurately assess the environmental and health effects of the use of biofuels such as ethanol, biodiesel or raw vegetable oils in vehicle engines. At present, is difficult to conclude what will change with biofuels in terms of toxicity. There are few research activities with the aim to study the toxicological effectiveness of biofuels or their emissions, even though this topic is of great relevance. Furthermore, the results of the available studies could fluctuate widely. Several findings on acute and mechanism-specific toxicity indicate less or comparable effects induced by biofuels in comparison to fossil fuels. However, indications for negative impacts that are induced both by the biofuels themselves and their emissions have been reported. Based on the data available, human health risks associated with spills or the use of biofuels currently cannot be ruled out. Therefore, additional experimental studies are necessary to provide a more comprehensive dataset for the identification of new alternative fuels which could have lower issue impacts for the environment.

Acknowledgments

The authors wish to thank the SENAI CIMATEC and Brazilian agencies: CNPq, FAPESB, CAPES, FINEP and INCT for Energy & Environment. In special to "Christina M. Batteate" proofread the English to ensure the accuracy and improve the language quality.

Author details

Lílian Lefol Nani Guarieiro[1,2] and Aline Lefol Nani Guarieiro[2]

1 SENAI-CIMATEC, Salvador, BA, Brazil

2 INCT de Energia e Ambiente, UFBA, Salvador, BA, Brazil

References

[1] Alves C. Aerossóis atmosféricos: perspectiva histórica, fontes, processos químicos de formação e composição orgânica. Química Nova 2005; 28 (5): 859-870.

[2] Akimoto Z. Major concerns and research needs for our understanding of the chemistry of the atmosphere (Technical Report - IUPAC). Pure and Applied Chemistry 1995; 67 (12): 2057-2064.

[3] Miguel AH. Environmental Polluition Research in South América.Environmental Science & Technology 1991; 25: 590-594.

[4] Daisey JM, Miguel AH, Andrade JB, Pereira P, Tanner RL. An Overview of the rio de Janeiro Aerosol Characterization Study. Journal of Air Pollution Control Association 1987; 37: 15-23.

[5] Tanner RL, Miguel AH, Andrade JB, Gaffney JS, Streit GE. Atmospheric Chemistry of Aldehydes: Enhanced Peroxyacetyl Nitrate Formation From Ethanol-Fueled Vehicular Emissions. Environmental Science &Technology 1988; 22: 1026-1034.

[6] Andrade MV, Pinheiro HLC, de Andrade JB. Ambient Levels of Formaldehyde and Acetaldehyde and Their Correlation with the Vehicular fleet in Salvador, BA, Brazil.Reg. Conf. On Global Change, p.151-152, dez, 1995.

[7] De Andrade JB, Macedo MA, Korn M, Oliveira E, Gennari RF. A comparison Study of Aerosol Emission Sources in Two Receptor Sites in Salvador, Brazil.Toxicology Environment Chemistry 1996; 54: 23-28.

[8] Bünger J, Krahl J, Munack A, Ruschel Y, Schröder O, Emmert B, et al. Strong mutagenic effects of diesel engine emissions using vegetable oil as fuel. ArchToxicol 2007;81:599–603.

[9] Costa Neto PR, Rossi LFS, Zagonel GF, Ramos LP. Produção de Biocombustível Alternativo ao Óleo Diesel Através da Transesterificação de óleo de Soja Usado em Frituras. Química Nova 2000; 23 (4): 531-537.

[10] Companhia de Tecnologia de Saneamento Ambiental, CETESB, Brazil. Emissões veiculares. Available at:http://www.cetesb.sp.gov.br/Ar/emissoes/introducao.asp (accessed 30 May 2012).

[11] Burtscher H. Physical characterization of particulate emissions from diesel engines: A review. Journal of Aerosol Science 2005; 36: 896–932.

[12] Oberdörster G, Maynard A, Donaldson K, Castranova V, Fitzpatrick J, Ausman K, et al. Principles for characterizing the potential human health effects from exposure to nanomaterials: Elements of a screening strategy. Part Fibre Toxicology 2005; 2: 1–35.

[13] Martins LD, Martins JA, Freitas ED, Rocha CRM, Gonçalves FLT, Ynoue RY, et al. Potential health impact of ultrafine particles under clean and polluted urban atmospheric conditions: A model-based study. Air Quality, Atmosphere& Health 2010; 3: 29–39.

[14] Szwarc A. Controle das emissões de poluentesporveículosautomotores no Brasil. Relatório de pesquisapara a Fundação COPPETEC/MMA. Mimeo, ADS Tecnologia e DesenvolvimentoSustentável, SP, 2001.

[15] Zervas E, Montagne X, Lahaye J. Emission of specific pollutants from a compression ignition engine. Influence of fuel hydrotreatment and fuel/air equivalence ratio. Atmospheric Environment 2001; 35: 1301-1306.

[16] Graham, L. Chemical characterization of emissions from advanced technology light-duty vehicles. Atmospheric Environment 2005; 39: 2385-2398.

[17] Nam, EK, Jensen TE, Wallington TJ. Methane emissions from vehicles. Environment Science Technology 2004; 38: 2005-2010.

[18] Caplain I, Cazierb F, Noualib H, Mercierb A, De'chauxa JC, Nolleta V, Joumardc R, Andre JM, Vidon R. Emissions of unregulated pollutants from European gasoline and diesel passenger cars. Atmospheric Environment 2006; 40: 5954–5966.

[19] Guarieiro LLN, Torres EA, de Andrade JB. Energia Verde. In: Alicia Ivanissevich e Angelo da Cunha Pinto. (Org.). Química Hoje. Química Hoje. 1 ed .Rio de Janeiro: Instituto Ciência Hoje, 2012; 1: 118-125.

[20] Tran DQ, Charanjit R. A kinetic model for pyrolysis of Douglas fir bark.Fuel 1978; 57: 293–8.

[21] Bridgwater AV. Renewable fuels and chemicals by thermal processing of biomass.ChemicalEngineeringJournal 2003; 91: 87–102.

[22] Pinto AC, Guarieiro LLN, Rezende MJC, Ribeiro NM, Torres EA, Lopes WA, Pereira PAP, de Andrade JB. Biodiesel: an Overview. Journal of the Brazilian Chemical Society 2005; 16: 1313.

[23] Sciencedirect, 2012. http://www.sciencedirect.com/(accessed 27 July 2012).

[24] Timilsina GR, Shrestha A. How much hope should we have for biofuels? Energy 2011; 36: 2055-2069.

[25] Kim S, Dale BE. Life cycle assessment of various cropping systems utilized for producing biofuels: bio-ethanol and biodiesel. Biomass Bioenergy 2005; 29: 426-39.

[26] Winther M, Møller F, Jensen TC. Emission consequences of introducing bio ethanol as a fuel for gasoline cars. Atmospheric Environment 2012; 55: 144-153.

[27] Agarwal AK. Biofuels (alcohols and biodiesel) applications as fuels for internal combustion engines Progress in Energy and Combustion Science 2007; 33: 233–271.

[28] Taylor AB, Mocan DP, Bell AJ, Hodgson NG, Myburgh IS, Botha JJ. Gasoline/alcohol blends: exhaust emission, performance and Burn-rate in multi-valve production engine, SAE paper no. 961988, 1996.

[29] de Melo TCC, Machado GB, Belchior CRP, Colaço MJ, Barros JEM, de Oliveira EJ, de Oliveira DG. Hydrous ethanol–gasoline blends – Combustion and emission investigations on a Flex-Fuel engine. Fuel 2012; 97: 796-804.

[30] Neimark A, Kholmer V, Sher E. The effect of oxygenates in motor fuel blends on the reduction of exhaust gas toxicity. SAE paper no. 940311, 1994.

[31] Smokers R, Smith R. Compatibility of pure and blended biofuels with respect to engine performance, durability and emission. A literature review, report 2GVAE04.01.Dutch ministry for spatial planning, 2004. p. 1–70.

[32] Merritt PM, Ulmet V, McCormick RL, Mitchell WE, Baumgard KJ. Regulated and unregulated exhaust emissions comparison for three tier II non-road diesel engines operating on ethanol- diesel blends. SAE paper no. 2005-01-2193, 2005.

[33] Guarieiro LLN, Vasconcellos PC, Solci MC. Poluentes Atmosféricos Provenientes da Queima de Combustíveis Fósseis e Biocombustíveis: Uma Breve Revisão. Revista Virtual de Quimica 2011; 3 (5): 434-445.

[34] EPA. A Comprehensive Analysis of Biodiesel Impacts on Exhaust Emissions. Draft Technical Report: EPA420-P-02-001; October 2002. http://www.epa.gov/otaq/models/analysis/biodsl/p02001.pdf.(accessed 01 August 2012).

[35] Schumacher LG, Borgelt SC, Fosseen D, Goetz W, Hires CWG. Heavy-duty engine exhaust emission tests using methyl ester soybean oil/diesel fuel blends. Bioresource Technology 1996; 57: 31-36.

[36] Alton R, Cetinkaya S, Yucesu HS. The potential of using vegetable oil fuels as fuel for diesel engines. Energy Convers Manage 2001;42:529–38.

[37] Agarwal AK, Das LM. Biodiesel development and characterization for use as a fuel in compression ignition engine. J Eng Gas Turbines Power 2001;123:440–7.

[38] Dorado MP, Ballesteros E, Arnal JM, Gomez J, Lopez FJ. Exhaust emissions from a diesel engine fueled with transesterified waste olive oil. Fuel 2003; 82: 1311-1315.

[39] Pang X, Mu Y, Yuan J, He H. Carbonyl emission from ethanolblended gasoline and biodiesel–ethanol–diesel used in engines. Atmospheric Environment 2008; 42: 1349-1358.

[40] Shi X, Pang X, Mu Y, He H, Shuai S, Wang J, Chen H, Li R. Emission reduction potential of using ethanol–biodiesel–diesel fuel blend on a heavy-duty diesel engine. Atmospheric Environment 2006; 40: 2567-2574.

[41] Guarieiro LLN, Pereira PAP, Torres EA, da Rocha GO, de Andrade JB. Carbonyl compounds emitted by a diesel engine filled with diesel and biodiesel–diesel blends: sampling optimization and emissions profile. Atmospheric Environment 2008; 42: 8211–8218.

[42] Rodrigues MC, Guarieiro LLN, Cardoso MP, Carvalho LS, da Rocha GO, de Andrade JB. Acetaldehyde and formaldehyde concentrations from sites impacted by heavy-duty diesel vehicles and their correlation with the fuel composition: Diesel and diesel/biodiesel blends. Fuel 2012; 92: 258–263.

[43] Bikas G, Zervas E. Nonregulated Pollutants emitted fromEuro 3 Diesel Vehicles as a function of their mileage. Energy and Fuels 2007; 21: 2731-2736.

[44] Turrio-Baldassarri L, Battistelli CL, Conti L, Crebelli R, De Berardis B, Iamiceli AL, Gambino M, Iannaccone S. Emission comparison of urban bus engine fueled with diesel oil and biodiesel blend. Science of the Total Environment 2004; 327: 147-162.

[45] Correa SM, Arbilla G. Carbonyl emissions in diesel and biodiesel exhaust. Atmospheric Environment 2008; 42: 769-775.

[46] Lev-On M, LeTavec C, Uihlein J, Kimura K, Gautam M, Thompson GJ, Wayne WS, Clark N, Zielinska B, Chatterjee S. Speciation of organic compounds from the ex-

haust of trucks and buses: effect of fuel and after-treatment on vehicle emission profiles. SAE Technical Paper SAE 2002, Warrendale, PA, 01, 2873.

[47] Durbin T, Collins JR, Norbeck J, Smith MR. Evaluation of the effects of alternative diesel fuel formulations on exhaust emission rates and reactivity – final report for the south coast air quality management district under contract 1999; no. 98102, South Coast Air Quality Management District, DiamondBar, CA, 68 pp.

[48] Lin Y, Lee W, Wu T, Wang C. Comparison of PAH and regulated harmful matter emissions from biodiesel blends and paraffinic fuel blends on engine accumulated mileage test. Fuel 2006; 85: 2516–2523.

[49] Martins LD, Silva Júnior CR, Solci MC, Pinto JP, Souza DZ, Vasconcellos P, Guarieiro ALN, Guarieiro LLN, Sousa ET, de Andrade JB. Particle emission from heavy-duty engine fuelled with blended diesel and biodiesel. Environmental Monitoring and Assessment 2012; 184: 2663-2676.

[50] Guarieiro LLN, De Souza, AF, Torres EA, De Andrade JB.Emission profile of 18 carbonyl compounds, CO, CO2, and NOx emitted by a dieselengine fuelled with diesel and ternary blends containing diesel, ethanol and biodiesel or vegetable oils Atmospheric Environment 2009; 43: 2754-2761.

[51] Caro PS, Mouloungui Z, Vaitilingom G, BergeJCh. Interest of combining an additive with diesel–ethanol blends for use in diesel engines. Fuel 2001; 80 (4): 565-574.

[52] Fernando S, Hanna M. Development of a Novel Biofuel Blend Using Ethanol–Biodiesel–Diesel Microemulsions: EB-Diesel. Energy Fuels 2004; 18 (6): 1695-1703.

[53] Pang X, Mu Y, Yuan J, He H. Carbonyl emission from ethanol-blended gasoline and biodiesel–ethanol–diesel used in engines. Atmospheric Environment 2008; 42: 1349-1358.

[54] Shi X, Yu Y, He H, Shuai S, Wang J, Li R. Emission characteristics using methyl soyate–ethanol–diesel fuel blends on a diesel engine. Fuel 2005; 84: 1543-1549.

[55] WHO – IARC: Diesel Engine Exhaust Carcinogenic, Lyon, France, June 12, 2012. http://new.paho.org/hq/index.php?option=com_content&task=view&id=6903&Itemid=259 (accessed 02 August 2012).

[56] WHO – Health in the Green economy: Co-benefits to health of climate change mitigation. Preliminary Finding – initial review.www.who.int/phe/en/ www.who.int/hia/green_economy/en/index.html(accessed 10 August 2012).

[57] Lapuerta M, Armas O, Rodrigues-Fernadez J. Effects of biodiesel fuel on diesel emissions. Progress in Energy and Combustion science 2008; 34: 198-223.

[58] Niven RK. Ethanol in gasoline: environmental impacts and sustainability review article. Renewable & Sustainable Energy Reviews 2005; 9: 535-555.

[59] Yang H-H, Chien S-M, Lo M-Y, Lan JC-W, Lu W-C, Ku Y-Y. Effects of biodiesel on emissions of regulated air pollutants and polycyclic aromatic hydrocarbons under engine durability testing.Atmopheric Environment 2007; 41: 7232-7240.

[60] Karavalakis G, Stournas S, Bakeas E. Effects of diesel/biodiesel blends on regulated and unregulated pollutants from a passenger vehicle operated over the European and Athens driving cycles. Atmospheric Environment 2009; 43: 1745-1752.

[61] Chien S-M, Huang Y-J, Chuang S-C, Yang H-H. Effects of Biodiesel Blending on Particulate and Polycyclic Aromatic Hydrocarbon Emissions in Nano/Ultrafine/Fine/Coarse Ranges from Diesel Engine. Aerosol and Air Quality Research 2009; 9 (1): 18-31.

[62] Bakeas E, Karavalakis G, Fontaras G, Stournas S. An experimental study on the impact of biodiesel origin on the regulated and PAH emissions from a Euro 4 light-duty vehicle. Fuel 2011; 90: 3200-3208.

[63] Song J, Ye SH. Study on the mutagenicity of diesel exhaust particles. Biomedicine. Environment. Science 1995; 8:240–245.

[64] Song CL. Zhou YC, Huang R-Ji, Wang Y-Q, Huang Q-F, Lü G, Liu K-M. Influence of ethanol–diesel blended fuels on diesel exhaust emissions and mutagenic and genotoxic activities of particulate extracts. Journal of Hazardous Materials 2007; 149: 355–363.

[65] Karavalakis G, Bakeas E, Fontaras G, Stournas S. Effect of biodiesel origin on regulated and particle-bound PAH (polycyclic aromatic hydrocarbon) emissions from a Euro 4 passenger car. Energy 2011; 36: 5328-5337.

[66] Wang B, Lee SC, Ho KF. Characteristics of carbonyls: concentrations and source strengths for indoor and outdoor residential microenvironments in China. Atmospheric Environment 2007; 41:2851–61.

[67] IARC, World Health Organization. International Agency for Research on Cancer. Summary of data reported and evaluation, vol. 88; 2006. p. 2–3. <http://www.inchem.org/documents/iarc/vol88/volume88.pdf>. (accessed 20 January 2012).

[68] McQueen CA. Comprehensive Toxicology. Second Edition. Copyright 2010 Elsevier; ISBN 978-0-08-046884-6.

[69] Madden MC, Bhavaraju L and Kodavanti UP. Toxicology of Biodiesel Combustion Products.Chapter 13.Biodiesel – Quality, Emissions and By-Products. November 2011. ISBN 978-953-307-784-0. Copyright In tech.

[70] Dutcher DD, Stolzenburg MR, Thomspon SL, Medrano JM, Gross DS, Kittelson DB, McMurry PH. Emissions from Ethanol-Gasoline Blends: A Single Particle Perspective. Atmosphere 2011; 2: 182-200.

[71] Agarwala AK, Guptab T and Kotharia A. Particulate emissions from biodiesel vs diesel fuelled compression ignition engine. Renewable and Sustainable Energy Reviews 2011; 15: 3278–3300.

[72] Hinds WC. Aerseol Technology: Properties, behavior, and measurement of airborne particles, Second Edition. John Wiley & sons, inc.1998.

[73] Wang S and Zhao Y. Air Pollution and Lung Cancer Risks.Tsinghua University, Beijing China.2011; 26-38.

[74] Diesel exhausts do cause cancer, says WHOhttp://www.bbc.co.uk/news/health-18415532 (accessed 02 August 2012).

[75] Holgate ST, SandströmT,FrewAJ,StenforsN,NördenhallC,SalviS,Blomberg A, Helleday R, Söderberg M. Health effects of acute exposure to air pollution. Part I: Healthy and asthmatic subjects exposed to diesel exhaust. Research report Health Effects Institute2003; Dec (112):1-30: discussion 51-67.

[76] Diaz-Sanchez D,TsienA,FlemingJ,Saxon A. Combined diesel exhaust particulate and ragweed allergen challenge markedly enhances human in vivo nasal ragweed-specific IgE and skews cytokine production to a T helper cell 2-type pattern. Journal Immunology 1997; 158:2406-13.

[77] Ghio AJ, Soukup JM, Case M, Dailey LA, Richards J, Berntsen J, Devlin RB, Stone S, Rappold A. Exposure to wood smoke particles produces inflammation in healthy volunteers.Occupotional Environmental Medicine; 2011: Jun 30.

[78] Vineis P, Forastiere F, Hoek G, Lipsett M. Outdoor air pollution and lung câncer: recent epidemiologic evidence. International Journal of Câncer 2004; 111 (5): 647-652.

[79] Che W, Zhang Z, Zhang H, Wu M, Liang Y, Liu F, Shu Y, Li N. Compositions and oxidative damage of condensate, particulate and semivolatile organic compounds from gasoline exhausts. Environmntal Toxicology and Pharmacology, 2007; 24 (1): 11-18.

[80] De Kok TMCM, Driece HAL, Hogervorst JGF, Briedé JJ. Toxicological assessment of ambient and traffic-related particulate matter: A review of recent studies. Mutation Research 2006; 613: 103-122.

[81] Ames BN, Lee FD and Durson WE. An improved bacterial test system for the detection and classification of mutagens and carcinogens.Proceedings of the National Academy of Sciences 1973; 70: 782-786.

[82] Bünger J, Muller MM, Krahl J, Baum K, Weigel A, Hallier E, Schulz TG. Mutagenicity of diesel exhaust particles from two fossil and two plant oil fuels. Mutagenesis 2000; 15: 391-397.

[83] Clark CR, Dutcher JS, McClellan RO, Naman,t TM, Seizingert DE. Influence of Ethanol and Methanol Gasoline Blends on the Mutagenicity of Particulate Exhaust Extracts . Arch. Environmental Contaminante Toxicology 1983; 12: 311-317.

[84] Bünger J, Krahl J, Franke HU, Munack A, Hallier E. Mutagenic and cytotoxic effects of exhaust particulate matter of biodiesel compared to fossil diesel fuel Mutat. Res., Genet.Toxicol. Environ. Mutagen 1998; 415: 13-23.

[85] Eckl PM, Leikermoser P, Wörgetter M, Prankl H, Wurst F. in Plant Oils as Fuels: Present State of Science and Future Developments, eds. N. Martini and J. S. Schell, Springer, Berlin, Germany 1998; 123-140.

[86] Bünger J, Krahl J, Baum K, Schroder O, Muller M, Westphal G, Ruhnau P, Schulz TG, Hallier E. Cytotoxic and mutagenic effects, particle size and concentration analysis of diesel engine emissions using biodiesel and petrol diesel as fuel. Archives of Toxicology 2000; 74: 490-498.

[87] Gagnon ML and White PA. The mutagenicity and dioxin-like activity of biodiesel emissions, Ottawa, Ontario, 2008.

[88] Jalava PI, Tapanainen M, Kuuspalo K, Markkanen A, Hakulinen P, Happo MS, Pennanen AS, Ihalainen M, Yli-Pirila P, Makkonen U, Teinila K, Maki-Paakkanen J, Salonen RO, Jokiniemi J, Hirvonen MR. Toxicological effects of emission particles from fossil- and biodiesel-fueled diesel engine with and without DOC/POC catalytic converter.Inhalation Toxicology 2010; 22: 48-58.

[89] Finch GL, Hobbs CH, Blair LF, Barr EB, Hahn FF, Jaramillo RJ, Kubatko JE, March TH, White RK, Krone JM, Menache MG, Nikula KJ, Mauderly JL, Van Gerpen J, Merceica MD, Zielinska B, Stankowski L, Burling K, Howell S. Effects Of Subchronic Inhalation Exposure Of Rats To Emissions From A Diesel Engine Burning Soybean Oil-Derived Biodiesel Fuel. Inhalation Toxicology 2002; 14(10):1017-1048.

[90] Krahl J, Munack A, Grope N, Ruschel Y, Schroder O, Bunger J. Biodiesel, Rapeseed Oil, Gas-To-Liquid, and a Premium Diesel Fuel in Heavy Duty Diesel Engines: Endurance, Emissions and Health Effects. Clean - Soil, Air, Water, 2007; 35 (5): 417-426.

[91] Bünger J, Krahl J, Munack A, Ruschel Y, Schroder O, Emmert B, Westphal G, Muller MM, Hallier E and Bruning T, Strong mutagenic effects of diesel engine emissions using vegetable oil as fuel. Archives of Toxicology 2007; 81(8): 599-603.

[92] Kado NY and Kuzmicky PA. Bioassay analyses of particulate matter from a diesel bus engine using various biodiesel feedstock fuels: Final report; report 3 in a series of 6, University of California, Department of Environmental Toxicology Report NREL/SR-510-31463, National Renewable Energy Laboratory, Golden, USA, 2003.

[93] Kooter IM, van Vugt MATM, Jedynska AD, Tromp PC, Houtzager MMG, Verbeek RP, Kadijk G, Mulderij M, Krul CAM. Toxicological characterization of diesel engine emissions using biodiesel and a closed soot filter. Atmospheric Environment 2011; 45: 1574-1580.

[94] Krahl J, Knothe G, Munack A, Ruschel Y, Schroder O, Hallier E, Westphal G, Bunger J. Comparison of exhaust emissions and their mutagenicity from the combustion of biodiesel, vegetable oil, gas-to-liquid and petrodiesel fuel. Fuel 2009; 88: 1064-1069.

[95] Li Q, Wyatt A, Kamens RM. Oxidant generation and toxicity enhancement of aged-diesel exhaust. Atmospheric Environment 2009; 43: 1037-1042.

[96] Kelly FJ. Oxidative Stress: Its role in air pollution and adverse health effects. Occupational Environmental Medicine 2003; 60: 612-616.

[97] Li N, Xia T, Nel AE. The role of oxidative stress in ambient particulate matter-induced lung diseases and its implications in the toxicity of engineered nanoparticles. Free Radical Biology & Medicine 2008; 44: 1689-1699.

[98] Li N, Sioutas C, Cho A, Schimitz D, Misra C, Sempf J, Wang M, Oberley T, Froines J, Nel A. Ultrafine Particulate Pollutants Induce Oxidative Stress and Mitochondrial Damage. Environmental Health Perspectives 2003; 111 (4): 455-460.

[99] Liu YY, Lin TC, Wang YJ, Ho WL. Carbonyl Compounds and Toxicity Assessments of Emissions from a Diesel Engine Running on Biodiesels. Journal Air & Waste Management Association 2009; 59: 163-171.

[100] Zhang ZZ, Che WJ, Liang Y, Wu M, Li N, Shu Y, Liu F, Wu DS. Comparison of cytotoxicity and genotoxicity induced by the extracts of methanol and gasoline engine exhausts Toxicology in Vitro 2007; 21: 1058-1065.